THE BUTTERFLIES OF BRITAIN & IRELAND

NEW REVISED EDITION

JEREMY THOMAS & RICHARD LEWINGTON

For our daughters,
Emily and Anna Thomas and
Alexandra Lewington

BLOOMSBURY WILDLIFE
Bloomsbury Publishing Plc
50 Bedford Square, London, WC1B 3DP, UK

BLOOMSBURY, BLOOMSBURY WILDLIFE and the Diana logo are trademarks of Bloomsbury Publishing Plc

First published in 1991 in the United Kingdom by Dorling Kindersley Ltd
This revised edition published 2016

A catalogue record for this book is available from the British Library

ISBN: 978-1-4729-6719-0

2 4 6 8 10 9 7 5 3

Printed and bound in China by C&C Offset Printing Co., Ltd.

Contents

Lycaenidae

Riodinidae

Nymphalidae

Introduction

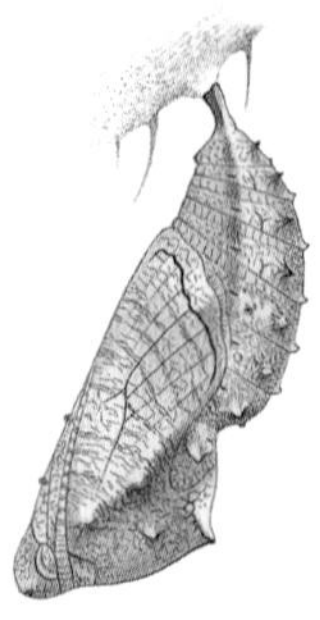

Red Admiral chrysalis

The aim of this book is to share some of the pleasure that butterflies have given us during the course of our careers. Its origin goes back to 1983 when, as a young artist, Richard found a wild chrysalis of a Red Admiral and realised that no book, past or present, had done justice to its subtle beauty. This feeling grew as he reared other species from egg to adult, and with it came the ambition to illustrate every stage in the life cycle of all British butterflies, emulating the great work of F W Frohawk a century ago.

As Richard embarked on his marathon, new details about the behaviour and ecology of butterflies were being discovered, together with ever more accurate data about their distributions and population changes. Most information was available only in unpublished reports or in obscure journals written for fellow academics. I attempted to present this in a more digestible form, by combining it with many personal observations and by drawing on the wealth of anecdotes left by Victorian and earlier naturalists. The resulting essays – 70 in all – provided the accompaniment for Richard's incomparable illustrations, and were published as *The Butterflies of Britain & Ireland* by Dorling Kindersley in 1991.

Our book was kindly received, winning the *Book of the Year* award for natural history the following year. However, although our publisher had produced a beautiful edition, long-term low-volume sales were not part of the business plan. So when the initial 20,000 copies had sold within three years, there was no further printing. Even the authors were taken by surprise: I possess two battered copies.

New research

As the years passed, Richard and I were regularly invited to revive our short-lived volume. I, however, hesitated, realising what a mammoth undertaking it would be to revise texts that were increasingly out of date. Even then I underestimated the task, little realising that there had been a sevenfold increase in the number of major authorities (922) that it would be appropriate to cite compared with 20 years ago.

For not only have two new species, Cryptic Wood White and the Geranium Bronze, been described from the British Isles, but there has been a burgeoning volume of research into almost every aspect of butterfly biology. This includes using modern techniques in genetics to throw light on the history and origins of our species; understanding how butterfly wing patterns are formed and why they have evolved such a magnificent array of refractive and pigmented colours; how adult butterflies manage to fly – for, like bumblebees, flight in several species cannot be explained by conventional forms of aerodynamics; how migrants such as the Painted Lady, Red Admiral and Monarch sweep across continents every year, whilst their offspring return in single flights of up to 2,000km, locating with pin-point accuracy the same few acres of ancestral over-wintering grounds (in the case of the Monarch) by using clock-compasses housed within their antennae for navigation. Other breakthroughs include how the caterpillars of some browns sequester obnoxious chemicals into their bodies by eating grasses infected by a fungus, which in turn makes the adult butterfly poisonous to enemies; and how the caterpillars and chrysalises of most blues and hairstreaks interact with ants to an astonishing degree, not least in the Silver-studded Blue, which is now known to have a relationship with black *Lasius* ants that is as intimate, and in some respects closer, than that between the Large Blue and red *Myrmica* ants. A further area of growing interest concerns the highly specialised parasitic wasps and flies whose grubs live inside the body of a caterpillar or chrysalis, gradually consuming its tissues. Twenty years ago, these parasitoids were commonly dismissed as ugly maggots that killed 'our' butterflies: today, they are rightly regarded as fascinating, and often beautiful, members of each butterfly's community, and equally deserving of conservation.

Mercifully for our readers, British Wildlife Publishing's new edition of *The Butterflies of Britain & Ireland* (updated in 2014) is not seven times larger, but it does contain 64 extra pages, allowing the average species account to be 40% longer and Richard to add about 100 new illustrations, including an extra picture of the adults in their natural settings, the life cycles of Cryptic Wood White and the Geranium Bronze, and about 30 more images depicting butterfly behaviour, their young stages and parasitoids.

Winners and losers

The past 25 years has seen an upheaval in the status and distribution of nearly all our butterflies, which have been monitored in unprecedented detail thanks to the mapping schemes and transect counts organised by Butterfly Conservation and The Centre for Ecology and Hydrology. Over 10,000 skilled amateurs have submitted several million records to these organisations since 1990, making the butterflies of the British Isles the most rigorously monitored group of insects anywhere in the world.

The results have identified winners and losers, of which the latter, sadly, outnumber the former by three to one. High on the list of losers are the species that traditionally bred in fresh clearings, on sunlit woodland floors. The most notable casualties are the Duke of Burgundy and the High Brown and Pearl-bordered Fritillaries. Despite some local conservation successes, it is a moot point whether the High Brown Fritillary or the Duke of Burgundy is the next most probable butterfly to become extinct in the British Isles. We chart their breathtaking declines in grisly detail.

There have been notable conservation successes, too, during the past two decades. Thirty years ago, the Adonis Blue, Silver-spotted Skipper and Heath Fritillary were approaching extinction, and the tide had only just turned a decade later. Today, the first two species have made remarkable recoveries, thanks to targeted conservation management, helped by the return of grazing rabbits and a series of warm summers. The Heath Fritillary has had a bumpier ride, but current prospects look good. Meanwhile, the Black Hairstreak is perhaps at an all-time high, thanks to careful conservation measures. Finally, the Large Blue, which in 1991 had yet to be reintroduced to Somerset, was spluttering on a single small site. Today, its reintroduction has been hugely successful, with more than 30 colonies in several counties.

High Brown Fritillary

Among the winners, there has been no more dramatic success than that of the Comma, which has spread northwards by more than 200km since 1990. This represents an astonishing change in fortunes compared with a century ago, when the Comma was on the verge of extinction and largely confined to the Welsh Borders. Fewer than half-a-dozen sightings of the butterfly were made in Suffolk, Surrey, Sussex and Dorset in the 100 years up to 1929, whereas today one expects to find it in every wood.

Climate-warming is one reason why the Comma and some other butterflies have spread north, although a medley of other factors may also be involved. As a relatively mobile butterfly, the Comma was quick to colonise an improving environment. The same is true of our rare and regular migrants, which now arrive with a frequency matched only during the years of the Second World War. For example, roughly ten times more sightings have been made of the Camberwell Beauty than of the Large Tortoiseshell in recent decades, and five times as many reports were received of the Monarch. For this reason, and because some extinct butterflies may be re-introduced in the foreseeable future, it no longer made sense to relegate the accounts of 11 rare vagrants or extinct species to a postscript at the end of the book. We have now promoted them to their rightful places among the more familiar species.

Species of dubious status

In addition to the 72 butterflies whose natural history is described in this book, are a further 40 species that have been reported from the British Isles, but which on various grounds we have excluded. Whereas the Short-tailed Blue and, more dubiously, the Geranium Bronze just make the cut, qualifying as a result of a few tens of sightings and evidence of breeding, the others consist mainly of single or suspect records. Several stem from genuine records, substantiated by photographs or specimens in museums, but the others include frauds, mistakes and pure wishful thinking. For butterfly enthusiasts are no less susceptible than anglers to embroidering their tales.

There is genuine uncertainty about a number of curiosities that feature in the earliest accounts of British butterflies. A few are simple mistakes that resulted from the mixing of records or specimens. Other species may well have resided here once, but became extinct before being properly documented. These include three species of copper – the Scarce, Purple-shot, and Purple-edged – as well as the so-called Arran Brown *Erebia ligea*. The case for believing the coppers to have been British is not strong, but many give the Arran Brown the benefit of the doubt. The merits of each case are admirably put, along with delightful anecdotes, in P B M Allen's series of *A Moth Hunter's Gossip* and in Michael Salmon's *The Aurelian Legacy*.

There is no doubt that the reported sightings and capture of many species of butterfly are fraudulent. The origin of an individual butterfly can be extremely difficult to verify,

▲ **American Painted Lady** *Vanessa virginiensis* Since its discovery in Pembrokeshire in 1808, about 20 genuine records have been made of this North American immigrant. It is, however, permanently established in Spain.

▲ **Arran Brown** *Erebia ligea* This handsome butterfly was reputedly discovered by Sir Patrick Walker on the Isle of Arran in 1803. The most recent specimen was caught by a schoolboy in 1969 on the western edge of Rannoch Moor.

▲ **Large Chequered Skipper** *Heteropterus morpho* This is believed to have been accidentally introduced to Jersey in hay during the wartime occupation. After 40 years' persistence in a single colony, it became extinct in 1996.

although molecular (DNA fingerprinting) techniques should solve the problem in the near future. I had personal experience with the Large Blue, when dead Continental specimens were planted in the field and claimed as English. Frauds are, in fact, less common now than in previous centuries, when there were genuine new species to be discovered in remote regions, and when collectors were prepared to pay high prices for British rarities. There have, of course, also been many innocent introductions of foreign specimens to Britain, as well as captive butterflies that have escaped. This has become more common because of the proliferation of butterfly farms; it is not unusual nowadays to see a gaudy South American species gliding around gardens in the vicinity of a rearing house. Other species that have been caught in the British Isles once or twice probably owe their origin to being accidentally imported as chrysalises or caterpillars, sometimes literally in a bunch of bananas. Others, such as the American or Hunter's Painted Lady *Vanessa virginiensis*, are vagrants from distant continents that crop up too rarely to merit a full account.

As the climate warms, we can expect additional species from central and southern Europe to colonise the British Isles. There has already been a polewards shift in the distributions of many Continental butterflies, akin to the expansions of the Comma, Speckled Wood and several grass-feeding skippers and browns within the British Isles, although, sadly, the English Channel has provided an insuperable obstacle to date. The Scarce Swallowtail *Iphiclides podalirius* is an obvious candidate to colonise southern England in the foreseeable future. Another is the European Map *Araschnia levana*, which briefly became established in the Forest of Dean after its release there in about 1912. It is widely believed that it was exterminated by a collector, A B Farn, who detested the idea of foreign butterflies in Britain. This seems unlikely to be the entire reason for this butterfly's demise; it is more probable that its habitat and the climate were only marginally suitable. In recent decades, the Map has spread hundreds of kilometres north and west across Europe, and we might expect this charming nettle-feeder to colonise our Isles naturally before long.

The illustrations in this book

For most of the 72 species described in this book, the upperside and underside of the male and female adult butterflies are illustrated, except in the few cases where the sexes look the same. Some species vary with the time of year, or in different parts of the British Isles, and these forms are shown as well. Also included are some of the strange varieties or aberrations in wing markings that periodically crop up in a number of species.

It should be remembered that, in the wild, butterflies only rarely rest with their wings held wide open. So, complementing the formal diagnostic portraits are two or more pictures of the living butterfly shown in a typical resting position.

Apart from the introductory illustration for each species, all the paintings of the adult butterflies are life-size. The eggs, caterpillars and chrysalises, however, are enlarged by varying amounts (indicated in each caption) to display the full beauty of their shapes and markings. In general, only the full-grown caterpillar has been painted, exceptions being where the young caterpillar looks markedly different from the older one.

The life of a butterfly

Every butterfly goes through four very different phases in its life: egg, caterpillar, chrysalis and adult. In many species, the first three of these stages can be just as interesting as the last, and they are often quite easy to find in the wild.

Most butterflies take one year to complete their life cycle, although certain species fit in two, three, or even four generations between spring and autumn. A chart at the base

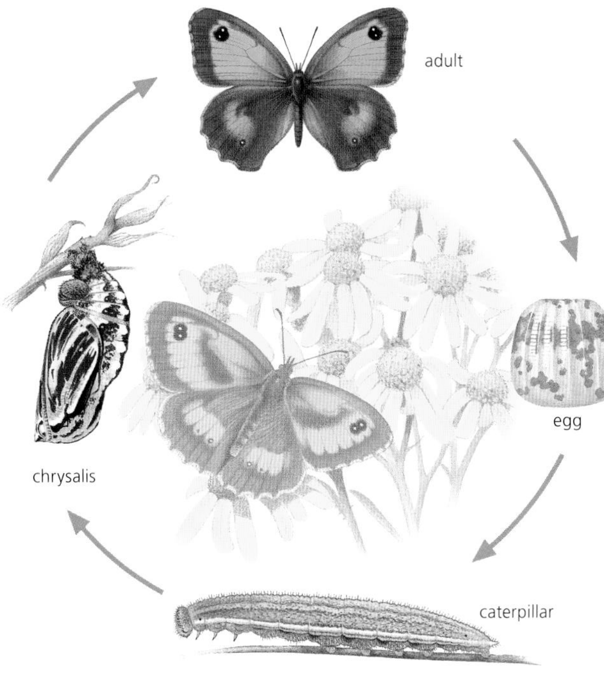

Life cycle of a Gatekeeper

of each species' illustrations shows the typical dates when each stage can be found. In some cases, this varies with the warmth of the season and, to a lesser extent, in different parts of the butterfly's range. In general, emergence dates are later, and there are fewer generations, in the north.

Like the egg of a bird, that of a butterfly represents a period of transition, in which a single microscopic embryo develops and grows into a little caterpillar by absorbing nutritious liquids deposited within the eggshell by its mother. Although little larger than a pinhead, the eggshells of most butterflies have intricate patterns when seen close up, with radiating ribs, spines, protrusions and sculpturing, each topped by a small depression, the micropyle, through which air penetrates to the embryo. We illustrate the egg of every species magnified 15-22 times – about the size they appear when observed through a strong hand lens. It is well worth searching for butterfly eggs in the wild: many of them have a rare beauty, and although tiny, they are generally 25-50 times more numerous than the adult butterfly. In a few species, notably the Brown Hairstreak, they are the easiest stage of the life cycle to find.

The caterpillar is perhaps the most interesting of the three immature stages. It is the only period in the life cycle in which growth occurs, and the necessity to expose itself when feeding makes many caterpillars vulnerable to enemies. Much of the behaviour and appearance of the caterpillars of different species is due to their different ways of avoiding predators. We have placed particular emphasis on illustrating and describing the interactions of ants with the caterpillars of blues and some hairstreaks, for this is an area of growing fascination that is also of the utmost importance for the survival of roughly a sixth of all our species.

Most caterpillars live singly, and are wonderfully camouflaged against their foodplants. Others live communally and are generally dark, allowing their dusky bodies to warm up *en masse* in the sunshine, reaching temperatures that would be unobtainable by single individuals, and permitting them to feed in cool weather. All caterpillars are tiny when they emerge from the egg, but grow rapidly, shedding their skin several times before pupating into a chrysalis.

For many naturalists, the chrysalis is the most beautiful stage in the life cycle, and the detailed paintings of these are one of the features of this book. Apart from those lycaenid chrysalises that are found in ant nests, all exhibit some form of camouflage. This ranges from mimics of dead, dew-spangled leaves in the case of the fritillaries, to living leaves in many browns, and to a bird-dropping in the case of the Black Hairstreak.

The adult butterfly

It is easy to imagine that an adult butterfly is a purposeless creature that lives out its short life by flitting aimlessly between flowers. Nothing could be further from the truth. Almost every action has a function, and it adds considerably to the pleasure of watching these insects if one can interpret what they are doing.

Having emerged from a chrysalis and inflated its wings, the main goal in a male butterfly's life is to pair with as many females as possible. The female, on the other hand, usually mates shortly after emergence, and spends the rest of her life avoiding males and searching for places to lay eggs. But before they can do anything, both sexes must raise their body temperatures to around 32°C.

Butterflies have the reputation of being cold-blooded animals that are active only in sunshine. In fact, they regulate their body temperatures to remarkably narrow limits by adjusting their wings to absorb or reflect the sun's heat. Temperature regulation occurs throughout the day, often at the same time as other activities. However, it dominates a butterfly's behaviour in marginal weather, especially first thing in the morning or late afternoon; these are often the only times when many species fully open their wings, in order to warm up.

Courtship and mating

Once warm, the males search for mates. Many fly to a particular spot where virgin females also gather. This may simply be around a distinctive part of their site, as in the case of the Small Heath, or it may be some distance away. A good example of the latter is shown by the Purple Emperor, which breeds at very low densities over considerable areas of countryside. If the sexes are to have any chance of meeting, it is necessary for them to gather at the highest point in the neighbourhood, a process known as 'hilltopping'.

Having arrived at a suitable area, most male butterflies hunt their mates in one of two ways. Some, such as the Duke of Burgundy, merely perch in a prominent place and launch themselves after any insect passing by. Others, such as the two wood whites, fly ceaselessly through their sites, searching every nook for hidden females. Some species behave in both ways, depending on the time of day, the season or the habitat.

Once an unmated female is discovered, there follows a courtship ritual which is prolonged and complex in species such as the Grayling and Silver-washed Fritillary, but brief and simple in others. The act of mating usually lasts about 30 minutes. During this time, a packet of sperm is transferred from the male to the female's body, usually sufficient to fertilise every egg that is laid. Many males also transfer a 'nuptial gift' of soluble foods alongside the sperm sac. The females of few species mate more than once, and when they do it is generally to obtain additional nuptial gifts rather than a fresh supply of sperm.

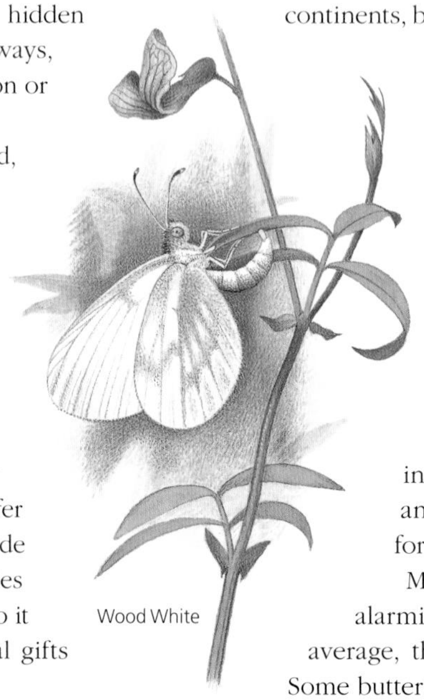

Wood White

Foodplants, habitats and distribution

The females of most species take enormous care during egg-laying, placing their offspring in situations where they are best adapted to survive. We describe this in detail for most species because it is usually the key to why certain species are rare, restricted or declining, whereas others are widespread and common. In most cases, the eggs are laid on a particular growth form of the one (or few) species of plant(s) that are eaten by the caterpillar. Many species, such as the Adonis Blue and Lulworth Skipper, restrict breeding to extremely warm parts of their sites. Colonies occur only where these specific breeding conditions exist, and, if there are many plants suitable for egg-laying present, the odds are that the colony of that butterfly will be large.

Many species of butterfly are surprisingly sedentary, and live in small, discrete areas from which they seldom stray. Others have no fixed abode, but roam the countryside selecting different places for different activities. A few butterflies, such as the Painted Lady, even migrate across continents, breeding in the British Isles during the summer, with their more numerous offspring migrating to North Africa in autumn.

The distribution of British butterflies has been recorded with unprecedented precision in recent years, although there are still gaps to fill in parts of Ireland. We include a map showing the current distribution of every species, based on the latest (post 2006) unpublished atlas data provided by Butterfly Conservation. Resident species and regular breeding migrants are shown in green, rare migrants are shown in blue, and those species that are extinct have their former distributions shown in orange.

Many species of butterfly have declined alarmingly in recent years, at a steeper rate, on average, than our native plants or breeding birds. Some butterflies now depend wholly on conservation bodies for their survival. The main causes of decline are described in detail in this book, in the hope that this will help the owners and managers of breeding sites to rectify some of these changes. Gardens can often be improved for butterflies, but unfortunately these attract only the most common and mobile species. The most effective way to help the survival of our butterflies is to join voluntary conservation bodies. Foremost among these are Butterfly Conservation, The National Trust, and the many county Wildlife Trusts. Details of these and other useful organisations are given at the end of our book (see page 284).

Butterfly names and families

The Latin names of the 72 species of butterfly that regularly occur in the British Isles are under constant revision as modern genetic techniques reveal unexpected levels of relatedness, especially between genera. At the time of writing, there is little consensus in the names for several species. Here, we follow names given in the December 2013 *Checklist of the Lepidoptera of the British Isles* by Agassiz, Beavan and Heckford. There have been changes, too, to the number of major families into which species are grouped, and again there are some disagreements. The consensus, at present, is that the British species belong to six separate families of closely related Lepidoptera. The individual members in each group possess similar family characteristics, which are briefly summarised here.

Family HESPERIIDAE: The skippers

These are the most moth-like and primitive of the families of butterflies. All are small, lively insects, with broad, hairy bodies and an unusually large gap between the bases of the two antennae. There are eight resident species in Britain. Five of these – the 'golden' skippers – sit in a characteristic pose, with the fore- and hindwings held at different angles.

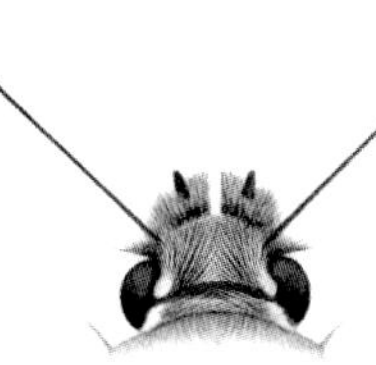

▼ The heads of skipper butterflies have a large gap between the bases of the antennae.

Family PAPILIONIDAE: The swallowtails

This magnificent family of butterflies includes the swallowtails, apollos and the wonderful birdwings of Australasia and the Far East. The family has just one representative in the British Isles, the Swallowtail. Like most of its relatives, this is a large, striking insect with distinctive long wing-tails.

Family PIERIDAE: The whites and yellows

These are medium-sized butterflies, with predominantly white or yellow wings. Most species roam widely through the countryside, rather than living in identifiable colonies. Their eggs are invariably tall and cylindrical, a characteristic most easily seen in those familiar pests of vegetable gardens, the Large and Small ('Cabbage') Whites.

Family LYCAENIDAE: The coppers, hairstreaks and blues

About a third of the world's butterfly species belong to this family. They are small, active butterflies with a metallic sheen to their wings. Twenty species have been recorded from the British Isles. During their development, many have an intimate relationship with ants. These provide protection for the caterpillar or chrysalis in return for a nutritious, sugary secretion. Lycaenid butterflies have small, bun-shaped eggs, slug-like caterpillars and short, stumpy chrysalises.

Family RIODINIDAE: The metalmarks

This beautiful family of butterflies reaches its greatest magnificence and diversity in Central America. There is just one European species, the Duke of Burgundy, which is one of the family's less distinguished members. After a period of relegation to being a subfamily of the Lycaenidae, the Riodinidae are now reinstated as a separate family, following modern phylogenetic studies.

Family NYMPHALIDAE: The brush- or four-footed butterflies

This already sizeable family of butterflies has been much enlarged in recent years by the inclusion of the browns, danaids, and some additional exotic families as members of this group. The diagnostic feature is that the first pair of legs on the adult is vestigial, the butterfly walking on two pairs alone. Worldwide, the Nymphalidae are subdivided into between eight and 15 subfamilies, depending on which authority one accepts. For convenience, we treat the British species as three main groups (clades):

◀ The body of a Red Admiral. Nymphalids have only two pairs of functional legs.

The 'Aristocrats' – emperors, admirals, vanessids and fritillaries Contains some of the biggest and gaudiest species in the world. Most of the 18 species in the British Isles possess caterpillars that are protected by sharp spines.

The browns Formerly considered to be a separate family (the Satyridae), this group of 11 British species consists of medium-sized or small butterflies, with small, gleaming eye-spots towards the outer margins of the wings. Most species have a brown coloration.

The Monarch Previously classed as belonging to a separate family (the Danaidae), the Monarch is a distinctive and magnificent migrant that occasionally reaches our shores from North America.

Chequered Skipper

Carterocephalus palaemon

This beautiful dappled skipper is now restricted to Scotland, having mysteriously disappeared from its last English sites in about 1975. Although always considered a rarity, its loss in England came as a jolt to conservationists, for in the 1960s there had been fine colonies on at least four east Midland nature reserves. Its extinction is lamentable because the Chequered Skipper was the most attractive of the English skippers, seen at its best in late spring, darting around Bugle flowers in sunny woodland rides.

Like all skippers, the Chequered Skipper flies in rapid dashes, zigzagging a few centimetres above the ground and changing direction – even reversing – at the slightest obstacle. It can be difficult to follow in full flight, when the wing markings blur into an orange-brown haze. Hopeful beginners sometimes misidentify the Duke of Burgundy for this rarity, but there is no mistaking the real thing. At rest, the Chequered Skipper's wings are much more angular than those of the 'Duke'. It also lacks spots along the base of its lower forewings, and the pale patches on the undersides form a jumbled mosaic with the darker areas, rather than occurring as two neat bands.

Regional differences

It is claimed that English and Scottish Chequered Skippers belong to distinct races, with the former being slightly larger and lighter, and possessing underwings that are more yellow-brown, with creamier yellow spots. We depict specimens from both countries, but in truth this is a variable butterfly and any regional distinction is slight. Nevertheless, with its disjunct distribution and regional differences in ecology as well as, perhaps, in appearance, it has been suggested that British Chequered Skippers derive from two separate colonisations after the last Ice Age, making the former English populations more closely related to colonies in northern France. That was the rationale for using livestock from Belgium woodlands when an attempt was made (unsuccessfully) to reintroduce it to Lincolnshire in 1995. In fact, recent genetic studies by D A Joyce and Andrew Pullin provide no support for the 'two colonisations' theory.

The Chequered Skipper is a handsome insect in all its forms. It emerges in one generation a year, peaking in late

Distribution
Locally common in Argyllshire and Inverness-shire. In England, last seen in the east Midlands in about 1975.

▲ **Male**
English specimen from former colony in Northamptonshire.

▲ **Female**
Northamptonshire specimen; females are generally slightly larger than males.

▲ **Male**
Rare aberrant form from Brigstock, Northamptonshire. Collected in 1933.

▲ **Male**
Typical specimen from Scottish colonies.

▲ **Resting adult**
English female at rest on grass blade.

◀ **Egg** [× 15]
Laid singly on underside of a grass blade.

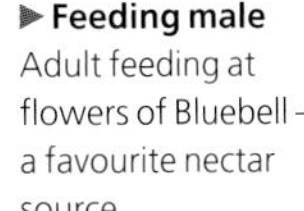

► **Feeding male**
Adult feeding at flowers of Bluebell – a favourite nectar source.

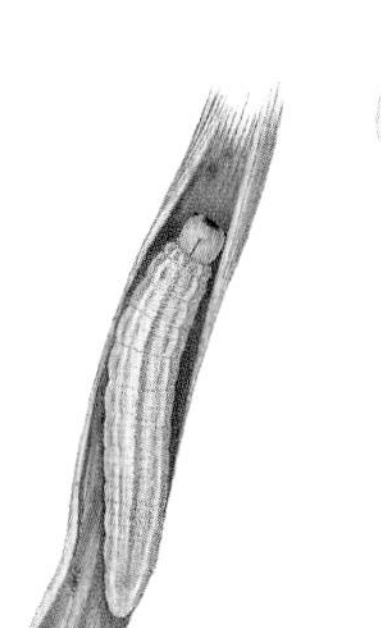

◀ **Spring caterpillar**
Straw-coloured caterpillar in spring, prior to forming chrysalis.

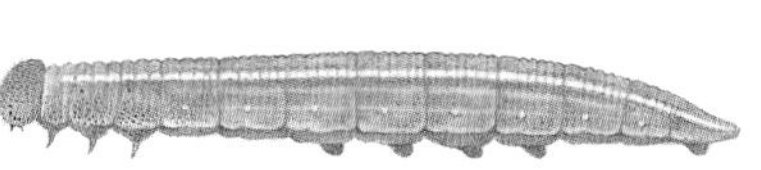

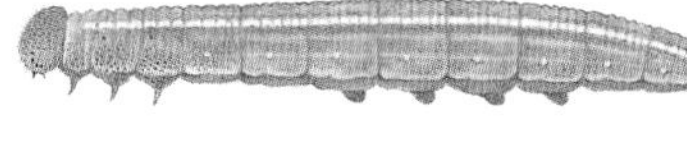

▲ **Summer/autumn caterpillar** [× 2¼]
Caterpillar remains green until hibernation.

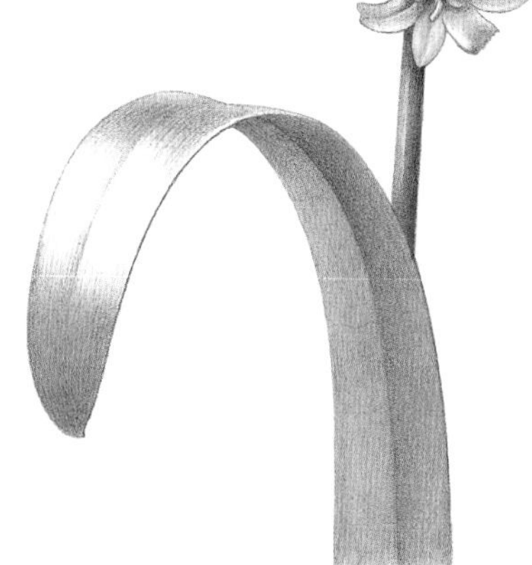

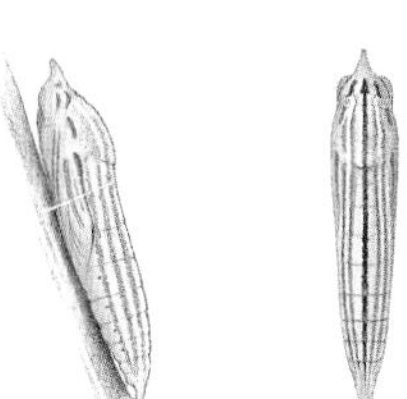

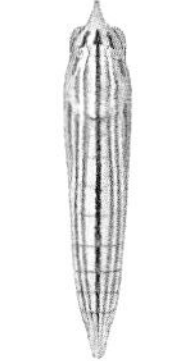

◀ **Chrysalis** [× 1½]
Formed inside a nest of dead leaves; camouflaged to resemble a dead grass blade.

Side view   View from above

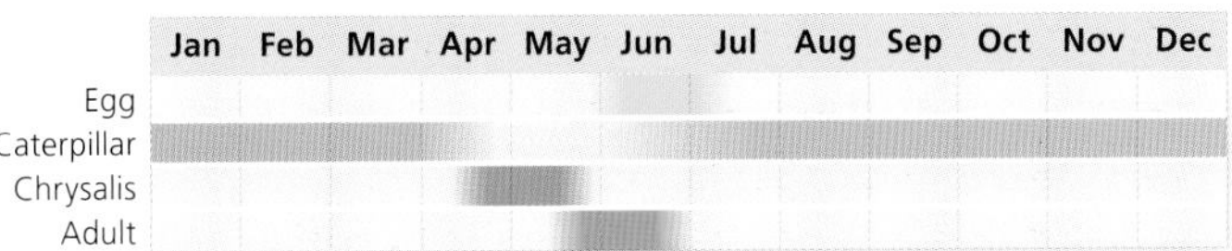

May or early June, depending on the weather. If anything, the Scottish butterflies emerge slightly earlier, but they can be frustrating to find because of the wind and rain that frequently lashes western Scotland in late spring.

Territorial males

The behaviour of this charming butterfly has been well studied by Neil Ravenscroft in Scotland and by Ray Collier in England. English Chequered Skippers were highly colonial, with the adults restricted to small areas of a ride or woodside for the whole flight period. In Scotland they are more free-ranging. One famous colony occurs more or less continuously along an 8km stretch of roadside, and another extends even further beside a loch. This is not to say that all Scottish populations are large. Although Ravenscroft found colonies consisting of hundreds of adults, others contain no more than a few dozen, at least in poor years for the species.

The male is the easier sex to find, for he establishes a territory in the sunshine, perched on prominent vegetation, especially sapling leaves, with wings draped back and open, darting out to intercept every passing intruder in rapid buzzing flights. Although fiercely territorial, males congregate in favourite places for perching, typically sunny nooks beside sheltered wood edges, particularly in south-facing dips in undulating terrain, and within these depressions at spots where the ground is bare or covered by dry leaf litter. These sites provide a comparatively warm micro-climate at a time of year when the weather can be cool and unpredictable.

Except when visiting the males' leks, the females tend to live separately, spreading over wide areas to seek the right grasses for egg-laying. Every naturalist familiar with the butterfly will have noted how the females love to visit flowers. They concentrate on blue species such as Bugle and Bluebell, which were also far and away the favourite nectar source of butterflies in the old English colonies.

A slow development

The Chequered Skipper's egg is a large pale sphere, laid singly on the underside of tall, coarse grass blades. Before they died out, the butterflies in English colonies bred mainly on False Brome in woodland rides and glades, with Tor-grass sometimes used in adjoining grassland. In Scotland, Purple Moor-grass is exclusively used. Purple Moor-grass is one of the most widespread and dominant plants of northern moorland but, as with so many species, the female Chequered Skipper is exceedingly particular over which tussocks she chooses for her offspring, limiting colonies to a very few sites.

Although its Scottish habitat is generally described as peaty bogs and mires, the female Chequered Skipper lays eggs at very low densities on grass growing in warm spots sheltered by scattered birches, Bracken and Bog-myrtle, on ground midway between the drier woodland (where males congregate) and the waterlogged peatlands that dominate the landscape. Purple Moor-grass grows in notably more luxuriant clumps in these intermediate zones, and the females choose the most vigorous specimens, selecting tussocks that remain lush for long enough in autumn for the slow-growing caterpillars to develop. Caterpillar survival is especially high on large, persistent, non-flowering Purple Moor-grass that has a high concentration of nitrogen in the leaves.

Like those of related skippers, the caterpillar moves sluggishly and lives in a tube of grass made by folding a blade double, with the edges secured by little cords of silk. It emerges to feed both above and below the tube, first making characteristic V-shaped notches, then eating more of the blade until the tube is left isolated on the midrib. The Chequered Skipper's caterpillar is one of the slowest growing of any British butterfly. Not until October, after 100 days of feeding, is it fully grown. It then changes from being a rather undistinguished pale green grub to a straw-coloured one, which blends with the hibernation nest that it constructs from dead grass and silk. There the caterpillar remains until the following spring, sheltered from the ferocity of Scottish winters. It reappears in mid-April and, after about a week without feeding, spins another nest of dead leaves where it forms a chrysalis. This also looks much like a withered grass blade.

Discovery and decline

It seems probable that English Chequered Skippers had equally particular breeding requirements to their Scottish counterparts, but their exact habitat may never now be known. The species was always scarce, and there is much uncertainty about the earliest recorded sightings, which were confused by misidentifications and fraud.

A Revd Dr Abbot is usually credited with the first genuine report, in 1798 in Clapham Park Wood, Bedfordshire, and he found another colony at Gamlingay, near Cambridge, a few years later. However, there is some doubt about these records. In his delightful reminiscences, P B M Allen showed that there was a lively trade in false English rarities in the late 18th century, and Dr Abbot himself was duped on at least one occasion. Moreover, he had a reputation for introducing rarities and aliens to Clapham Park Wood, only to 'discover' them soon after.

Despite uncertainties about the Chequered Skipper's original range, English colonies were confined for most of the 19th century to a band of woods and associated limestone grassland that stretched from Oxford to south Lincolnshire, with another group of colonies further north in Lincolnshire.

Most were fairly wet sites, some more or less waterlogged. The Chequered Skipper lived mainly in rides and glades, and one by one these colonies disappeared until, by 1950, the butterfly was largely restricted to Lincolnshire and to the woods of Rockingham Forest in Northamptonshire and Rutland.

There then followed a brief expansion when the Chequered Skipper recolonised some former sites, including Monks Wood and Woodwalton Fen in Cambridgeshire. It remained plentiful on the best sites, so much so that in 1961 R E M Pilcher wrote: 'Castor Hanglands [near Peterborough] has always been the metropolis of this insect. It is still very common ... but no longer enjoys its former abundance. An insect in no apparent danger of extinction but well worth some trouble to maintain in good numbers.' It was, alas, to become extinct on that nature reserve ten years later, having already disappeared from almost every other English site in the decade. Recent attempts to reintroduce the Chequered Skipper failed, and today its monument is the Chequered Skipper pub at Ashton, near Oundle, named half a century ago by Miriam Rothschild to celebrate the greatest prize on her estate.

Scottish stronghold

The situation in Scotland is more encouraging. The butterfly was unknown there until 1942, when Lt-Col Mackworth-Praed astonished the entomological world by announcing that he had found specimens near Fort William, roughly 480km north of the nearest known colony. In fact, a Miss Evans had discovered the butterfly nearby three years earlier, but this had remained a secret.

These two sites held the only known Scottish colonies for 30 years, until an amateur naturalist, J E H Blackie, informed us at Monks Wood that he had found some in Argyll. Surveys were quickly organised, notably by the Scottish Wildlife Trust, and soon about 40 colonies were located. Today, about 50 populations are known, most breeding on slopes beneath open broadleaved woodland, below 200m altitude, and often bordering lochs or rivers.

Despite these recent discoveries, no-one suggests that the butterfly has spread, merely that it had been overlooked in a region that few entomologists visited. It is, however, a heartening chapter in the history of this beautiful skipper, the more so since many Scottish colonies breed in conservation areas, where management regimes have been influenced by Ravenscroft's perceptive studies.

Small Skipper

Thymelicus sylvestris

At least four British skippers are smaller than this species, whose unimaginative name was intended simply to distinguish it from the only other skipper, the Large Skipper, that is really common in the south. It would have been better, in my opinion, to have stuck by Petiver's (1704) names of 'Streaked Golden Hog' for Small Skipper and 'Streakt Cloudy Hog' or 'Chequered Hog' for Large Skipper, but these disappeared in 1766 when Moses Harris published *The Aurelian*.

Although diminutive, this is a creature of considerable charm that weaves between grass heads, flashing gold in the sunshine of high summer. July is the main month to see it, but emergence starts in mid-June in a warm year, and old, faded individuals last well into August in a cool one. There is one generation each year.

Small Skippers live in self-contained colonies, seldom wandering far from the patches of tall grassland that serve as breeding grounds. They are, nonetheless, seen quite often in southern country gardens, for a colony can be supported by just a few square metres of rough grass nearby. Typical colonies contain a few dozen adults, but this can rise to thousands in extensive areas of Yorkshire-fog, the favoured foodplant of the caterpillar.

Flying with a purpose

Small Skippers tend to be secretive, but when they do take to the air their golden wings flash conspicuously in the sunlight. The male is the more active sex, and his powers of manoeuvre are remarkable. At one moment he will weave slowly between tall grass stems, then zip away sideways, only to stop almost instantaneously in mid-air to hover above another clump. Females, by contrast, are sedentary, and perch for long hours among the tussocks, sitting with the two forewings pointing upwards in a V, while the hindwings are held parallel to the ground. The need to feed and to lay eggs eventually forces them into the open. Like many butterflies, each activity is performed separately on different excursions: you will rarely see a female stop to take nectar during an egg-laying flight.

Both sexes, when feeding, visit a wide variety of flowers, but are nevertheless selective about which individual blooms they choose. In one grassland, Dave Goulson noted that they drank nectar readily from ten different species,

Distribution
A common skipper of rough grassland and woods, that has expanded 100km northwards in range since the 1980s.

▲**Male**
Males have a curved black sex-brand on each forewing; markings of both sexes are similar.

▲**Female**
Black sex-brand is absent from the forewings; inconspicuous, spending much time on the ground.

▲**Egg** [× 15]
Laid in groups usually of 3-5. Initially white, the egg then becomes yellow.

▲**Antenna**
Underside of antenna tip is orange, distinguishing it from the Essex Skipper.

◀**Hibernating cocoons**
Newly hatched caterpillars spin cocoons inside grass sheaths, in which they hibernate.

▲**Caterpillar** [× 2¼]
After emerging from hibernation, the caterpillar lives within a tube formed out of a single grass blade.

▼**Chrysalis** [× 1½]
Formed near the ground. Head of the chrysalis has a short, blunt 'beak'.

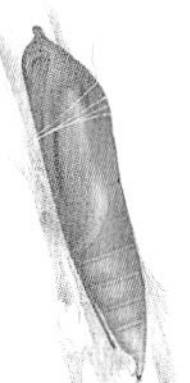

Side view View from above

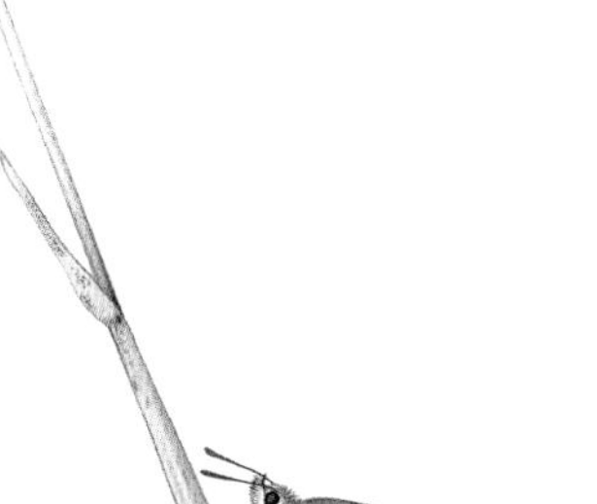

▲**Egg-laying**
Female inserts her abdomen into a grass sheath and lays her eggs in a row with a pumping action.

▲**Feeding male**
Male drinking at a favourite foodplant, Greater Knapweed.

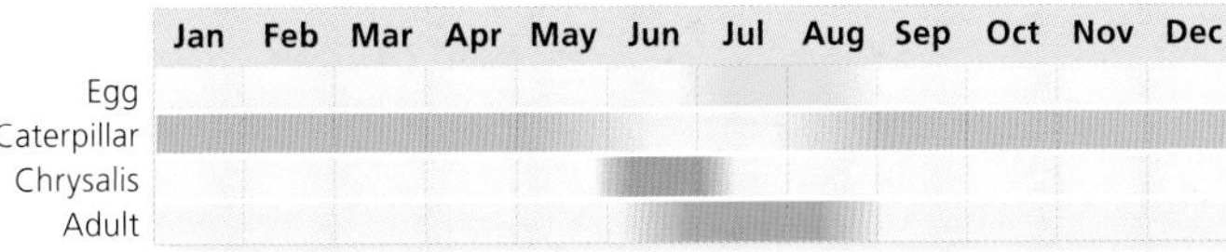

including clovers, Common Bird's-foot-trefoil, restharrows, knapweeds, thistles, brambles and hawkbits. However, they rejected nearly one in 12 of the available blooms, and had a strong tendency to feed on the same plant species several times in succession, displaying an accurate short-term memory for productive flowers. Typical flights contained about eight separate stops to drink, with the butterfly switching to a different plant species only when it encountered a patch where its previous favourite was scarce. Thus, in the long term, Small Skippers utilised most of the available nectar sources on a site.

A specialist of Yorkshire-fog

Female Small Skippers fly rapidly when seeking nectar, but their egg-laying flights are slow, deliberate, and quite easy to spot, as they hover around the tall sheltered clumps of Yorkshire-fog on which they prefer to lay. Each shuffles backwards down the stem, revolving and probing the furled grass sheath with an abdomen swollen with eggs, yet somehow bent double. As she descends, her ovipositor slips inside the sheath, then after a short adjustment to find the perfect position, she lowers both antennae, stiffens, and pumps three to five eggs into the sheath.

The egg is white, turning primrose-yellow, and is much rounder and less robust than the otherwise similar egg of the Essex Skipper. It hatches in August. The caterpillar eats most of its shell and then spins a tiny silk cocoon around itself, still inside the grass sheath. Each row of cocooned caterpillars remains in its stem until April, before emerging to nibble the springtime growth of tender new leaves. From then on the caterpillars live singly, each constructing progressively larger tubes by spinning a Yorkshire-fog leaf double. By mid-June the caterpillar is fully fed, and descends to the base of its grass clump to spin a loose tent of leaves in which the chrysalis is formed. This stage lasts for about a fortnight.

Moving north

Yorkshire-fog is abundant on all types of soil, and Small Skipper colonies occur almost everywhere in England and Wales where this grass is regularly left to grow as tall clumps. On the whole, the Small Skipper is found in taller, lusher grassland than the Essex Skipper, and in more open places than the Large Skipper. It is, nonetheless, common within its range in woodland glades and sunny rides, as well as in a host of other grassland habitats, ranging from overgrown downland and grassy heaths to road verges, cliffs and undercliffs. Although it remains one of the commoner butterflies throughout the south, there is little doubt that this species has suffered much through the elimination of wild grasses on most agricultural land during the past half century; indeed, its trend on monitored sites has been of unremitting decline since systematic recording began in 1976. On the other hand, there was always an abrupt climatic limit to its northern edge of range in Britain, and as the seasons have become warmer, this charming skipper has spread in a continuous block to about 100km north of its traditional boundary during the past 25 years. Today, its range extends just a few miles across the border into Scotland.

Essex Skipper

Thymelicus lineola

This comparatively common southern butterfly was first discovered in Germany as long ago as 1808, but was overlooked by British naturalists, who mistook it for the Small Skipper. Another 80 years were to pass before Mr F W Hawes of Essex realised that three specimens in his collection were in fact members of a different species. So it was that our penultimate resident butterfly was added to the British list, almost exactly a century before the Cryptic Wood White was recognised in Ireland. In a low-key letter to *The Entomologist,* Hawes announced his discovery and astutely predicted that 'If all collections in this country were carefully examined, I have little doubt that other *T. lineola* would be found hidden away in some unexpected corners.' And so, indeed, they were. Even today, many experienced lepidopterists are unable to distinguish between Essex and Small Skippers in the field, and overlook the former. This is surprising, for the distinction between the two is relatively simple to make, and requires no more than persistence and reasonable eyesight.

Points of identification

The trick in identifying the Essex Skipper is to examine the curved tips of the antennae, in order to see the colour of their undersurfaces. This is best done in the early evening, when the skippers bask or roost communally, but it is also perfectly possible during the heat of the day. Adults frequently settle on grass heads or stems with their bodies pointing upwards at 45°. All that is necessary is to creep up to them on all fours, until you are head-on and can look upwards at the antennae. The difference then becomes obvious: the Small Skipper has antennae that are dull orange or brown underneath the tips, whereas those of the Essex Skipper are glossy black, looking rather as if the tips have been pressed into an ink pad.

Many naturalists prefer to net these two species for identification. This is perfectly harmless, for both are tough little insects. The trapped butterfly is extracted from the net and gently held between thumb and forefinger by the thorax, with the wings pressed firmly enough to prevent them from flapping. Some people find that they can also distinguish the two species by the slightly more pointed wings of the Essex Skipper and, in the case of males, from the black sex-brands in the

Distribution
Common in grassland in the south-eastern half of its range; much more local elsewhere, but expanding west and north, and probably much overlooked.

middle of the golden forewings. These are very fine and run parallel to the leading edge of the wing on the Essex Skipper, whereas they are bolder and at a slight angle on the Small Skipper. The differences are, however, slight and variable.

Subtle differences

It is worth describing these differences at length, for our knowledge of the behaviour and distribution of the Essex Skipper is still poor, simply because many entomologists fail to distinguish between the species. We can confidently state, though, that the Essex Skipper lives in sedentary colonies, with the annual emergence of adults occurring throughout July, peaking a good week later than that of the Small Skipper. There is considerable overlap, however, in both flight period and habitat of the two species, which can be seen flying together in many places at the height of summer. Their behaviour is also similar, although many claim that the Essex Skipper's flight is the slower and more purposeful of the two, and that it generally flies closer to the ground. The only exception is with the courting male, which makes a series of steep ascents and dives before pairing with a female back-to-back on a grass blade.

Essex Skipper colonies range enormously in size, from a few individuals that breed on small, rough banksides to the many thousands found on large embankments and dykes. They gather by the score in the late afternoon on good sites, to bask gregariously in the fading sunlight before settling to roost in groups of four or five per grass stem. Next morning they separate to lead solitary lives, the females flying slowly just above ground level through a tall, sparse sward of grasses as they prepare to lay their burden of eggs.

Surviving the winter

Each female is as particular as the Small Skipper in choosing her egg-laying sites. She takes great care to select a firm, tightly furled sheath or hard, dead stem, usually of Cock's-foot or Creeping Soft-grass. Timothy, Tor-grass and False Brome are also occasionally used, but Essex Skippers avoid the Small Skipper's favourite grass, Yorkshire-fog, probably because its sheaths are insufficiently compact.

The egg has a thick, flattened shell, well designed both for insertion into a tight grass sheath and for survival through the winter. The fully formed caterpillar remains in the shell for eight months, unlike the Small Skipper which hatches after two or three weeks. Essex Skipper eggs are laid in strings of four or five, and are primrose-yellow at first but gradually clear to a pearly opaqueness.

The thick, protective shells of the eggs of all butterflies that hibernate are too tough to provide food for their occupants, so the emerging caterpillar merely nibbles a neat exit hole in one end and immediately starts feeding on grass blades. It eats only the tenderest spring growth, and soon spins the two edges of a blade around itself, forming the characteristic tube of a grass-feeding skipper. These tubes, the distinctive V-shaped feeding damage on leaf blades, and the caterpillar itself, can be found by patiently examining Cock's-foot on good sites in June. Apart from having different foodplants, Essex and Small Skipper caterpillars can be distinguished by the stripes and colour of their heads. The head of this species is yellowish with brown lines, whereas the Small Skipper has a pale green head. The adult emerges from the chrysalis after about three weeks.

Habitat preferences

No-one quite knows which habitat this golden skipper prefers. I usually find it in well-drained, wild grassland that has been left to grow fairly tall, but which is not so lush or dense as that occupied by Small Skippers. Typical sites are grassy sea walls, dyke sides, and embankments, especially but not exclusively on chalky or sandy soils.

Despite this preference for dry spots, the Essex Skipper's overwintering eggs can withstand prolonged submersion, as occurs regularly on the coastal marshes of East Anglia and north Kent, where it abounds. It was, indeed, originally thought to be confined to these habitats, but more recent surveys have revealed that it is very much more widespread. Colonies can be found in tall chalk grassland, on grassy banks, in woodland rides, rough corners, and even on strips of road verge throughout most of the south-east quarter of England. They are much more localised in south-west England, but are probably often overlooked. It is always worth checking what appear to be Small Skippers anywhere in the southern half of the country.

Spreading along the roads

There is no doubt, also, that the Essex Skipper has been spreading to many places where it was formerly absent or scarce during the past half century. This expansion was first seen in Sussex and Hampshire, where it was perhaps aided by the development of tall, sparse grassland on chalk downland and verges following the disappearance of rabbits after the mid-1950s.

This skipper has also benefited from the steep embankments and cuttings that border many of our new motorways and trunk roads. For example, in Dorset, where there are many new colonies, the county council was persuaded in 1982 to seed the Bere Regis bypass – a deep cutting through 'improved' arable land – with fine grasses and wild vetches. Cock's-foot and other plants arrived unaided and, after five years, an enormous population of Essex Skippers had developed. Today, they dart in their thousands up to the verge of this busy trunk road.

▲Male
The sex-brands on the forewings are shorter and straighter than those on the Small Skipper.

▲Female
Slightly smaller than the Small Skipper; ground-colour more dull, with darker veins.

▲Colour variant
Adult male, rare *fulva* form.

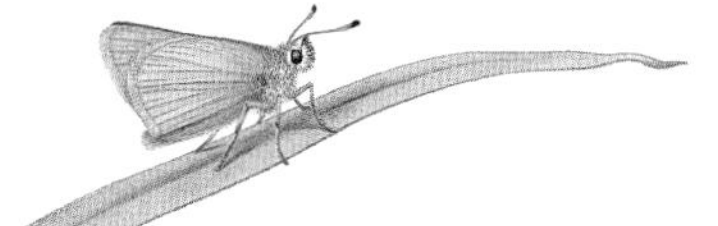

▲Perching male
Similar to Small and Lulworth Skippers; black antenna tips can just be seen at close range.

▲Antenna
Underside of antenna tip is glossy black.

▼Egg [x 15]
Primrose-yellow when laid, turning pearly opaque; the tiny caterpillar develops within the egg, hatching in spring.

Newly laid

Overwintering

▼Feeding female
Adults visit a wide variety of flowers, including thistles.

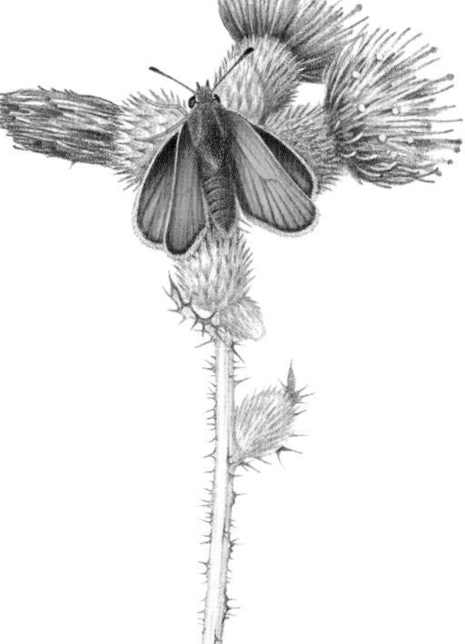

▼Caterpillar [× 2¼]
The brown-striped head capsule of the caterpillar distinguishes it from the Small Skipper.

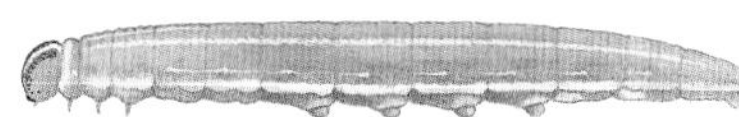

Head capsule [x 4½]

◄Chrysalis [× 1½]
Formed at the base of the foodplant; 'beak' of chrysalis is long, with a rounded tip.

Side view View from above

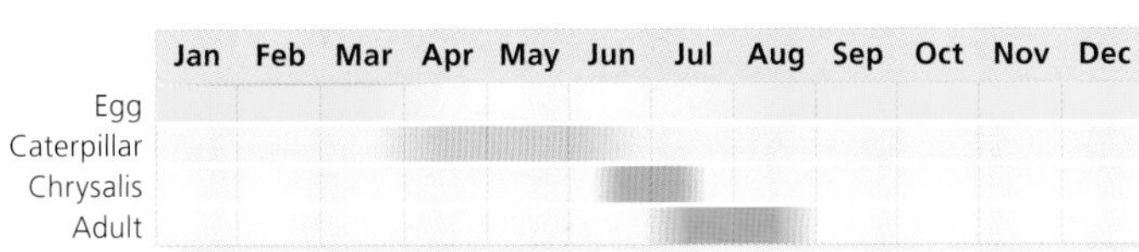

Falling off the back of a lorry

This little skipper's ability to colonise new road embankments has probably also helped its spread in North America, where it was introduced and is known as the European Skipper. It was first recorded at London, Ontario, in 1910, just 21 years after its discovery in Britain. Over the decades it has become widespread and common, so much so that it is now regarded as a pest of Timothy grass. Although less sedentary as an adult than the Small Skipper, the most likely means of its dispersal is through the transportation of hay crops to distant states. In one study, about 5,000 hibernating eggs were found per bale of hay.

It is likely that the same thing happens in Britain. As farms increasingly become more specialised, the shipping of hay has become commonplace, often over considerable distances. This happened on an exceptional scale in the drought years of 1975 and 1976, when convoys of hay were transported from the heart of the Essex Skipper's domain – East Anglia – to the parched dairy farms of the south-west. As the hay lorries rumbled down southern bypasses, it is easy to imagine the butterfly's eggs being scattered along the grass verges, establishing new colonies along the way.

Whatever the mechanism, this skipper's southern and westward expansion has continued in Britain, colonising most of Dorset and Somerset in recent years and reaching Wales in 2000. Remarkably, it was discovered in Co Wexford, in 2006, and has subsequently been found in large numbers on several neighbouring sites: how and when it arrived there is again unknown, but it is heartening that, in an age of declining populations, Ireland's list of resident butterflies should have increased by three species in the past 15 years (the Comma and Cryptic Wood White being the others). There has also been a substantial movement north on its eastern flank during the past 25 years, from its former boundary on the Wash to the Humber today. This, no doubt, results partly from the warming climate, and helps to make this one of the few British butterflies that has genuinely increased in recent years, despite the intensified land use in our country.

Lulworth Skipper

Thymelicus acteon

The tiny Lulworth Skipper is found only on sheltered downs, cliff-tops, and crumbling undercliffs in the extreme south of England. Today, no more than 60 colonies are known, a decline of about 25% from the high-water mark in the last century, when numbers soared for three decades after grazing more or less ceased on most of its rough grassland sites, before gradually being re-imposed. Nevertheless, its late 20th-century heyday apart, there is good reason to believe that the species is more abundant than at any time since its discovery at Durdle Door in 1832 by J C Dale, squire of Glanvilles Wootton, in north Dorset.

Any increase by a British butterfly is encouraging these days, but it is particularly pleasing in this case, for although the Lulworth Skipper is common enough on southern European limestone, it is a local and severely declining species across the northern half of the continent, to an extent that it is now classed as being Vulnerable in Europe as a whole. It is already extinct in the Netherlands.

This is both the smallest and darkest of our golden skippers. It is also one of the latest to emerge. Apart from its small size, the male Lulworth Skipper can be distinguished from other golden skippers by its dark, dun-coloured wings that are tinged with olive-brown. The tiny female has one distinctive circle of golden marks on each forewing, a pattern that has been aptly likened to the rays around the eye of a peacock's feather. The same marks are sometimes faintly visible on the male.

Painstaking research

The young stages of this attractive skipper were painstakingly studied by F W Frohawk at the turn of the last century, but little was known of the natural history of the butterfly in the wild. In the late 1970s I had the pleasure of filling some of the gaps, helped by Robert Smith, David Simcox, and Chris Thomas, when we surveyed all but a handful of the colonies in Britain. A second survey, led by Nigel Bourn 20 years later, focused on the population structure of the skipper and the quality of habitat on surviving or unoccupied patches. A third, in 2010, again by Bourn and colleagues at Butterfly Conservation, provided state-of-the-art information.

Distribution
Almost entirely confined to south-facing chalk and limestone hills, cliffs and undercliffs between Weymouth and Swanage.

During the first study, we analysed the habitat of each colony to see why some contained hundreds of thousands of adults whereas others supported fewer than 50. We spent long summer days catching the adults in order to paint different combinations of coloured spots on their wings, so that every individual could be distinguished at a later date. By doing this, it was possible to determine the numbers present, the lifespan of the adults, and whether they migrated between breeding sites along the coast or remained in self-contained colonies – all information that is vital for the conservation of any butterfly.

By marking over 400 butterflies, we found that Lulworth Skippers live in discrete colonies, with little or no detectable interchange between neighbouring sites. Recent genetic work on central European colonies confirmed that this is a considerably more sedentary species than the Small or Essex Skippers, its close relatives, and that Lulworth Skipper populations are apt to become inbred.

Careful selection

The adult skippers fly only in sunshine, but are then extremely active, darting and weaving just above the grass or making rapid hops from one flower to the next. Wild Marjoram is a great favourite, and so small is this skipper that five or six individuals can often be seen jostling together at the same nectar-rich flowerhead. Typical adults live just seven to ten days – a normal lifespan for any butterfly that does not hibernate at this stage. No elaborate courtship was observed. Indeed, their behaviour is unremarkable in all aspects except when egg-laying, over which the females take immense care.

The egg-laying female Lulworth Skipper fusses around tall grass clumps like a broody hen. She invariably selects Tor-grass, but lays only on large, flowering tussocks that are growing in sunny, sheltered nooks. I have never found eggs or caterpillars on any clump that was under 10cm high, and their preferred grasses are 30-50cm tall. Having found a suitable tussock, the female crawls backwards down a flower sheath, probing the crease in the same way as the Small Skipper. She inserts a row of up to 15 eggs within the sheath. The eggs hatch after about three weeks and the newly emerged caterpillars almost immediately hibernate, each spinning a little cocoon around both itself and the remnants of the eggshell. These white silken beads can be found by splitting open the old brown stalks on suitable plants at any time in autumn and winter.

In spring, the little caterpillar bores through both the cocoon and plant sheath to feed on tender young grass blades. Each lives separately within a tube made by fastening two edges of a blade together with stout cords of silk. These are quite easy to find in May and June on good sites, and can be spotted by their feeding damage both above and below the tube, which start as V-shaped notches in the blade and end up as an isolated tube on the midrib. The caterpillar constructs fresh tubes as it grows larger, but often rests exposed along a blade in its final days. The chrysalis is much harder to find, being formed inside a loose nest of grass and silk spun deep within a tussock of Tor-grass. It lasts for about a fortnight.

Warmth and shelter

The temperature in a grass clump is coolest near the ground and warmest halfway up, below the windswept tips. Lulworth Skipper caterpillars live in the warmest zone.

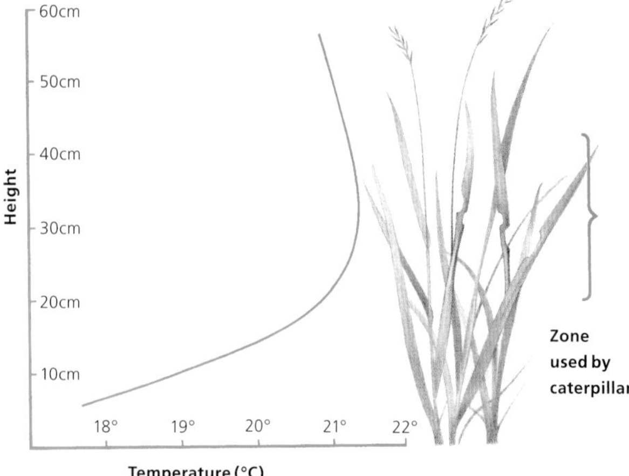

A Dorset speciality

The Lulworth Skipper reaches the northern limit of its range in England, and can live only in the warmest and most sheltered places near the south coast. Unlike other golden skippers, it has not shifted north during the warmer springs and summers of recent decades. Most of its sites are steep, south-facing downland or cliffs, and all contain Tor-grass in great abundance. As expected from its egg-laying preferences, this skipper is by no means ubiquitous in places where the turf is dominated by Tor-grass, even in the Isle of Purbeck. Colonies are restricted to stretches where the grass is left to grow into the tall clumps needed for egg-laying. Small colonies exist where there is just a scattering of suitable plants, but vast numbers develop if the vegetation becomes overgrown or dense. One colony that I studied increased by 20-fold in just four years after the farmer reduced grazing on it. This dependency was confirmed by Nigel Bourn's recent surveys, which found a strong correlation between the density of Lulworth Skippers supported by a site and the quality of the habitat, represented not by the abundance of Tor-grass present but by the number of plants growing in the optimum form for egg-laying. Isolated sites were also less likely to contain a colony.

It is changes in grazing regimes that are responsible for

▲ Male
This is the smallest and darkest of the 'golden' skippers. Underwings are tinged with olive-brown.

▲ Female
Each forewing has a characteristic circle of golden marks. Female's underwings are less green than those of the male.

▲ Colour variant
Female form *alba* – a rare example of variation.

▲ Resting male
Male resting on grass, with wings held together.

◄ Egg [× 15]
Laid in rows in flower sheaths of Tor-grass; hatching takes place after three weeks.

▼ Caterpillar [× 2¼]
Feeds only on Tor-grass, retreating to the safety of a grass blade tube when not eating.

Head capsule [x 4½]

◄ Caterpillar tube
Constructed by fastening the edges of grass blades together with silk; fresh tubes are made as the caterpillar grows.

◄ Chrysalis [× 1½]
Concealed deep in a tussock of grass; has a long and pointed 'beak'.

Side view View from above

	Jan	Feb	Mar	Apr	May	Jun	Jul	Aug	Sep	Oct	Nov	Dec
Egg												
Caterpillar												
Chrysalis												
Adult												

the welcome increase and recent decline of this skipper. In the 19th century, most known colonies bred on unfarmed undercliffs, where periodic falls created a moonscape overgrown by giant tussocks of Tor-grass among a mosaic of scrub and shorter plants. A favourite Victorian collecting ground was the Burning Cliff in Ringstead Bay, Dorset, where oil and gas released by a fall in 1826 had burnt for three years. Thirty years later, a Mr Pretor boasted of collecting 80 Lulworth Skippers in an hour's hard netting: it would be easy (although illegal) to do so today, for these slumps continue to support large numbers of butterflies down to the shoreline. However, little of the downland that adjoins the Dorset coast contained colonies in Victoria's reign, almost certainly because most was heavily stocked with sheep. Much of Dorset's sheepwalk was abandoned in the past century, for it was generally unprofitable to 'improve' the pasture on these steep slopes by ploughing or spraying. Nevertheless, in many areas large numbers of rabbits kept the turf closely cropped, to the great advantage of most butterflies and plants, but to the distinct disadvantage of the Lulworth Skipper.

Improving fortunes

A dramatic shift in fortunes occurred in the 1950s when rabbits were all but eliminated by myxomatosis. Not only did Tor-grass spread unchecked, at the expense of less vigorous grasses and herbs, but it was now free to grow into the large tussocks preferred by the Lulworth Skipper for egg-laying. From then on, this skipper both increased in numbers and spread to adjoining land. However, it is not known to have colonised any new region of Britain where tall Tor-grass is abundant, probably because it is too poor a migrant to reach distant sites. For example, large stretches of the North Downs near the Kent coast now look suitable.

At its peak in the 1980s, there were about 90 colonies of the Lulworth Skipper, whereas fewer than 40 were known before the Second World War. Today, with the return of rabbits and domestic grazing, perhaps 60 colonies exist, including some that have recently colonised Portland. As is clear from the distribution map, the vast majority are in south-east Dorset, confined to uncultivated stretches of chalk and limestone. During the first survey we found three apparent exceptions – on acid heathland and heavy clays – but these all proved to be butterflies that were breeding on narrow strips of chalk rubble that had been laid as ballast for the small quarry railways that once criss-crossed these soils.

The finest colonies of Lulworth Skipper can be found on or very near to the coast between Weymouth and Swanage. The largest are just east of Lulworth Cove itself, where vast acreages of ideal grassland have developed on the extensive army ranges. During the 1970s, nearly 400,000 adults emerged annually on Bindon Hill, with perhaps a million individuals on the ranges as a whole. Today, only a tenth of those densities occur, yet they still make a beautiful sight in high summer, monopolising every Marjoram bloom that is available. On a few coastal sites, for example at Dungy Head, between Durdle Door and Lulworth, adult skippers may be seen as early as late May, more than a month before their appearance elsewhere.

A second string of colonies runs parallel to the Purbeck coast, but a few kilometres inland. Most lie on the southern scarp of the Purbeck Hills, and most contain a few tens or hundreds of adults. Elsewhere in Britain there are two colonies near Burton Bradstock, west of Weymouth, but the butterfly is now extinct on former sites on the Devon coast.

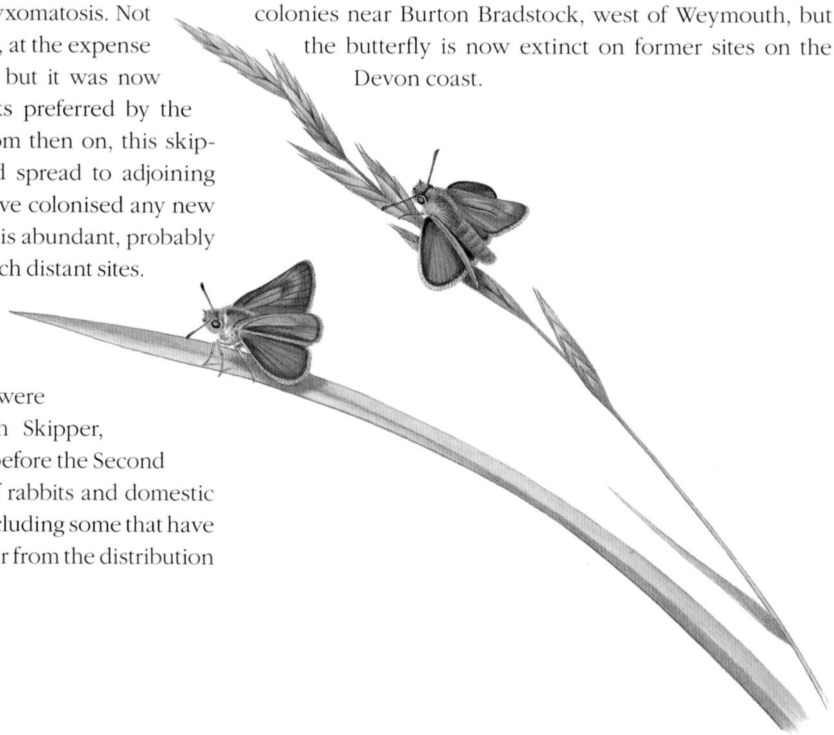

Silver-spotted Skipper

Hesperia comma

This handsome skipper is common on limestone throughout central southern Europe, but is rare in Britain, where it is confined to the warmest southern downs. Yet where it does occur, it can be abundant. The adults make a charming sight in high summer, darting between flowers above the sparse, springy turf, or basking in the August sunshine on crumbly chalk scree. This is one of our latest butterflies to emerge. The adults are first seen in late July following a warm summer, but more often appear in the second week of August. They peak towards the end of the month and then rapidly decline, although a few tattered individuals may last late into September.

A classic ecological study

With Chris Thomas and David Simcox, I studied the British colonies during the 1970s and early 1980s, at a time when this skipper had declined to about 20 downland sites, supporting 68 mostly small populations, and seemed destined for extinction in our country. We aimed to understand its biology and habitat requirements sufficiently well to reverse the decline. Our initial results were much extended in subsequent years by Rob Wilson, Jane Hill, Owen Lewis and Chris Thomas, who used the Silver-spotted Skipper to explore new questions in landscape and climate-change biology, making this butterfly one of the classic study-systems in European ecology.

Adult Silver-spotted Skippers live about six days on average, and are inactive in cloudy weather or when the air temperature falls below 20°C. They require warmer conditions than any other British butterfly and, even on hot days, spend most of their lives basking in hoof-prints, along paths, or on patches of scree, in fact anywhere where the sun can bake the ground. Like all golden skippers, they sit with the fore- and hindwings held at different planes.

Distribution
A rare species that has recovered some lost ground in the centre and east of England's chalk regions during the past 25 years.

Courtship and egg-laying

The adults' flight is rapid and buzzing, and like all skippers they have the ability to dart forwards, sideways or backwards at high speed. They make numerous visits to late-summer flowers, especially concentrating on Dwarf Thistle, knapweeds, yellow members of the daisy family, and Autumn Gentian. They also fly during courtship and egg-laying, although males do not actively seek the females, but instead perch in sunlit spots waiting for a mate to pass by. There is then a rapid, tumbling aerial courtship before the male drives the female to the ground. He crash-lands beside her, and for a minute or two there is a curious scene when the two butterflies stand alongside each other, the male twitching and slightly to the rear, before mating takes place.

Once mated, the female makes slow, investigative flights a few centimetres above the turf as she searches for spots in which to lay. She deposits eggs singly on Sheep's-fescue, the caterpillar's only foodplant, but is extremely fussy over the choice of plants. First, she alights in a warm patch of bare ground. Then after waiting a few seconds, she starts a jerky walk that takes her around or just into the surrounding vegetation, where she lays her curious pudding-basin of an egg. In the late 1970s we analysed the exact position of over 800 Silver-spotted Skipper eggs. We found that females rejected the vast majority of Sheep's-fescue plants, restricting themselves to a small subset growing as vigorous fine tufts, ideally no more than 2.5cm in diameter, and with about three-quarters of their edges abutting onto bare ground or scree. These conditions prevail on sites that have been heavily grazed in the recent past. But the females also avoid any hummocks of Sheep's-fescue that are under 1cm tall and that have had their nutritious growing tips nibbled off, choosing plants that have been left to grow up without grazing during the summer months.

The ideal habitat

We defined this butterfly's idea of heaven as a thin-soiled, south-facing southern down that had a shortish, sparse sward in which about 40% of the ground was bare, 45% consisted of little Sheep's-fescue plants, and the remaining 15% was composed of flowers. We explained this curious constraint on breeding by the skipper's need for the hottest, least shaded fescue-swards available in Britain, knowing that under warmer climates in southern France colonies were common in much more overgrown turf. And, indeed, during the warmer years of the past two decades, Chris Thomas and co-workers found that the butterfly became less particular when egg-laying in Britain, using fescues on less southerly aspects and plants with as little as 20% of their edges abutting bare ground.

This greater tolerance of breeding habitat is a bonus because it has roughly doubled the areas of grassland available to egg-laying Silver-spotted Skippers in recent years, over and above the extensive restoration of its habitat. During the cooler decades of my youth, sites needed regular heavy grazing or perturbations. Unfortunately, most of the butterfly's traditional breeding sites had largely been abandoned by farmers in the early 20th century, causing a spate of extinctions from the mid-1950s onwards when rabbits also disappeared as a result of myxomatosis. Ironically, Sheep's-fescue increased on many sites, but without rabbits the small, sparse plants grew into dense clumps, which were too cool for egg-laying. From the early 1980s onwards, however, we knew how to advise conservation bodies and farmers to manage the few surviving sites, and from then on numbers increased in the core populations. Crucially, the colonisation of restored patches at last outnumbered extinctions. The return of rabbits was an unexpected boon, and by the 1990s there was adequate grazing on many former sites.

Regardless of climate, the eggs of the Silver-spotted Skipper are easy to find once you know which type of Sheep's-fescue the butterfly prefers. They are large, off-white and conspicuous, and are attached to the fine, wiry leaves of the grass. Up to ten eggs can be found on the same little tuft on the best sites, but it is wise to search in early autumn, for many eggs drop off to complete their hibernation on the ground.

The caterpillar bores out through the tough eggshell in March, and immediately constructs a small nest of silk and tender fescue leaves. It lives in this and similar nests for the next 14 weeks, usually alone but sometimes with up to three other caterpillars. Nests are not particularly easy to spot, and are disappointing when found, for they contain one of the ugliest caterpillars of all – a brown-green wrinkled maggot that stretches out of its nest to browse on the surrounding leaf-tips. When gently blown upon, the caterpillar darts back into its nest to press against the bare ground – a useful escape mechanism for a creature whose habitat is subject to heavy grazing.

When fully grown, the caterpillar abandons its nest and starts wandering in search of a denser tussock in which to pupate. It spins a coarse cocoon at ground level in a grass tussock, and lives as a chrysalis for about a fortnight.

A reluctant disperser

The status of this skipper has caused great concern over the years. It appears always to have been scarce, although Richard South, a leading lepidopterist at the start of the 20th century, considered that colonies then bred 'on most of our chalk hills'. However, earlier entomologists – who knew it as the 'Pearl Skipper' or 'August Skipper' – list few sites, and there is doubt over some, for it was frequently confused with the Large Skipper.

There is no doubt that this butterfly was much more widespread in the past. Occasional colonies were known from as far north as Yorkshire, whereas today the Chilterns mark its northern limit. At present, there are probably many hundreds more patches of chalk or limestone grassland that

▲**Male**
Each forewing has a black sex-brand made up of scent scales.

▲**Female**
The forewings lack the black sex-brand, otherwise similar to male.

◄**Colour variant**
Undersides are deeper green in some adults.

►**Resting female**
Female on leaf of Salad Burnet, a typical plant of the butterfly's downland habitat.

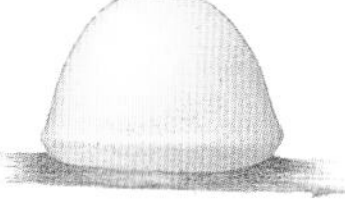

▲**Egg** [× 15]
Laid singly on a blade of Sheep's-fescue; no other grass is used.

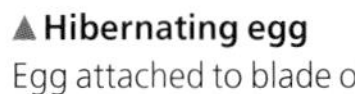

▲**Hibernating egg**
Egg attached to blade of Sheep's-fescue.

▲**Caterpillar** [× 1½]
Lives within a nest of silk and grass blades, protruding its head to feed, and retreating if disturbed.

▲**Chrysalis** [× 1½]
Formed at ground level in a grass tussock; adult emerges after about two weeks.

▲**Basking adults**
Male (right) and female warming their bodies by basking in sunshine amongst chalk scree.

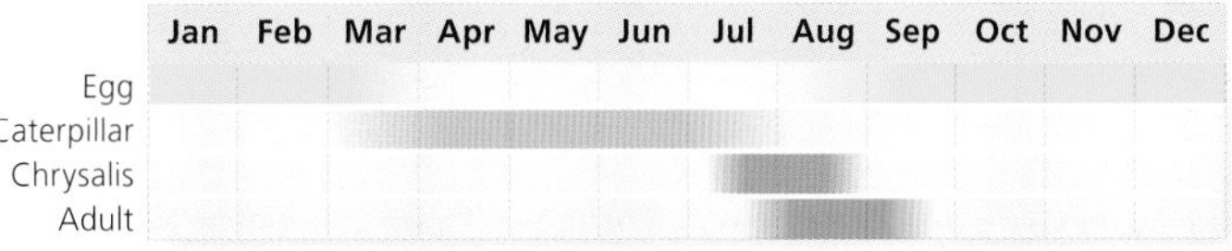

could support this warmth-loving butterfly, thanks to the triple benefit of a warmer climate, rabbits and agri-environmental grazing schemes. The fact that few do is explained by the Silver-spotted Skipper's second constraint: its reluctance to disperse far from current colonies.

Early studies showed that Silver-spotted Skippers seldom emigrate far from isolated downs, although there was significant mixing between nearby sites on the same escarpment, for example along the North Downs west of Dorking. Yet it still had great difficulty in colonising new sites if this meant flying 500m or more over intensively farmed land. The recovery of the Silver-spotted Skipper has thus been painfully slow. By the 1990s, only those sites that were very close to an existing population had been recolonised. To take an extreme example, after suitable habitat had been restored to Old Winchester Hill, in Hampshire, it took 17 years for the butterfly to cross the Meon Valley – a distance of just 3km – which separated it from Beacon Hill, where more than a thousand adults flew every year.

Stepping-stones to recovery

The slow stepping-stone expansion of the Silver-spotted Skipper has been carefully monitored by Zoe Davies and colleagues in recent decades. They found that the skippers from East Sussex, where egg-laying spots were few and far between, had larger thoraxes in relation to their abdomens than those from the North Downs, where females need only to fly short distances to find suitable habitat. Broad thoraxes are associated with more free-flying individuals, for example in the Swallowtail and Large Blue, and it is no coincidence that this skipper has spread greater distances to reach restored sites in East Sussex than it has in the North Downs. The furthest single 'jump' to found a new colony in Sussex was 9km, but this is exceptional. Overall, about 90% of recent British colonisations have been to downs which lie within 10km of one of the 1982 refuge colonies, and more than 80% of the apparently suitable grassland that lies between 5km and 15km from a 1982 population has yet to be colonised.

Nevertheless, the outlook for this butterfly is encouraging. Today, there are more than 250 colonies of Silver-spotted Skipper in England, roughly five times as many as during its nadir of the 1970s, and several sites support populations of a few thousand adults. Current strongholds are in the Chilterns, Hampshire, the North Downs between Guildford and Reigate, East Sussex to Beachy Head, and south-east Kent.

Large Skipper

Ochlodes sylvanus

The Large Skipper is a common butterfly in England and Wales, and can be found in almost any sheltered patch of grassland that contains tall clumps of native coarse grasses. But it becomes a rarity as one travels further north, extending no further than the Scottish borderlands in the north-east and Ayrshire in the north-west. It has never reliably been recorded in Ireland.

Like all our golden skippers, the adult adopts a characteristic basking stance, with the forewings and hindwings held apart at different angles. It is easy to identify, being the only common skipper to have mottled rather than clear golden wings, and the markings are more or less constant throughout the butterfly's range; it was indeed called 'The Chequered Hog' or 'Streakt Cloudy Hog' – pronounced 'og' – by early collectors. Hopeful beginners sometimes mistake old, faded adults for the Silver-spotted Skipper or the Lulworth Skipper, but both have different colours or patterns, and are restricted to a few sites in southern England. If in doubt, it is always safest to assume that any mottled golden skipper is a Large Skipper.

A long summer emergence

There is one generation of Large Skippers a year, which lasts for most of the summer. Given a warm spring, the first males emerge in mid-May, building up to a peak in mid- to late June and flying throughout July. Small numbers usually survive into August, and a few battered individuals linger on until mid-September, but these are pale and hardly recognisable from their scaleless wings.

The Large Skipper is one of the few butterflies that are regularly seen in gardens. This generally means that there is a colony in an adjoining field or verge, for the adults seldom fly far from their self-contained breeding sites and rarely breed in gardens. Typical colonies contain just a few dozen adults. However, much larger numbers can develop under ideal conditions, such as during the first five years after a large area of woodland has been cleared, replanted, and left to grow wild.

Distribution
Common in rough grassland and woods throughout its range. Scarcer in the north, and absent from most islands.

Patrollers and perchers

For many naturalists, the image of this skipper is of a burly little butterfly, darting in golden flashes around shrubs and tall grass heads in the summer sunshine. A more scientific account is provided by Roger Dennis, who made an excellent study of a colony in Cheshire. Dennis confirmed that the males were active only in hot weather, and showed that they divided their time between feeding and locating females, which they found either by patrolling back and forth through their habitat or, at other times, by perching close to the ground, waiting for virgin females to fly past.

The males patrol mainly from 10am to noon, although it is possible to find the odd one doing this at any time of day. A patrol involves slow, extended flights, hovering just above the ground, or weaving around clumps of grass, scanning each for the presence of a female. The final detection is by scent. Dennis watched one male spend 15 minutes weaving in and out of the branches of a seedling tree before it eventually located the female that this contained. These slow patrolling flights are punctuated by swifter dashes, covering several metres a second, as the male abandons one small area to search in another further off.

Patrolling flights give way to perching from noon until about 4pm. Each male selects a position in the sun, usually where shrubs, a wood edge or a hedgerow form some sort of boundary along which females like to fly. Open junctions in sunny woods are particular favourites, and are much fought over, for each male tries to exclude rivals from his territory.

Having selected a suitable spot, a male then chooses a flat leaf that is sturdy enough to act as a launching pad for his many forays. This is usually just above ground level and is invariably in the sun. Here he sits with wings apart, basking and waiting. He challenges any passing insect with extraordinary force, while rival males are attacked and pursued in violent aerial conflicts. These are so rapid that the combatants are hard to see, but so fierce that you can hear the clashing of their wings. The intruder is usually banished, whereupon the victor returns to the same leaf in his territory. If a female Large Skipper flies by, she triggers a rapid courtship flight, culminating in both insects landing 3-4m up in a tree or bush top. An intricate courtship ensues before pairing.

Searching for the right grass

Once mated, a typical female spends most of the day basking or resting, interrupted by short periods of egg-laying or feeding. The egg-laying flight is gentle and investigative, like that of a patrolling male. She circles around and between tall tussocks, examining each before laying single white eggs on the under-surfaces of suitable blades. On most soils egg-laying is restricted to Cock's-foot, large clumps growing in open ground being particular favourites. However, Purple Moor-grass is usually the main foodplant on wet, acid soils, and there are reliable records of egg-laying on False Brome, Tor-grass and Wood Small-reed. The last three are unusual choices, and perhaps never support a colony in isolation.

Eggs are easily found on good sites, for each is comparatively large and lies exposed on its leaf blade. It hatches after about two weeks. The caterpillar immediately constructs a grass tube, drawing the two edges of a leaf blade together around it using silk cords. It lives inside this tube, periodically emerging to nibble the edges of its leaf. The grass tube is similar to that inhabited by other golden skippers, and would be unsanitary but for an ingenious comb-like device that enables these skippers to flick their droppings a good metre away from the tube.

The caterpillar hibernates after its skin has been shed for the fourth time, and resumes feeding in spring. It moults twice more before forming a chrysalis. The ugly grub-like caterpillar is quite easy to find in its grass tube in early May, and can be distinguished from other skippers by its blue-green body and brown and cream head. The chrysalis is also hidden in grass, in a nest of blades spun in the heart of a tussock. It lasts for about three weeks before the adult emerges.

Surviving and spreading north

The Large Skipper tends to be found in wetter and more wooded places than our other golden skippers, although there is a considerable overlap in habitats. Typical sites are sunny areas of grassland where Cock's-foot or Purple Moor-grass grow in large, ungrazed clumps. Very small patches of land, including road verges, hedge banks, and rough corners, can support a colony. Most sites are sheltered by scattered shrubs or wood edges, and there can be few southern woodland glades that do not contain a colony.

Although Large Skipper numbers have been stable for the past 40 years on most well-monitored sites (which are often protected), numerous other sites have been destroyed because of the intensification of agriculture and the general tidying up of the countryside. Nevertheless, this remains one of the commonest butterflies in lowland Wales, the Midlands and southern England. It is common even in East Anglia, where a high proportion of other butterflies has disappeared. Colonies are distinctly scarcer further north, being confined to warm, sheltered, low-lying sites. Yet here it has spread in recent decades, thanks to climate warming, from around Durham to near the Scottish border in the east and from Dumfries into Ayrshire in the west.

The Large Skipper is absent from virtually all the smaller British islands, apart from Anglesey and the Isle of Wight, suggesting that it was a late coloniser as the glaciers of the last Ice Age retreated. This would also account for its surprising absence from Ireland, where there exist innumerable places in which one might expect it to thrive.

▲**Male**
Male has a prominent black sex-brand on each forewing.

▲**Female**
Markings, like those of the male, are constant throughout the range.

▲**Egg** [x 15]
Laid singly, hatching after about two weeks.

▲**Egg on grass blade**
Eggs are laid on the underside of grass blades; Cock's-foot is a favourite grass for laying.

▲**Perching male**
Male in a typically alert posture, waiting to challenge rival males or pursue prospective mates.

▼**Caterpillar** [x 1½]
Normally enclosed within larval tube; head capsule is clearly striped with cream.

Head capsule [x 3]

►**Larval tube**
Caterpillar shelters inside a tube which it forms by drawing together the edges of a grass blade.

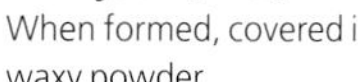

▼**Chrysalis** [x 1½]
When formed, covered in a white, waxy powder.

▲**Basking adults**
A male (right) and female basking in characteristic skipper stance, with wings angled and body positioned for rapid take-off.

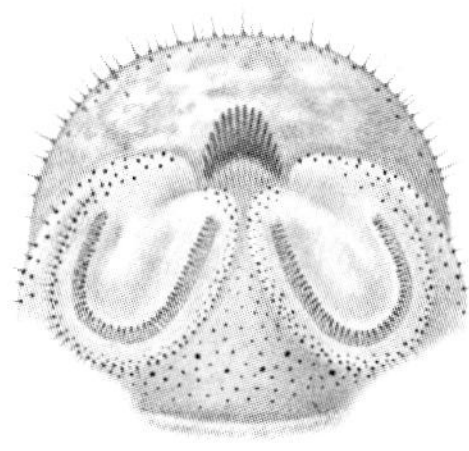

◄**Keeping clean** [x 9]
Those skippers that live in grass tubes have a comb-like device that is used to keep the tube clean. It springs upwards, flicking the caterpillar's droppings up to 1m away.

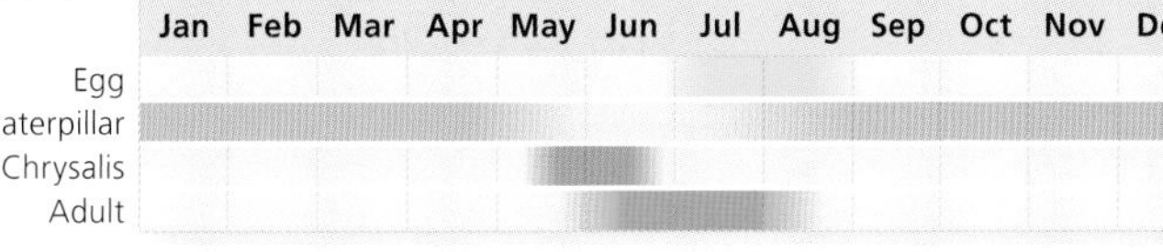

Grizzled Skipper

Pyrgus malvae

The Grizzled Skipper is a butterfly that appeals more to the specialist than the general naturalist. Its small size and moth-like appearance seldom invite a second look, which is a pity, for it is highly attractive when newly emerged and an insect of considerable character. It is best seen on a hot spring day, when the males bask nervously in sunlit spots, forever on the alert for rivals that have entered their domains. The dog-fights that follow are remarkable for their ferocity. The males circle, dive and soar, each trying to outmanoeuvre the other and flying so fast that the blurred wings become impossible to follow. Few butterflies can match these spectacular aeronautics.

Although a grey blur in flight, the black-and-white wings and fringes are distinctive when the adults do settle. Indeed, the chequered pattern has led to the butterfly being given a variety of names over the years, for the species was once much more common than today and was familiar to all early entomologists. The first, coined by James Petiver, were 'Our Marsh Fritillary' (1699) and 'Mr Dandridge's March Fritillary' (1704), a reminder that in those days any chequered animal or plant was called a fritillary, after the Latin for a chequer board. It then passed through several titles, including 'The Spotted Skipper' and, my favourite, 'The Grizzle'.

Colonies and egg-laying

Adult Grizzled Skippers live in self-contained colonies, of which the largest seldom comprise more than 150 individuals in a typical year. The males gather at the bases of hills, in sheltered hollows and along tracks, where they sit twitching on dead leaves or bare patches of ground, with their thoraxes tilted towards the sun and wings fully expanded to absorb maximum warmth. Every so often they dart into the sky, circling at high speed to court a female or fend off a rival. Females are seen less often, but are conspicuous enough when egg-laying, when they flutter back and forth investigating every nook and cranny.

Much of our knowledge about the ecology of this charming butterfly derives from an excellent study by Tom Brereton. The small, bun-shaped eggs are laid singly, generally on the undersurface of a leaf of Wild Strawberry, Agrimony or Creeping Cinquefoil. In addition, Tormentil, Barren Strawberry, Wood Avens, Salad Burnet, Dog-rose and the short

Distribution
Declining and now rare through most of its range, although common on a few southern downs that have escaped intensive agriculture.

▲**Male**
Distinctive patterning is blurred when in flight; spotting on wings in both sexes is variable.

▲**Female**
Patterning is very similar to male's; less conspicuous than male, except on egg-laying flights.

▲**Male**
The rare aberrant form *taras*, is seen regularly on a few sites. Originally described by William Lewin (1795) as a new species, 'The Scarce Spotted Skipper'.

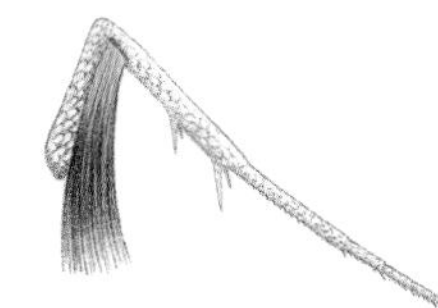

▲**Hair pencil**
Hindlegs have a tuft of black hairs that can be extended to release an aphrodisiac scent during courtship.

▲**Basking male**
Males perch for long periods on bare ground, absorbing the warmth of the sun.

▲**Egg** [× 22]
Laid singly on a variety of plants, hatching after about ten days.

◀**Roosting adults**
Grizzled Skippers may congregate on dead flowerheads in order to roost.

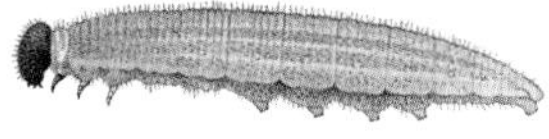

▲**Caterpillar** [× 2¼]
Lives within a tent made of leaves, bound together with silk, and emerges to feed.

▲**Chrysalis** [× 2¼]
The hibernating stage of the life cycle, formed near ground level.

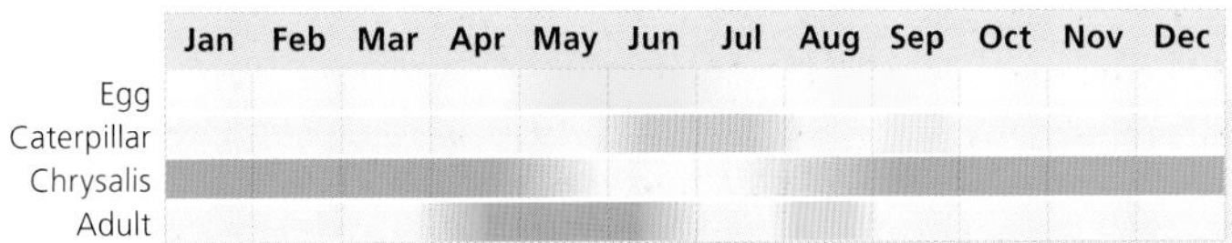

suckers of Bramble may also be used; Bramble sometimes supports an entire colony.

The egg hatches after about ten days, and at first the solitary caterpillar lives under a film of silk which it spins along the midrib on the top of its leaf. It emerges to nibble on the leaf surface around the edges of its shelter, leaving the leaf structurally intact and creating distinctive blotches that are quite easy to find. A slow grower, feeding mainly in the early morning or evening, it constructs a more substantial shelter after the first or second skin moult by folding and spinning the edges of its leaf together into a tent. The caterpillar makes fresh shelters over the next two months as it grows larger. This is a crucial period for survival, for when the weather is both warm and wet in July, the adult emergence tends to be much more numerous the following year, and *vice versa*. The caterpillar finally descends to the base of its foodplant or to rough vegetation nearby to spin a neat netting of silk in which the remarkably beautiful chrysalis is formed. It is in this stage, which lasts nine months, that it spends the winter. A few, however, pupate early after a warm spring, to form an occasional second brood of adults in August.

A range of habitats

Grizzled Skippers live in a variety of places where their foodplants grow in short, bushy growths, generally less than 10cm tall but not *very* short, situated among a heterogeneous structure of vegetation with both bare ground and taller plants nearby. Typical sites include the sunny edges of scrub on chalk and limestone downs, crumbling south-facing banks supporting sparse, scrambling vegetation, and sheltered but unshaded woodland rides and glades. They are also found in cuttings and on the edges of railway tracks, where Wild Strawberry encroaches onto the ballast as straggly runners that bake among the chippings. Other industrial sites where they may be found include disused mineral workings, spoil heaps and rubbish tips.

Despite this range of habitats, the Grizzled Skipper is far from common. Scattered colonies occur as far north as Yorkshire, but most have disappeared in recent years and the species is now rare outside central southern England. However, it is still a butterfly that one half expects to find in any suitable habitat from the Cotswolds and Chilterns southwards, especially in large, sunny woods and on unfertilised chalk and limestone downs.

Dingy Skipper

Erynnis tages

The Dingy Skipper was regarded as a common butterfly in much of England and Wales by the early entomologists, and as a rather local species in Scotland and Ireland. James Petiver described it first, as 'Handley's Brown Butterfly', in a catalogue of 1704, later revising it to 'Handley's Small Brown Butterfly'. Its current name, originally spelt Dingey Skipper, derives from Moses Harris's *The Aurelian*, published in 1766. On the Continent it is known as Dickkopffalter (Thick-headed Butterfly) in Germany and, more elegantly, as La Grisette (Little Grey One) in France.

The Dingy Skipper remained locally common in Britain until 50 years ago, but is nowadays distinctly scarce throughout its range – sufficiently so to be declared a 'conservation priority species'. Whilst in the Netherlands it verges on extinction, here it belongs to that large group of butterflies which one hopes, rather than expects, to find when exploring new, suitable-looking ground.

Beginners usually find the adult difficult to identify in flight, for its wings beat so quickly that they become a grey-brown blur, resembling those of many other butterflies or day-flying moths. It is much easier to distinguish at rest, although some find the Grizzled Skipper similar. The latter has a clear-cut pattern of small black-and-white squares on its wings, the fringes of which are also chequered. The Dingy Skipper, by comparison, has much browner wings, which can be quite variegated when fresh. Some Dingy Skippers also have an oily, iridescent sheen for the first few days of adulthood. However, they lose many scales with age, and increasingly become pale, drab and the embodiment of their name.

Distribution
A declining species that is rare through most of its range, but locally common on southern downs and coasts.

Close-knit colonies

The Dingy Skipper is essentially a springtime butterfly. The first males can be seen in mid-April in warm years, but the main emergence occurs throughout May, reaching a peak in the first week of June. It is over by the end of the month, except perhaps for a few faded stragglers that last for a week or two into July. There may then be a small second brood in August, but this occurs only after hot summers on warm southern sites.

A typical Dingy Skipper colony is quite small – indeed,

there are probably few that contain more than 100 adults, even in their better years. My brother, Chris Thomas, measured one of the largest known populations in 1978, a hazardous undertaking since the butterfly is a rapid flier, and the site consisted of about 3ha of crumbling undercliff perched near the top of a high cliff-face. By marking the adults with minute, coloured ink-spots, he was able to study individual movements and the size of the colony. He recorded about 300 butterflies present when emergence had reached its annual peak. This probably represented a total flight of about 900 adults that spring, for these butterflies are short-lived and, as a rough rule of thumb, no more than a third of the individuals that emerge in a year are alive on the peak day. Later, working with David Gutiérrez in north Wales, he found that none of the colonies studied held more than 200 adults.

Marking the adults also confirmed that Dingy Skippers essentially live in close-knit colonies, with butterflies breeding and flying on the same small sites year after year, although in Wales about a fifth of adults migrated between sites that were up to 200m apart. Within sites, long periods of the day are spent basking, with the wings held wide open and their bodies pressed hard against bare patches of ground in order to obtain maximum warmth from the spring sunshine. Males tend to gather towards the bases of hillsides, in gullies, or in any small depression where they can settle on the ground and wait for passing females to chase.

The flight of both sexes is extremely fast. They dart back and forth just above the short, sparse vegetation that typifies sites, and then climb at high speed into the sky like aircraft peeling off from a formation. Females fly widely over their breeding sites, but in late afternoon both sexes gather to roost in tall vegetation. They sit on grass heads or on dead flowers, heads uppermost and with their backs pointing towards the evening sun. All skippers have a moth-like appearance, but roosting Dingy Skippers look more like moths than most, as they drape their wings tightly around a flowerhead, with the upper surfaces exposed.

Finding eggs

On the large majority of sites, the females lay their eggs on Common Bird's-foot-trefoil, while on warm downs they prefer Horseshoe Vetch. On heavier soils, entire colonies breed on the taller, hairier leaves of Greater Bird's-foot-trefoil. The eggs are quite easy to find on good sites: search medium-sized plants with stems longer than 5cm, growing in warm sheltered nooks, part surrounded by bare soil; few, if any, eggs are laid on the tops of mounds or in close-cropped turf. Although each egg is laid singly, several may be found together on more suitable plants. Most will be on the youngest leaflets or, more often, in the groove where the stalk joins the leaves. Although pale greenish yellow when laid, they soon turn bright orange, making them especially conspicuous.

Each egg hatches after about a fortnight, and the young caterpillar immediately spins two or three leaflets together with silk threads, forming a tiny tent in which to live. It feeds on these leaves, periodically spinning a larger tent as it grows. This nest can be found with practice, but the effort is not well rewarded. The caterpillar concealed within is an ugly, grey-green grub with a purplish-black shiny head. It is unlikely to be confused with that of any other butterfly in Britain. The caterpillar is fully grown by August, when it spins and twists the leaves to make a more substantial nest. It lives in this for a further eight months, hibernating as a caterpillar then forming a chrysalis the following April. Another month passes before the adult butterfly emerges.

Sheltered warm habitats

Colonies of Dingy Skipper can be found in a wide range of habitats and on all soils. Most live on warm, south-facing downland, particularly where the turf has been disturbed by grazing stock, yet is not cropped very short. The species is especially common on dunes, cliffs and undercliffs along much of the coast, for Common Bird's-foot-trefoil flourishes wherever there is a regular supply of bare soil. These sites are also sheltered and warm, which is probably why, together with abandoned quarries, they support the highest densities of Dingy Skippers known in the British Isles. Smaller numbers can be found in the grassier areas of heathland, on embankments, and on wasteland. Many woods also contain small colonies, which breed either along the broad, rutted access rides or, in wetter areas, along ditches where they feed on Greater Bird's-foot trefoil.

Declining colonies

Unfortunately, numerous colonies have been lost in recent years, making this one of our most rapidly declining butterflies. The principal cause has been the 'improvement' of most ancient grassland for agriculture, which eliminates the butterfly's foodplants. In addition, much of the grassland that escaped improvement became unsuitable because it was seldom perturbed or grazed, or, after the return of rabbits, because it was too closely cropped. Like the Duke of Burgundy, this butterfly prefers an intermediate stage of grassland succession.

Until recently, undergrazing was the greater problem: both Common Bird's-foot-trefoil and Horseshoe Vetch are soon shaded out by the tall grasses that grow unchecked when grazing stops. In woods, coppicing once played a vital role in maintaining breeding grounds. Now that the practice has virtually ceased, many of these sites have also been

▲**Male**
Front margin of upper forewing contains scent scales, shown unfolded on the right wing. Wings of both sexes become more pale and drab as they age and lose scales.

▲**Female**
Female lacks scent scales; coloration of both sexes is otherwise similar. Underwings are slightly brighter than those of the male.

▲**Female, Irish form**
This is the only skipper found in Ireland.

◄**Resting adult**
At night, adults rest on grasses or flowerheads, adopting a characteristic moth-like pose.

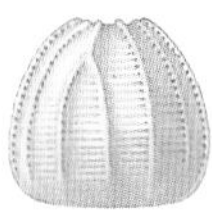

▲**Egg** [× 22]
Laid singly; initially pale greenish yellow, becoming a conspicuous orange after five days.

▼**Caterpillar** [× 2¼]
Hibernates when fully grown in a nest of trefoil or vetch leaves.

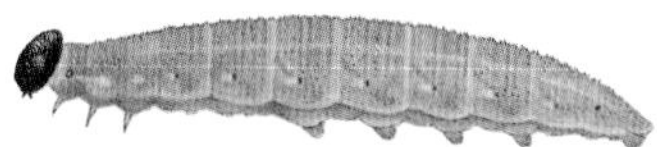

Head capsule [x 3]

▲**Basking female**
Adults often bask, between rapid bursts of flight, with their bodies pressed against bare ground.

▼**Chrysalis** [× 2¼]
Formed in spring; butterfly emerges after about a month.

	Jan	Feb	Mar	Apr	May	Jun	Jul	Aug	Sep	Oct	Nov	Dec
Egg												
Caterpillar												
Chrysalis												
Adult												

eliminated. Finally, many colonies have disappeared as a result of development on wasteland and the general tidying-up of the countryside.

A wide distribution

Despite these losses, this species remains our most widely distributed skipper. Apart from the newly discovered Essex Skipper colonies in Co Wexford, it is the only skipper in Ireland, but is not particularly common, being restricted to areas of limestone outcrop, such as the Burren. These western Irish butterflies have a slightly different appearance to other Dingy Skippers. For reasons yet to be explained, virtually no colonies occur on the Irish coast or on the smaller British islands, except for Anglesey and the Isle of Wight, even though Common Bird's-foot-trefoil is abundant in most of these localities.

In Scotland, colonies occur even further north than those of the Chequered Skipper, but again the butterfly is uncommon. It is restricted to two main areas: around Ross, Banff, Aberdeen and Inverness in the north, and along the warm south-west coastline.

Fine Dingy Skipper populations survive in most northern English counties, but they are few and far between. The butterfly becomes distinctly more common further south, with many colonies along the coasts of Wales and southern England. Inland, it has become much more localised. It was never abundant in East Anglia, and the declines there have been severe, leaving it restricted to one or two colonies per county. Its current strongholds are the central southern counties of England, particularly Wiltshire, Hampshire and Dorset, where scattered colonies survive on all soils. In Dorset alone there are several hundred colonies, making the region internationally important for this declining European butterfly.

Swallowtail

Papilio machaon

This is one of the rarest and most magnificent of our resident butterflies. Colonies are confined nowadays to the Norfolk Broads, although they once occurred in ancient fenlands from Lincolnshire to Cambridge. Occasional sightings are also made in a very different habitat – the chalk downs, from Kent to Dorset – but these are of a Continental subspecies that occasionally migrates to Britain and briefly establishes itself here.

Resident British Swallowtails belong to a unique subspecies called *britannicus*, which differs in several ways from the Continental form, *gorganus*. Although superficially similar, our Swallowtails are slightly smaller than their Continental counterparts, and have more extensive dark markings. They also behave differently: *britannicus* Swallowtails live in self-contained colonies and breed almost exclusively on Milk-parsley, whereas *gorganus* Swallowtails roam widely through the countryside, laying eggs on a range of umbellifers, including Wild Carrot and Fennel. Furthermore, British Swallowtails have one main generation a year, in late May and June, with no more than a partial second emergence in warm years, during August. The Continental subspecies has two full broods. Its charming vernacular name in France is 'Le Grande Porte-Queue', which roughly translates as 'The Great Tail-bearer'.

Distribution
The English subspecies is resident only in the Norfolk Broads; occasional Continental immigrants are seen on southern downs.

A speciality of the Broads

The British Swallowtail is seen at its best skimming over open water, from one Broad to the next. Although non-migratory, it is still a powerful flier that wanders between all the Broads adjoining the rivers Ant, Thurne, Bure and mid-Yare. This is reassuring for its future because it ensures that new patches of habitat are quickly colonised and, as genetic studies have recently shown, there is no detectable inbreeding in our surviving populations. It is probable that our Swallowtail roamed even more widely before its vast fenland breeding grounds were drained. Jack Dempster showed that the butterflies from a century ago were more robust insects, with broader thoraxes housing more powerful wing muscles than is the case today. I remember, as a student, assisting Dempster by throwing individual Swallowtails into the air at one end of his tunnel-like room, while he recorded the time each took to fly to a far window, demonstrating that the adults with wide thoraxes were more powerful fliers. It seems that as its breeding sites were destroyed and became fragmented, there was selection for a less mobile form of butterfly.

Adult Swallowtails feed mainly in the morning and late afternoon. They drink nectar from Ragged-Robin, Red Campion and other fenland flowers, slowly flapping their wings to support their heavy body while they feed. Males spend the rest of the day patrolling around prominent shrubs among the reedbeds, and virgin females fly there for mating. The encounters with females are spectacular. Both partners hover a little in the breeze before soaring high into the air and then descending to mate in a reedbed or on a shrub. They may remain together for several hours before the female departs in search of egg-sites.

A good deal is known about the natural history of this butterfly, thanks to the researches of Jack Dempster and Marney Hall. They noted how each female was highly selective when egg-laying, skimming a few centimetres above the vegetation to lay on large, prominent Milk-parsley plants, or on those regenerating in freshly mown areas. The shining, spherical egg is laid singly on the tenderest leaflets of Milk-parsley, and is easy to find once you know the type of plant the butterfly chooses.

Vulnerable caterpillars

Swallowtail eggs hatch after a week or two, and the minute caterpillars begin nibbling the upper surfaces of leaves. They show a wonderful mimicry in the period up to their third moult, resembling small bird-droppings encrusted on the Milk-parsley. This camouflage, however, does not fool spiders, and up to 65% of caterpillars may be killed before they undergo their first skin change. As they grow larger, fewer caterpillars are killed by spiders, but instead attract the attention of birds, which can eat a further two-thirds of the population. Reed Buntings are the main culprits, with Sedge Warblers and Bearded Tits also taking appreciable quantities.

Repelling predators
Looking like a bright orange snake's tongue, the osmeterium is flicked out for a few seconds whenever the caterpillar is alarmed. It emits an acrid smell reminiscent of rotting pineapple.

It is curious that the caterpillar should be so vulnerable to birds, for it has a strange, scented organ, the osmeterium, which it protrudes from its head when threatened, and is designed to deter enemies. Perhaps it just reduces bird attack, for in Sweden Christer Wiklund found that caged Great Tits, offered a choice between mealworms and Swallowtail caterpillars, initially pecked a caterpillar but seldom killed one, and soon leant to avoid their obnoxious striped bodies. The scent is certainly powerful. The Victorian entomologist, Barrett, noted how 'Fenmen always assert that they know by the scent when a large specimen... has fallen among the mown herbage, and this assists them to find it.' This, of course, was in the days when a high premium was paid by collectors for genuine British Swallowtails. The undisturbed caterpillar is also easy to find, for it advertises its deterrent by having conspicuous stripes on its body and by sitting high on a foodplant.

After feeding for roughly a month, the caterpillars desert the Milk-parsley to pupate on the stems of reeds and other fenland plants. The Swallowtail chrysalis has two colour forms, yellow-green and brown. Neither is easy to find.

Loss of the Fens

It is impossible to say just how widespread Swallowtails once were in British wetlands. Many breeding sites would have been inaccessible to early collectors, and most of the vast drainage schemes were completed before systematic records began. The butterfly clearly bred in several southern marshlands, including the Thames Valley, where one of the first authentic specimens was caught at St James's Palace in the 17th century. This was given to James Petiver, who christened it the 'Royal William'. Perhaps this was a stray from the then undeveloped marshes around London for, years later, Barrett records that its caterpillars were taken 'year after year in Osier beds in Battersea Fields'.

But the true home of the Swallowtail was the vast acreage of fenlands around the Wash. The butterfly was reported to be abundant at Whittlesea Mere, Yaxley and Burwell, in Cambridgeshire, before the great drains were dug. As the swamps were reclaimed it became increasingly confined to isolated fragments. The last of the fenland colonies survived at Wicken Fen, near Cambridge, where it was common until the 1940s. This relic population received a mortal blow during the Second World War when a substantial area was ploughed for potatoes. To make matters worse, the traditional cutting of sedges lapsed, smothering the butterfly's foodplants. The Swallowtail lingered on for a further decade at Wicken, before becoming extinct through the degradation of its habitat.

Similar problems afflicted colonies in the Broads, at least until the 1990s. An important factor here was the decline

▲ British male
The British Swallowtail (subspecies *P. m. britannicus*) is darker than its Continental counterpart. Markings of both sexes are similar.

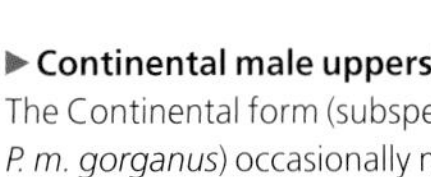

► Continental male upperside
The Continental form (subspecies *P. m. gorganus*) occasionally migrates to southern England.

▲ Egg [× 15]
After about a week the egg darkens as the caterpillar develops inside.

▼ Caterpillar [× 1½]
The young caterpillar resembles a bird-dropping, but later stages develop green and black stripes.

Young caterpillar

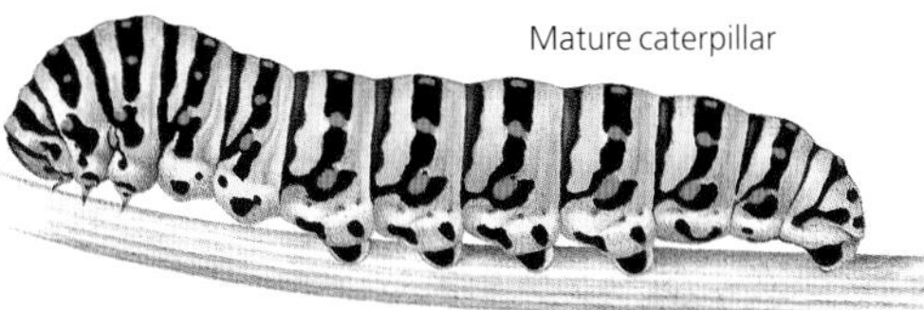

Mature caterpillar

► Chrysalis [× 1½]
Two distinct colour forms regularly occur, determined by light levels.

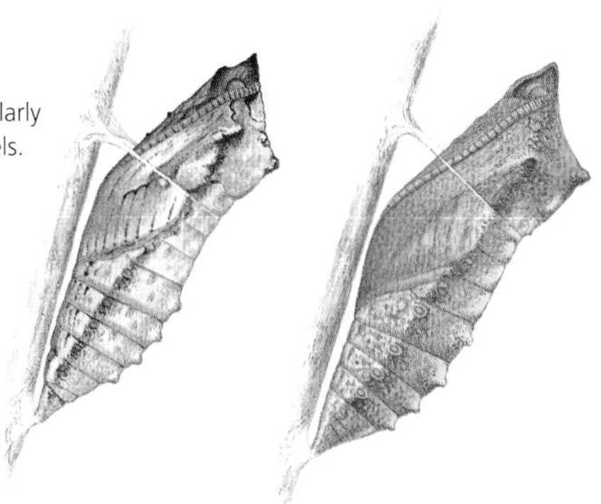

Brown form

Yellow-green form

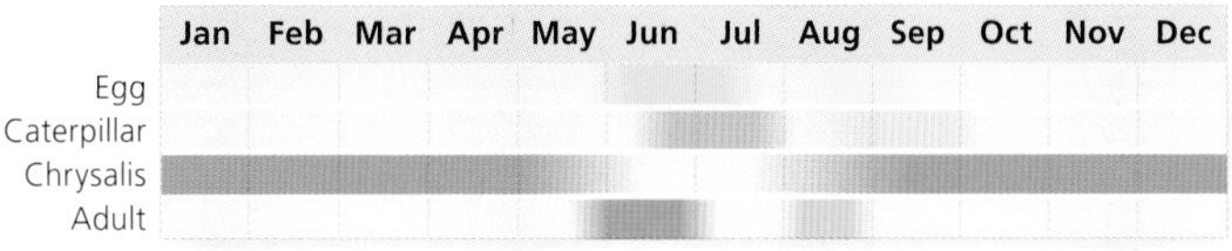

of fen management, especially in the areas with Great Fen-sedge and mixed fenland vegetation, where Milk-parsley grows at its greatest abundance. The Great Fen-sedge was traditionally cut on a three-year rotation to provide material to crown the ridges of thatched houses, while the mixed fen provided an annual crop of bedding for domestic animals. Reed-cutting, too, was in decline, but this was less serious for the Swallowtail because comparatively little Milk-parsley grows in the very wet areas where Reed proliferates. The worst period of neglect occurred during the Second World War, when there was no cutting at all for four years. Some management resumed in the post-war years, but on such a small and patchy scale that the butterfly declined considerably, and its survival was a cause of great concern in the mid-1970s.

Wetland recovery

Since then, there has been a revival in the Swallowtail's fortunes. For this, we are indebted to Dempster and Hall for providing knowledge of its exact requirements, and to the conservation organisations and volunteers that translated this advice into practice. We must be equally thankful for the socio-economic changes that once again made fen management feasible, and often profitable. For the restoration of cottages throughout England has ensured that fine Norfolk Great Fen-sedge and Reed are once more in demand. Large areas of overgrown fen have been opened up by the Broads Authority and are now being cut on a commercial basis; crucially, a new generation of sedge-cutters is being trained. In addition, several surviving fens are managed by conservation bodies, which also mow much of the mixed fenland vegetation. As a result, there has been a resurgence of vigorous Milk-parsley plants, and the Swallowtail has been more abundant in the Broads during the past 20-30 years than for the preceding four decades. It has yet to be successfully reintroduced to Wicken, but once the Great Fen Project is complete, it is reasonable to expect it to inhabit patches of Milk-parsley throughout the 3,700ha of wetlands currently being restored between Holme Fen, near Peterborough, and Woodwalton Fen, north of Huntingdon.

Arrivals from the Continent

The future of the butterfly now seems reasonably secure. There is also the probability, as our climate becomes warmer, that the Continental subspecies *gorganus* will establish itself in the south. It is a far commoner immigrant than is generally thought, although records are bedevilled by introductions and fraud.

There is no doubt that *gorganus* Swallowtails were well established on a few downs in Dorset and Kent in the early 19th century, only to disappear during the cold summer of 1816. Caterpillars have regularly been found eating Wild Carrot on southern chalk during the last century, and more often in gardens, feeding on carrot-tops as well as parsnip and rue. The entomologist J M Chalmers-Hunt calculated that there has, on average, been at least one sighting of a Continental Swallowtail in Kent every three years since 1850. The butterfly established itself again in the south during prolonged warm periods, for example near Deal in 1857-69, near Hythe in 1918-26, and during the mid-1940s in Dorset, Kent, south Hampshire and the Isle of Wight.

The most recent and largest recorded immigration of *gorganus* occurred in 2013, with reports from Kent, Sussex, Dorset and Somerset, and breeding confirmed in the last three counties. Colin Pratt and Neil Hulme have meticulously assembled records for Sussex, and report a minimum of seven adults being seen between June and August, exceeding the previous county record of five adults in the 1940s. Equally exciting, eggs and 11 full-grown larvae were reported in September on cultivated carrots and fennel in gardens and allotments in three Sussex locations, and it seems likely that others bred on Wild Carrot on the downs, leading to the possibility of a home-bred emergence in 2014. Although slightly less beautiful than our native subspecies, the permanent presence of Continental Swallowtails on southern downland would offer some recompense for our own Swallowtail's loss from the once extensive fenlands.

◄ **Resting adult**
In dull weather, the adults hide deep in fenland vegetation, with their large wings hanging downwards.

Wood White

Leptidea sinapis

The distribution of the Wood White in the British Isles was an enigma for a great many years. Widespread and quite common in Irish grasslands, it has always been much more localised in England and Wales, where it is confined to woods or occasional scrublands. The first clue to understanding this dichotomy was provided by Réal in 1988, who suggested that the Wood White in the Pyrenees consisted of two separate cryptic species that masqueraded as one, each with identical wing markings but possessing distinctive genitalia in both the female and the male that made mating between the two types impossible. Over the next two decades this exciting possibility was rigorously tested by taxonomists, geneticists, ecologists and students of butterfly behaviour in many parts of Europe: not only did they confirm the existence of a second species – appropriately christened Réal's Wood White – but in 2011, thanks to molecular DNA markers, Vlad Dincă and colleagues in Barcelona showed that a *third* sibling species existed in west Europe, including the majority that inhabit Ireland, which is today known as the Cryptic Wood White.

Separate distributions

The discovery that two species of Wood White occur in the British Isles was made in 2001, following a detailed examination of museum specimens by Brian Nelson and Maurice Hughes. They found that the butterfly in England and Wales consists exclusively of the 'true' Wood White, *Leptidea sinapis*, whereas in Ireland this species is found only in the Burren and neighbouring sites near the west coast. All of the many specimens that have been examined from other Irish localities were initially believed to be Réal's Wood White, *L. reali*, but in fact proved to be Cryptic Wood Whites, *L. juvernica*. Curiously, there appears to be a complete segregation of the two species: no Cryptic Wood White has been found in the Burren, and *vice versa* for the rest of Ireland; moreover, there is a 5-10km-wide buffer zone between their ranges where neither occurs.

Distribution
A declining rarity in English and Welsh woodlands. In Ireland, it is found only in the Burren.

The geographical separation of the two species in the British Isles is curious, because whilst each occupies a distinctive habitat – Cryptic Wood White preferring open grassland and the Wood White more wooded and warmer habitats – they frequently overlap in the northern half of the Continent and sometimes fly together on the same site. These habitat preferences help to explain why 'Wood Whites' were always quite common in Irish grasslands, yet were rare inhabitants of woods in England and Wales; but they also raise the mystifying question of why the Crypic Wood White does not occur in England and Wales. Although a few other Irish species are absent from Great Britain, this is the only butterfly with this pattern of distribution. The main differences that have been described to date between these close relatives are given under the Cryptic Wood White account.

The daintiest white

Along with the Cryptic Wood White, the Wood White is the smallest, daintiest, and by far the rarest of our British whites. It is quite easy to recognise even in flight, for it flaps its slender wings so slowly that the distinctive outline is clearly visible, as are the male's black wingtips. Sometimes a Green-veined White fluttering weakly on a cool spring day may be mistaken for the female, but at rest there should be no confusion. The delicacy of the wings, their oval shape, and the long, slender body all distinguish this from other whites, apart from the Cryptic. It differs, too, in that it always sits with closed wings, so that only the undersides are visible.

Unlike other pierid butterflies, the Wood White lives in self-contained colonies, although males may stray a kilometre or two from their breeding sites. Typical colonies contain a few dozen adults, but there are still places or years when it is a common springtime butterfly. Until quite recently, several thousand emerged each year in Salcey Forest, near Northampton, adding immense charm to an ancient woodland hideously disfigured by conifer plantations. Today, only the populations at Chiddingfold, Surrey, and Wigmore Rolls, in Herefordshire, can be classed as large.

Adult behaviour

The first adults are generally seen in late April, but emerge up to three weeks later in the east Midlands. On southern sites, numbers peak in May and last well into June before dwindling in the second half of the month. There is usually then a second brood, which has been as large as the first after recent warm springs on southern sites. These fly from mid-July to late August, and look slightly different, the males having smaller but darker black wingtips.

Martin Warren and Stephen Jeffcoate working in England, and Christer Wiklund in Sweden, have uncovered much about the life cycle of this butterfly. Given reasonable weather, the males spend most of their lives flying 50-100cm above ground level, slowly patrolling woodland rides and shrub edges in a continuous search for mates. They swerve to investigate any white object, and if this proves to be a freshly emerged female, they begin the curious head-to-head courtship shown on page 49. Virgin females mate quite quickly, but the courtship may last four minutes or longer if the female has already paired, before the male eventually tires of his quest.

Female Wood Whites fly only half as frequently as the males, and hence are seen less often. When spotted, most will be feeding from flowers: Bugle, Ragged-Robin and taller bird's-foot-trefoils are favourites in springtime, whereas trefoils and knapweeds are most commonly used in the second emergence. Males spend less time feeding on nectar, but supplement their diet with mineral salts through drinking from the muddy edges of puddles. 'Puddling' generally occurs in hot dry weather, and is more often seen further south in Europe. It is well worth looking for, as often many males will group together and become so preoccupied that they can be approached very closely before they gracefully flutter away in a white cloud of wings.

Egg-laying

An adult Wood White can live for up to three weeks, but most die of old age after a fortnight, and the average lifespan is eight days. By then, a typical female will have laid between 30 and 60 eggs, larger numbers being laid if the weather stays fine. Eggs are laid singly, usually on yellow Meadow Vetchling, Bitter-vetch, Tufted Vetch, Greater Bird's-foot-trefoil or Common Bird's-foot-trefoil. These are all quite common plants but, like most butterflies, female Wood Whites are distinctly careful over which specimens they choose. In the east Midlands, they prefer to lay on sheltered plants growing along ride edges that are shaded for at least a fifth, but generally not more than half, of the day. They typically choose rather prominent vetches that have clambered above the surrounding vegetation. Once the favoured growth-forms are known, the pale, bottle-shaped egg is quite easy to find on good sites during June. The female is also precise in where she places eggs. When searching Meadow Vetchling, look especially beneath the first and second pair of leaflets down from the tip and under the youngest bracts. On Bitter-vetch, most eggs are found beneath the second to fourth pair of leaflets from the tip.

In the Chiddingfold woods of Surrey, Jeffcoate noticed an interesting switch in egg-laying behaviour between the two generations, not dissimilar to that described for the Adonis Blue. The first generation of females, flying early in spring, lay eggs on foodplants growing in the warmest spots available to them: fresh clearings with short, open vegetation, where the eggs are laid in full sunshine and typically within 5cm of the ground. Although laid singly, small batches of eggs may be found on the small number of foodplants growing in these optimum conditions, regardless of plant species. But in the warmer weather of July, the second-generation females switch their egg-laying to the 'classic' semi-shaded rides, with taller vegetation, described for the east Midlands.

The eggs hatch after two weeks, but by then up to half

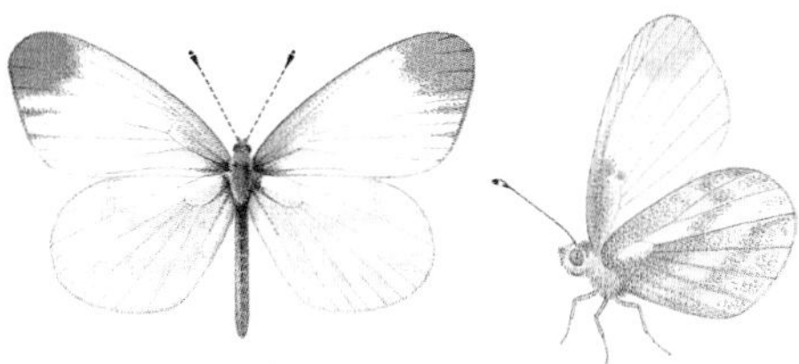

▲**English male, first brood**
Springtime male with larger wings, greyer than those of second brood.

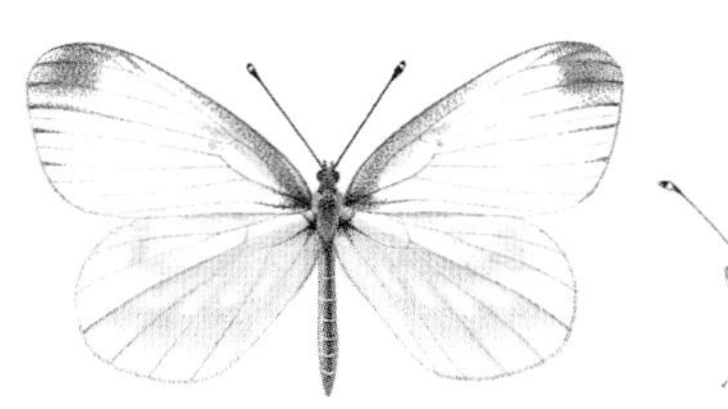

▲**English female, first brood**
Wings of the female are more rounded than those of the male, and black tips less intense.

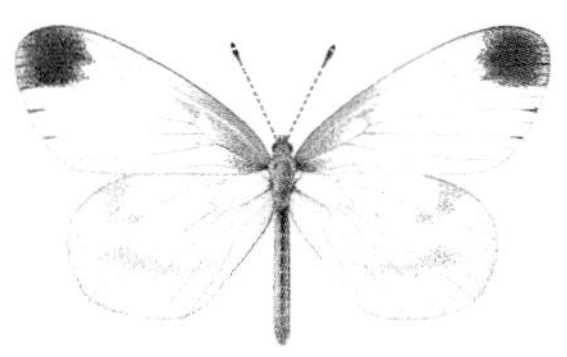

▲**English male, second brood**
Smaller and less numerous than males of the first brood.

▲**English female, second brood**
Smaller and less numerous than first-brood counterparts.

◀**Subspecies**
L. s. juvernica
Irish specimens from the Burren have greener undersides than those in England.

◀**Egg** [× 15]
Laid singly, on a variety of vetches and trefoils.

▼**Caterpillar** [× 2¼]
Perfect camouflage conceals the caterpillar when resting on the stem of its foodplant.

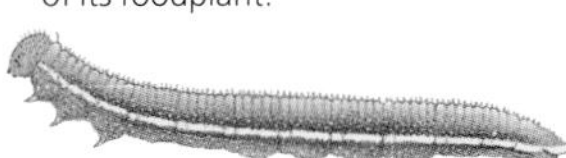

▼**Chrysalis** [× 2¼]
Formed in dense clumps of grass away from the foodplant.

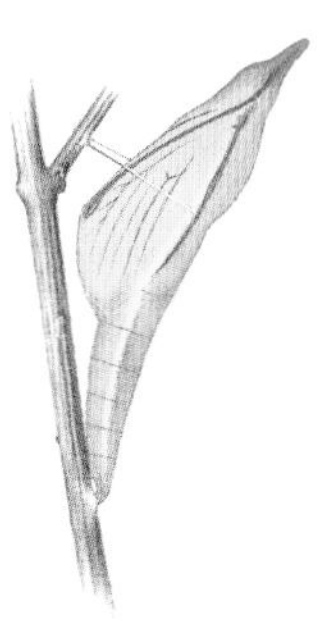

◀**Feeding adult**
English female on Marsh Thistle, with wings closed.

▲**Resting adult**
English male, second brood, in a typical resting position.

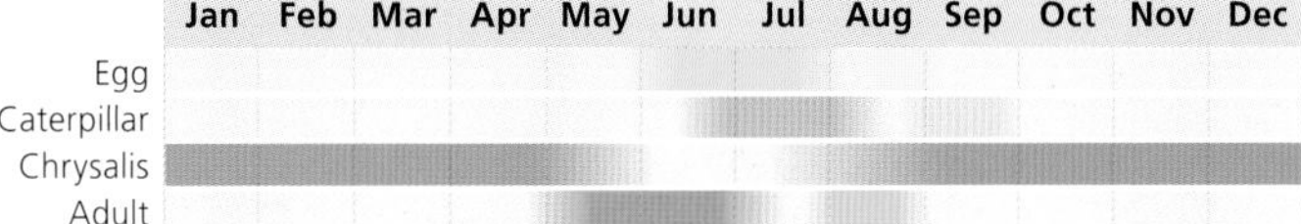

may have been killed by predators or by tiny parasitic *Trichogramma* wasps. The solitary caterpillar eats the vetch leaves, eating from the tip of a shoot down. This, too, is a vulnerable time, and many caterpillars are killed, mainly by birds. Although beautifully camouflaged, the caterpillar can be found, with practice, on suitable vetches in early August.

The chrysalis is the prettiest stage in the life cycle, with the veins and edges picked out in pink on a translucent green background. In Sweden, Magne Friberg showed that the stripe along the sheath that covers the antenna is a sharp, red line in this species, whereas in the Cryptic Wood White it is blurred and mingles with the white background. We confirm that this also applies to English and Irish specimens, and is currently the most reliable way of distinguishing between the two butterflies without killing them. Unfortunately, the chrysalis is hard to find in the wild, for the caterpillar leaves its vetch plant to pupate among dense clumps of grass. For most naturalists, the only realistic chance of seeing this chrysalis is to rear the butterfly in captivity, which is simple and enjoyable to do.

Fluctuating fortunes

The number of Wood Whites in any colony fluctuates considerably from one year to the next, depending chiefly on whether the females experienced sufficient sunshine the year before to lay large quantities of eggs. Over a longer period, numbers shift up or down in response to the number of suitable foodplants growing on a site; the largest colonies are not necessarily found where vetches and vetchling are most abundant, but rather where they grow in ideal situations for egg-laying. The species of legume may be less important than its growth-form: for example, in Surrey, Bitter-vetch and Common Bird's-foot-trefoil are more often used than Meadow Vetchling.

It appears that the Wood White requires warmer, more sheltered conditions in which to breed than is the case with the Cryptic Wood White. Suitable situations were quite common in earlier centuries, when most British woods were coppiced. In those days Wood Whites bred abundantly in the shrubby regrowth, thriving in most counties south of the Lake District. Its great decline occurred in the late 19th and early 20th centuries, when coppicing was largely abandoned. By the early 1900s it had disappeared from Cumbria, East Anglia, Kent, the New Forest and the Isle of Wight, and by 1920 it had been lost from Nottinghamshire, North Wales and Yorkshire.

The story was a somewhat happier one in the second half of the last century. With the widespread planting of conifers in former deciduous woods, suitable breeding conditions often existed along the bushy edges of access rides, between 20 and 40 years after the conifers were introduced. Many British plantations reached that stage in the last third of the 20th century, and the Wood White increased in several woods and colonised new ones, helped on occasion by the deliberate introduction of females. Today, most plantations are too shaded for peak conditions, and many have lost their colonies. The Wood White also benefited from the abandonment of railway lines after the cuts of the 1960s. For several decades, the scrubby regrowth in sheltered cuttings offered ideal conditions, and some were colonised and supported large colonies. Several of these, too, are now overgrown, and again the butterfly has been shaded out.

Conservation

Despite the many natural losses, efforts to conserve the Wood White have been made on a few nature reserves and in some Forestry Commission woodlands, where the optimum habitat conditions described by Warren have been deliberately generated through the maintenance of open, sunny rides and the regular cutting back of vegetation along their edges on three- to six-year rotations. Better still has been the regrowth that occurs after more extreme disturbances of the ground, following ride widening and ditch clearances. Fortunately, this is one of the few species that can be maintained without commercial loss in plantations, provided there are regular clearings and perturbations, and appropriate ride management. In Surrey, it has come back from the brink in the last decade, thanks to conservation management; similar advances have been achieved in Herefordshire and Shropshire.

Thirty years ago, there were perhaps 90 colonies of Wood White surviving in England and Wales; today, there may be nearer 50, and most populations are small and vulnerable. There are three main strongholds. The finest colonies occur on the clays of the east Midlands, especially in conifer plantations in Northamptonshire and Bedfordshire, and along some railway tracks. A second centre is in the heavily wooded Weald of West Sussex and Surrey, particularly around Chiddingfold. Wood Whites also occur locally in the Forests of Dean and Wyre, and in woods in Herefordshire and Worcestershire. Elsewhere, there are a few scattered colonies, including the scrubby undercliffs west of Lyme Regis.

A Burren speciality

In Ireland, the Wood White is restricted to the Burren, County Clare, and to the immediate north in County Galway. Here, it abounds in the more sheltered areas of limestone pavement, in sunny spots where Hazel scrub sprouts between the rocks. It can be found, often in large numbers, in any suitable habitat in this region, but is absent from the exposed expanses of limestone pavement and from the open grassland.

Cryptic Wood White

Leptidea juvernica

It is always a thrilling moment when a new British insect is found. This is a regular experience for students of beetles, flies and parasitic wasps, but the large majority of our butterflies were discovered before the 19th century, the Black Hairstreak (1828) and Lulworth Skipper (1832) being the last of the distinctive residents to be located. Novelty since then has been provided by the discovery that a few familiar butterflies actually consist of two very similar cryptic or sibling species: thus, the Essex Skipper was distinguished from the Small Skipper in 1888, and Berger's Clouded Yellow from the Pale Clouded Yellow in 1945. There was therefore much excitement when, in 2001, Brian Nelson and Maurice Hughes demonstrated that the wood whites of Ireland consist both of the named species and, apparently, the newly recognised Réal's Wood White described from the Pyrenees in 1988. We now know that nearly all Irish wood whites belong not to Réal's new butterfly but to a third, sibling species, the Cryptic Wood White, that was recognised as recently as 2011, with the true Wood White being confined to the Burren and adjoining areas of County Galway. No specimen of Cryptic Wood White has been found in England or Wales, nor does this now seem likely.

Distribution
Found only in Ireland, where it is widely distributed in grassland habitats. Pale green area shows the expected range. Absent from the Burren, in Ireland, and from Britain.

Separating the species

Much of our knowledge about the distinguishing features of our two wood whites comes from pioneering studies by Magne Friberg and Christer Wiklund in Sweden. Their results tally so closely with my casual observations in the British Isles, and with the taxonomic and ecological work of Nelson, Hughes and Stephen Jeffcoate, that I have drawn freely from them. Nevertheless, it is evident that some variation in behaviour, habitat use and the number of broods in each species exists in different parts of Europe, so this must be regarded as a preliminary and incomplete account of the wood whites of the British Isles, which I trust will be improved upon by future naturalists.

In the first place, there is no doubt that the three wood whites of western Europe are separate species. The results of genetic DNA sequencing in France, Sweden and Spain are unequivocal, whilst the original separation based on male and female genitalia also distinguishes *L. sinapis* from her two siblings. Since most of the known features that distinguish our two wood whites have evolved to ensure that each mates only with its own species on sites where both occur, an account of their courtship and mating is given first.

Courtship and wing clapping

The adults of both wood whites emerge on similar dates, and although Cryptics may be a few days earlier, this is insufficient to keep the two species apart. The males of both are patrollers *par excellence*. They flap in elegant slow motion along woodland rides, scrub edges and adjoining grassland in the case of the Wood White, or across more open ground in the case of the Cryptic, maintaining the steady, investigative flight a metre or so above the ground that will be familiar to every butterfly-watcher. The male dips to examine anything that is white, and if this proves to be a female the two soon settle head-to-head before the male starts the famous head-waving courtship first described by Christer Wiklund (see illustration on page 49). In Sweden, where our two species overlap on some sites, the male of neither wood white seems able to distinguish between his own female and that of the sibling species. Each will court the other's females with the same persistence as his own. It is the female that eventually chooses whether to accept her suitor, and she invariably selects her own species. She may, however, take seven or eight minutes to make her choice.

At first, the courting male of each species behaves in the same way. He waves his antennae and extended proboscis in sideways sweeps across the female's face, perhaps giving an early clue to his identity, for the conspicuous white patch on the tips of his antennae is noticeably duller, narrower and yet more elongated on the new Cryptic species. This is not, however, a safe way to distinguish between the two butterflies, for some antennal scales of Wood Whites fall off with age, leaving both males looking very similar.

After 20-30 seconds of head waving, the first clear-cut difference occurs. The male Wood White briefly flaps his wings open then shut again, and repeats this at frequent intervals throughout his three to four minute courtship. The Cryptic Wood White never claps or opens its wings during courtship. On present knowledge, this is the surest way of distinguishing between adults of these sibling species without catching, killing and dissecting them.

Choosing the right mate

We have seen that it is the female that decides whether a male is acceptable for mating. Two further cues may help her in her choice. Although identical to our eyes, the black upperside wingtips of the male Cryptic Wood White are darker in the ultraviolet range of wavelength than in the Wood White, and it is possible that females can detect this in wing-clapping suitors. The second, more probable, signal is provided by the chemicals emitted by males. The release of a distinctive aphrodisiac plays a large part in the way most butterflies recognise a suitable mate, and although Friberg and colleagues found many similarities in the cocktail of volatiles released by the two wood whites, the original *Leptidea sinapis* generally contains two unique components: dihydroisophorene and cyclonol.

The next stage in mating provides a foolproof way for naturalists to distinguish between *juvernica* and *sinapis*. As our illustration shows, the female signals her acceptance of a male by lowering her abdomen from between her wings and bending it towards the male. They then pair for about 30 minutes while the male transfers his spermatophore – a packet containing a lifetime's supply of sperm and a little 'nuptial gift' of nutritious fluid – into a chamber called the *bursa copulatrix* inside the tip of the female's abdomen. As a general rule, the hardened genitals of most butterfly and moth species are shaped to allow the male and female to couple using a 'lock-and-key' mechanism that, with few exceptions, provides a unique fit for each species. In the case of our two wood white species, the male 'key' only fits the 'lock' of his own female, not that of his sister species. We illustrate the distinctive genitalia of both the males and females of the two wood whites, based on photographs kindly supplied by Maurice Hughes. It can be clearly seen that the *aedeagus* (the male tube that penetrates the female body to deliver sperm) is longer in the Cryptic Wood White than in the Wood White, and that this matches a longer *ductus bursae* in Cryptic females, that is the mating tube that leads to her *bursa copulatrix*. The male's *saccus*, an articulated plate that locks onto the female genitalia to hold her tight during mating, is also significantly longer in Cryptic Wood Whites.

The technique of extracting the genitalia from dead adult Lepidoptera and measuring them under a microscope is familiar to students of moths, but may not appeal to the amateur butterfly-watcher. However, apart from the clue that the rare second-generation females of the Cryptic Wood White have a dusting of grey scales on the tips of their upper forewings, whereas the female Wood White is typically pure white, no other physical difference is known between the two species, except in its chrysalis stage. There, the pretty pink line that picks out the impression of the antenna is less sharp in the Cryptic Wood White.

A different geography

On current knowledge, a rough and ready guide towards distinguishing between the two wood whites in the British Isles is by their geographical location. As we have seen, the Wood White is the only species found on the British mainland, and in Ireland it is confined to the Burren and nearby sites, where the Cryptic Wood White is absent. The latter species appears to be the only wood white found across the rest of Ireland, where it is relatively common. The two butterflies also occupy rather different habitats, although both share a similar taste in foodplants. If anything, the female Cryptic Wood White lays an even greater proportion of her eggs on Meadow Vetchling, a favoured plant of the Wood White. She also lays on Common Bird's-foot-trefoil, but because of her habitat is less likely to encounter the other vetches used by Wood Whites. Stephen Jeffcoate reports that Common Bird's-foot-trefoil is the only foodplant used on the wonderful Murlough sand-dune sites in County Down.

The Cryptic Wood White's preference for sheltered, open

Courtship and mating

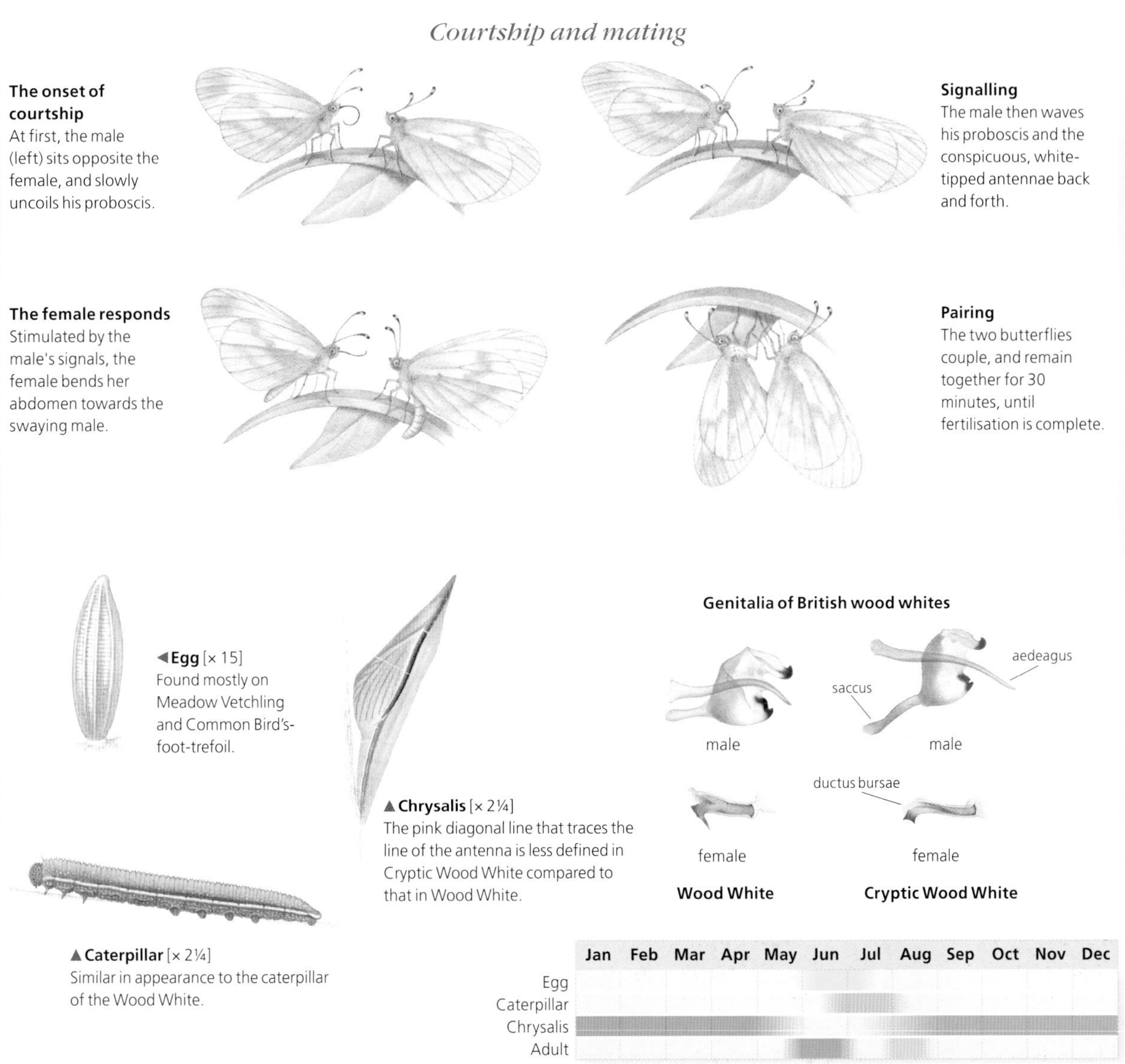

The onset of courtship
At first, the male (left) sits opposite the female, and slowly uncoils his proboscis.

Signalling
The male then waves his proboscis and the conspicuous, white-tipped antennae back and forth.

The female responds
Stimulated by the male's signals, the female bends her abdomen towards the swaying male.

Pairing
The two butterflies couple, and remain together for 30 minutes, until fertilisation is complete.

◀**Egg** [× 15]
Found mostly on Meadow Vetchling and Common Bird's-foot-trefoil.

▲**Chrysalis** [× 2¼]
The pink diagonal line that traces the line of the antenna is less defined in Cryptic Wood White compared to that in Wood White.

▲**Caterpillar** [× 2¼]
Similar in appearance to the caterpillar of the Wood White.

grassland was confirmed in a neat experiment when Friberg and Wiklund released marked adults of each species into the other's habitat in Sweden. All the Cryptic Wood Whites soon returned to grassland, although not every Wood White flew in the opposite direction to woodland.

In Ireland, the Cryptic Wood White is found locally, but not uncommonly, in sheltered, unfertilised grasslands across the whole island, other than in the Burren and the southern edge of County Galway. Look for it anywhere in mid-height to shortish flower-rich grassland, especially among scattered shrubs or other sheltered areas. Typical sites include coastal dunes, unimproved meadows, fen edges and recently cut bogs, and the wider grassy tracks and rides of sunny woods. Other colonies breed in disused quarries and along many railway and road verges. Indeed, before the new species was recognised, Henry Heal mused that in Ireland the term 'railway white' was more appropriate than 'wood white'.

Although still a widespread and under-recorded species – Butterfly Conservation's 2000-2005 survey located more than 100 'new' 10km squares containing the Cryptic Wood White – there is little doubt that many Irish colonies have been lost in recent years as a result of the intensification of agriculture following accession to the EU and its Common Agricultural Policy, and the general tidying-up of the countryside. On the other hand, this delicate butterfly expanded northwards to reach the north coast of Ireland in the 1970s and 1980s, presumably in response to climate warming.

Pale and Berger's Clouded Yellows

Colias hyale and *Colias alfacariensis*

These two clouded yellows are rare migrants to the British Isles and are extremely difficult to distinguish. Indeed, they were only recognised as separate species as recently as 1945, and there remains much confusion as to the true occurrence of each in Britain. To make matters worse, both also resemble the pale *helice* form of the female Clouded Yellow which, although scarce, has been considerably more common than either species in the British Isles in recent years.

Berger's Clouded Yellow

Identifying adults

In some parts of the Continent, a clue to the identity of these species may be gained from their location, as only Berger's Clouded Yellow reaches as far south as the Mediterranean countries. However in Britain, and in much of Europe, it is virtually impossible to distinguish between Pale, Berger's and *helice* Clouded Yellows on the wing. All three species settle with their wings closed, which obscures the main distinguishing mark of the *helice* form of the Clouded Yellow – a broader black edge to the wings, especially the hindwings, that extends much further around the corner towards the body, along both the lower edge of the forewing and the upper edge of the hindwing.

Pale and Berger's Clouded Yellows are best distinguished by the shape of the leading edge of their forewings. This is curved in the former and straight in the latter, giving the Pale Clouded Yellow's wings an altogether more pointed look. The male Berger's Clouded Yellow also has a slightly warmer and more intense yellow ground-colour, the orange spot on the underside is brighter, and the dark markings are less distinct. All these characteristics vary within the two species. Although the shape of the forewing is a fair guide in the field, for positive identification the butterflies must be killed and their detailed anatomy examined. Today, few naturalists have either the ability or inclination to do this, and so the identity of vagrants is often difficult to establish.

Distribution
The map shows the areas where Pale Clouded Yellow has been recorded. Berger's Clouded Yellow – the rarer of the two species – has not been recorded beyond southern England.

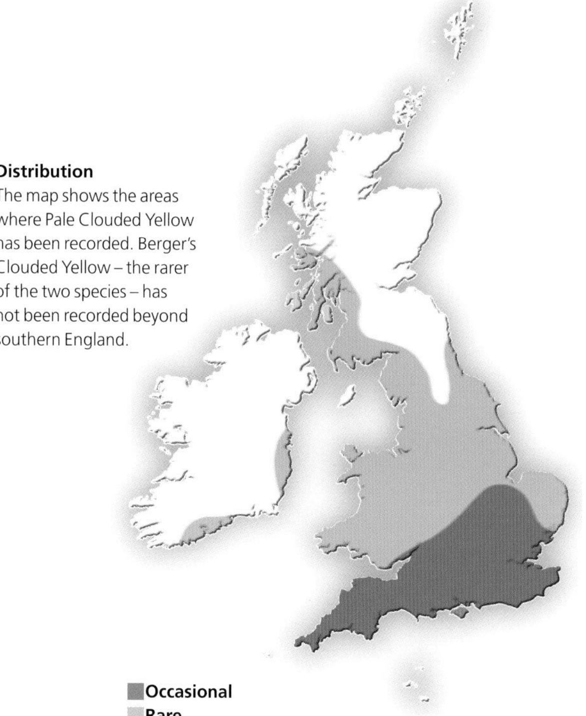

Distinctive caterpillars

Another way around the problem of identification is to breed the two species in captivity or, preferably, to watch egg-laying females in the wild. The Pale Clouded Yellow

Pale Clouded Yellow

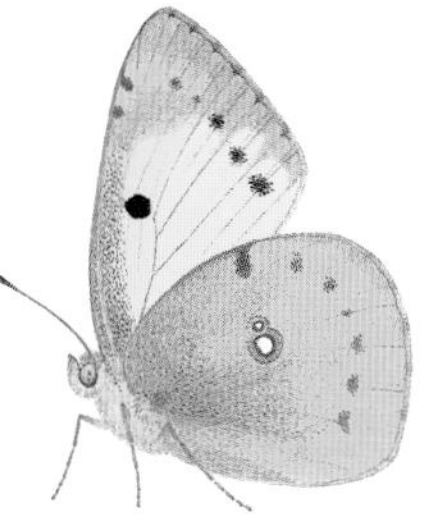

▲**Male**
Forewings slightly pointed; black edges to hindwings more extensive than those of Berger's Clouded Yellow.

▲**Female**
Black dusting at base of wings is heavy in both sexes.

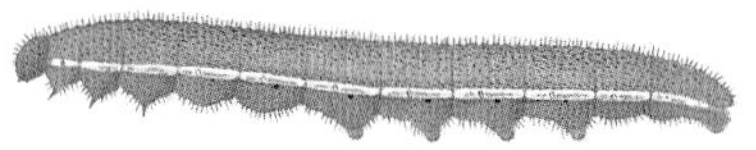

▲**Caterpillar** [× 1½]
Hairier than Clouded Yellow caterpillar, and with a more granular appearance.

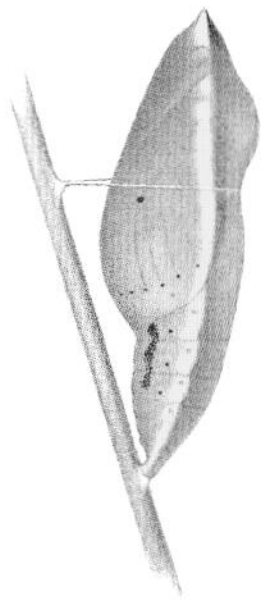

▲**Chrysalis** [× 1½]
Very similar to Clouded Yellow, but with a straighter head.

Berger's Clouded Yellow

▲**Male**
Forewings more rounded than those of Pale Clouded Yellow; brighter lemon-yellow coloration.

▲**Female**
Very similar to female Pale Clouded Yellow, but with paler and more rounded wings.

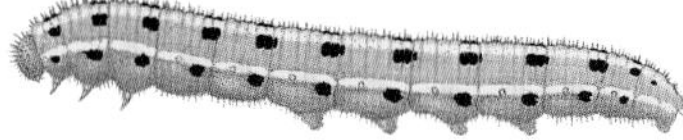

▲**Caterpillar** [× 1½]
Camouflaged to resemble its sole foodplant, Horseshoe Vetch.

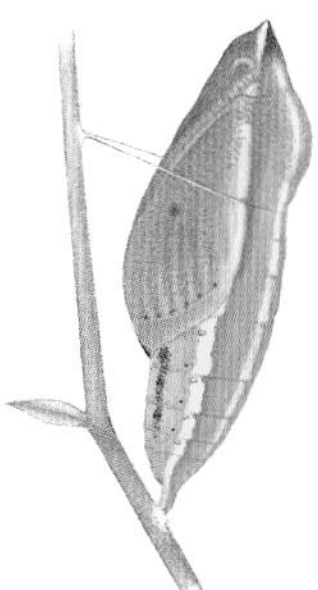

◄**Chrysalis** [× 1½]
Well concealed among low vegetation; adult emerges after 1 to 2 weeks.

	Jan	Feb	Mar	Apr	May	Jun	Jul	Aug	Sep	Oct	Nov	Dec
Egg												
Caterpillar												
Chrysalis												
Adult												

lays on Lucerne and probably other medicks and melilots as well as, reputedly, on clovers, whereas Berger's Clouded Yellow restricts its eggs to Horseshoe Vetch. Both lay the typical long, bottle-shaped eggs of all whites, and these turn orange or pink after a few days on the foodplant.

It is the caterpillar that is really distinctive, and confirms that these two similar butterflies are separate species. That of the Pale Clouded Yellow is similar to the Clouded Yellow's, in that it is pale green with a fine, pale line down either side linking the spiracles. The Clouded Yellow caterpillar, which shares the same foodplants, has a rather smoother and less granular appearance. The caterpillar of Berger's Clouded Yellow is totally different, and has the same basic colours of the Chalkhill and Adonis Blue caterpillars, which are likewise camouflaged to resemble Horseshoe Vetch. In this case, the cylindrical green caterpillar has bright yellow lines down its body and along both flanks, each punctuated by a pair of black blobs in every segment either side of each line. The chrysalises of all three clouded yellows are pale green, unlikely to be found, and almost impossible to distinguish.

The ecology and behaviour of these two butterflies is also rather different. Berger's Clouded Yellow is more likely to be seen on southern chalk downs where Horseshoe Vetch is abundant, whereas the Pale Clouded Yellow generally settles in clover and Lucerne fields. Of the two, the Pale Clouded Yellow is more docile, and spends much of the day hiding or feeding on Lucerne, making strong zigzag flights only in sunshine.

Patterns of immigration

Both these species are irregular migrants, most frequently seen in Kent, Sussex and the south-east. It is believed that most originate from central Europe, and having reached Britain they seldom wander far, even though they have a distinct urge to fly northwards in spring and south in autumn. Only a handful reach the Midlands, and both species are virtually unknown further north. Thus, neither species has anything like the migratory powers of the Clouded Yellow, which arrives every year from southern Europe. Yet these two, by contrast, are far more common in the Netherlands than the Clouded Yellow. Unfortunately, although both hibernate as caterpillars and can survive cold conditions, neither is able to withstand the damp winters of the British Isles.

Both Pale and Berger's Clouded Yellows are generally seen in August or early September, with a few sightings in May and June, and some as late as October. It is likely that all early sightings are of immigrants, but that most seen from August onwards are their offspring. Both species are known to breed in Britain. Their occurrence has been highly irregular, and the records are confused due to misidentifications of *helice* Clouded Yellows and the uncertain identity of these two species. So far as one can tell, Berger's Clouded Yellow, which has a more southerly distribution in Europe, is by far the rarer, and perhaps accounts for just one in ten sightings of these two species. Both tend to arrive in the same years, and were comparatively common on occasions in the years 1826 to 1950, when a total of 8,500 individuals was sighted – a seventh of the number of true Clouded Yellows seen in the same period. The ratio has been nearer to one in a thousand in recent decades, reflecting both the extremely low number of Pale and Berger's Clouded Yellows that have migrated here and the exceedingly high numbers of Clouded Yellows. The decline of Pale and Berger's Clouded Yellows probably reflects the intensification of agriculture and declining numbers, especially of Pale Clouded Yellows, in central Europe since the war.

20th-century records

Pale or Berger's Clouded Yellows were recorded in Britain as early as the 18th century. About a quarter of the 8,500 individuals seen up to 1950 occured in one extraordinary year, 1900, when they were comparatively common throughout the south. There was then a lull until that celebrated decade for immigrants, the 1940s. Unlike the Clouded Yellow, which was common throughout the decade, these two species had an exceptional span of five years, from 1945 to 1949, when they were locally common in all but one year, and when a total of 1,979 specimens was reported. It is tempting to think that this had something to do with the disruption of agriculture in the latter stages of the war, which allowed Pale Clouded Yellows in particular to breed on the one- to two-year-old Lucerne plants that they prefer. This, however, may be coincidence, for weather patterns were unusual during this period.

There have been very few sightings of these butterflies since the 1940s, although I have occasionally seen both on the southern Dorset downs. In most years no reliable record has been made of either species, and the last record of any numbers occurred in 1991, on the Isle of Portland. There, A S Harmer found four chrysalises and 15 adult Berger's Clouded Yellows in September, all evidently having bred on the abundant Horseshoe Vetch that clothes the abandoned limestone quarries and undercliffs around the peninsular. The report brought the same excitement to Dorset entomologists that the grand invasion of Monarchs was to do, four years later, at the same locality. But there was widespread disgust when it transpired that a well-known moth hunter had encamped on Portland and taken 16 adult Berger's Clouded Yellows for his collection. To the best of my knowledge, not a single butterfly was seen after this depletion.

Clouded Yellow

Colias croceus

Although the Clouded Yellow is seen every year in the British Isles, this is one of three butterflies that seldom survive our cool damp winters. Its existence this far north depends on the arrival of fresh immigrants each spring from southern Europe. But unlike the Painted Lady and Red Admiral, this species arrives in irregular numbers. On average, it is scarce in nine years out of ten, making the great 'Clouded Yellow years' all the more memorable.

The Clouded Yellow is one of the great migratory insects of the Western Palaearctic. Permanent populations occur in the southern half of Europe and in north Africa, where the butterflies breed continuously on Lucerne, clovers and other leguminous plants. These populations give rise to large migrations every spring, during which Clouded Yellows teem in strong, purposeful flights northwards through Europe. It is very much a one-way flight at this time of year, and the sea is clearly no obstacle. Indeed, Britain receives a great many more Clouded Yellows than the Netherlands, and this is one of only six species of European butterfly to have colonised the Azores, 1,450km from the nearest colonies on the Spanish coast.

Annual arrivals

In Britain, the first bands of immigrants usually reach the south coast in May and June, although there are occasional arrivals as early as February. The number to reach us varies greatly each year. Although normally seen in ones and twos, the major immigrations can be astounding. F W Frohawk quotes one famous account from the 19th century by the Revd D Percy Harrison:

> 'My greatest experience was in Cornwall as far back as 1868, when I was only 11, and sat on the cliff near Marazion, and saw a yellow patch out at sea, which as it came nearer showed itself to be composed of thousands of Clouded Yellows, which approached flying close over the water, and rising and falling over every wave till they reached the cliffs, when I was surrounded by clouds of *C. edusa* [= *croceus*], which settled on every flower... They swarmed in the district for a space of some three weeks and were good specimens when they arrived.'

E B Ford, in *Butterflies* (1945), makes the same point that immigrants often arrive in mint condition having flown

Distribution
Seen every year on southern coasts; occasionally abundant, spreading in diminishing numbers north to central Scotland.

perhaps hundreds of kilometres. This is my experience, too. It is when butterflies pursue, court or reject mates among tangled vegetation, or when females scrabble around for egg-laying sites, that the scales fly and adults lose their pristine appearance.

Many entomologists have waxed lyrical over the spectacular arrival of Clouded Yellows to our shores. Michael Salmon quotes several in his delightful book, *The Aurelian Legacy*, including Coleman's wonderment at Folkestone at the 'constant flutter of orange specks', and L Hugh Newman's experience of 'an occasion during the Second World War when the Clouded Yellow migrated in such numbers that military observers saw them approach the coast in the form of a great golden ball, which they thought at first to be a cloud of poison gas drifting over the water'. These and other descriptions indicate that the Clouded Yellow, like its fellow pierids the Small and Large Whites, often migrates in discrete high-density swarms, unlike the Red Admiral and Painted Lady which fly in ones or twos in an apparently endless stream of butterflies across a European front that may stretch from the Atlantic coast to Poland.

A home-grown generation

In Dorset, which with Devon receives more Clouded Yellows than any other county, the adults immediately strike inland on arrival and then fan out through the county, settling particularly in clover fields and on chalk downs. They lay their eggs mainly on the leaves of clovers, Lucerne and, less often, on Common Bird's-foot-trefoil and other native legumes. There is, therefore, no shortage of habitat, and it is interesting that this is the only species of butterfly in the British Isles capable of breeding on modern improved grasslands, in which sown clover is often an important component.

The bottle-shaped eggs are laid openly on the tops of leaves, and are easy to find in Clouded Yellow years. Initially white, they soon turn pinkish orange and hatch after about a week. The caterpillar greedily devours the leaves of its foodplant and may be fully grown within a month. The chrysalis, which I have yet to find in the wild, lasts a further two to three weeks.

The offspring of the first batch of spring immigrants generally emerge in mid-August. These are far more plentiful than their parents, and their numbers may also be boosted by fresh arrivals from the Continent. I saw this myself when making weekly counts of butterflies on the downs near Swanage during the Clouded Yellow year of 1983. In the first half of June, only three Clouded Yellows occurred within the narrow boundaries of my transects route, although odd adults could be seen egg-laying throughout the area. But in August there was a magnificent emergence; I counted 109 Clouded Yellows along the transect, with 20 to 30 individuals always visible elsewhere on the down at any one time. There is often a further brood that can produce enormous numbers in September and October, although this failed to materialise in 1983 owing, probably, to unseasonably bad weather.

Years of mass immigrations

There is perhaps no year in which a few Clouded Yellows do not reach Britain, but immigrants are generally few and far between, and seldom penetrate beyond the southern English counties. Indeed, they usually breed near the coast, especially that of the Isle of Wight, Dorset, Devon and Cornwall. Here, one can expect to see the odd individual in August every year, and about half-a-dozen on a coastal walk every five years or so. But about ten times a century there is an immigration on an altogether grander scale.

Clouded Yellow years do not come regularly once a decade, but are erratic and unpredictable. Thus there were six in the period between 1941 and 1950, followed by a lean period of 33 years before the next in 1983.

The 1983 immigration came as a complete surprise to a generation of entomologists – myself included – who had never experienced one of these events. It had, indeed, been suspected that mass immigrations might be a thing of the past, on the grounds that vast migratory swarms were now unlikely to build up owing to the modernisation of agriculture in the Clouded Yellow's permanent breeding grounds. Happily, that has proved not to be the case. Nor are long gaps between Clouded Yellow years new; in the 19th century there were three periods of between 15 and 22 years separating the major immigrations.

Since 1990, there has been an unprecedented number of Clouded Yellow years, notably in 1992, 1996, 2000 (when numbers exceeded those of 1983), in five consecutive years from 2002 to 2006, and again in 2013. True to form, interspersed with these years were seasons when few or no Clouded Yellows were recorded, for example in 1997, 1999, 2001 and 2007 to 2008.

In most Clouded Yellow years, the butterflies spread north in diminishing numbers, reaching as far as the southern half of Scotland, and in the summers of 1983 and 2000 they were more common than Brimstones in many parts of southern England. These two butterflies have a superficial similarity, but are easy to distinguish because their tones are entirely different. The Clouded Yellow is a rich, sulphurous yellow, which led to the early English name of the 'Saffron Butterfly' and its French name 'Le Souci' (the Marigold). The Brimstone, by contrast, has a paler, clearer, almost luminous tint. It is well to note, however, that the full colours of these pierids are apparent only when they fly, for both settle and

▲ **Male**
Markings are fairly constant throughout their range; male upperwing margins are solid black.

▲ **Female**
Normal colour form; black margins of wings are broken by ragged yellow spots, unlike those of male.

▲ **Female**
Pale coloured form *helice*; colour ranges from white to grey. There is no male equivalent of this form.

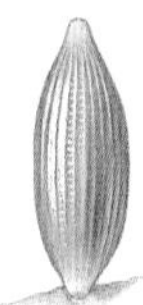

◀ **Egg** [× 15]
Laid singly, quickly changing from white to pinkish orange.

▼ **Caterpillar** [× 1½]
Very similar to Pale Clouded Yellow caterpillar, but less heavily speckled on back.

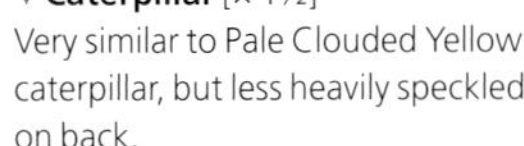

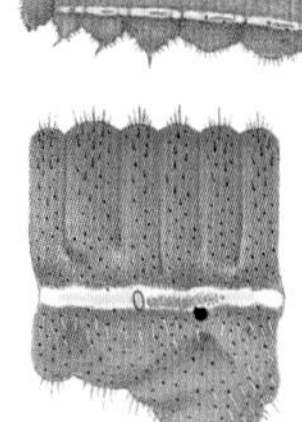

Individual segment

▶ **Feeding adult**
Clovers and Lucerne are both caterpillar foodplants and nectar sources for the adults.

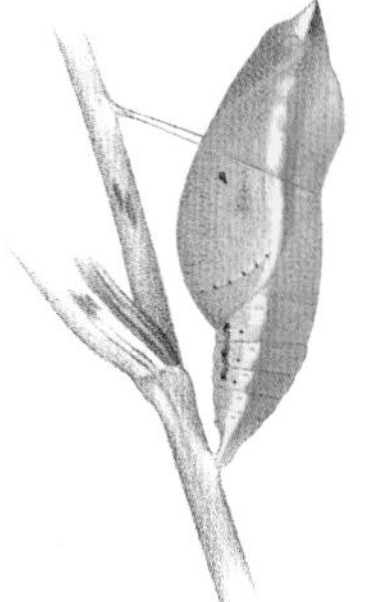

◀ **Chrysalis** [× 1½]
Head slightly upturned; attached to foodplant by a silk girdle.

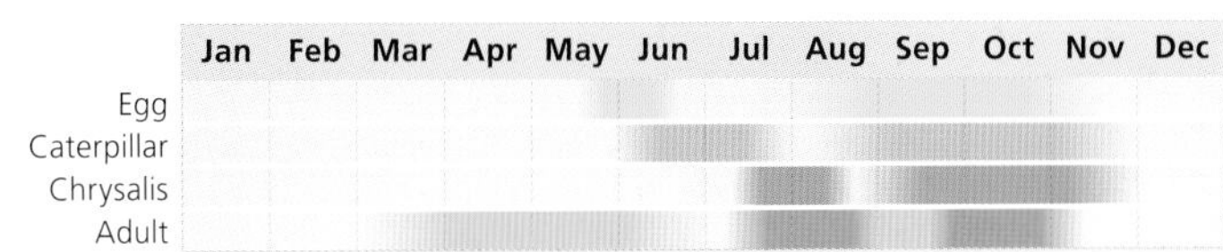

bask with their wings firmly closed, so that only the paler undersides are visible.

Many naturalists are unaware that a beautiful pale-coloured form of the female Clouded Yellow exists, called *helice*. These account for up to one tenth of all females, and appear in various tones ranging from white to grey. Some are easily confused with the much rarer Pale or Berger's Clouded Yellows. They look almost white on the wing, and in the great Clouded Yellow years can be more common than Small, Large or Green-veined Whites on southern English downs.

Clouded Yellow years often coincide with large immigrations of Red Admirals and Painted Ladies. They should be enjoyed when they occur, for it is as likely that once the spectacle is over an exceptionally poor year will follow. This was the case after 1947, which was by some margin the greatest Clouded Yellow year of the last century, with 36,000 sightings of the butterfly, out-numbering every other species along the southern coast in high summer. It also occurred following the great immigration of 2000. Although accounts of both years are peppered with superlatives, these immigrations were clearly surpassed by that of 1877, although as there were few entomologists in those days, exact comparisons are impossible.

Heading south

Clouded Yellow years end as quickly as they begin. A few individuals may survive the mildest winters, as Mike Skelton has recently proved at Bournemouth, but the vast majority migrate south. Southern migrations occur regularly throughout Europe in the autumn, and the butterflies can be seen making their way through passes in the Pyrenees. Return migrations are seldom observed in Britain, but they can be spectacular. In *Insect Migration* (1958), C B Williams recounts the experience of J Blake, who witnessed the end of the *annus mirabilis* of 1947 aboard a steamer sailing up the English Channel between Ushant and Start Point, on 14th October:

> 'For many miles he saw Clouded Yellows over the sea moving steadily to the SSW. He considered that the flight was on a front of about fifty miles and that there must have been well over a hundred thousand butterflies taking part.'

Where they made a landfall, nobody knows. Perhaps they simply pressed on into the Atlantic and perished, for 1948 was one of the poorest years on record.

Brimstone

Gonepteryx rhamni

The Brimstone is a conspicuous nomadic butterfly that will be familiar to every naturalist who lives within a few kilometres of its foodplants, the Buckthorn and Alder Buckthorn. As one of our longest-lived species, it can be seen in almost every month of the year. However, there are two distinct peaks: one on the first warm days of spring, when the butterfly flutters along hedgerows and around woods, the other in high summer, when it gorges on nectar on any site with suitable flowers, laying down reserves before its winter rest.

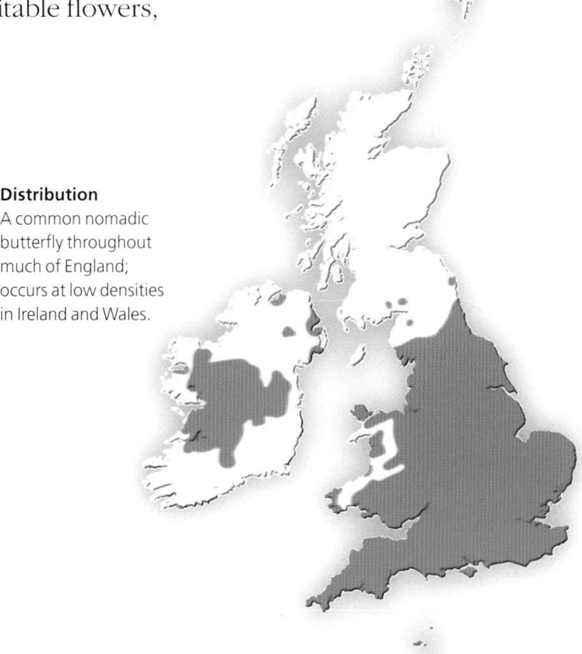

Distribution
A common nomadic butterfly throughout much of England; occurs at low densities in Ireland and Wales.

Favourite flowers

Brimstones fly during the heat of the day but tend to roost early, usually disappearing between 3pm and 4pm to settle upside-down beneath leaves in shrubs. There is one generation a year, which emerges over several weeks from early July onwards. Young adults spend much of the day feeding, concentrating on purple flowers such as Wild Teasel, buddleia, Purple-loosestrife and thistles. They quickly learn which plant provides the strongest nectar in any locality, and thereafter focus on this.

The Brimstone also selects flowers for anatomical reasons, as I found when I once counted the butterflies visiting different flowers along a woodland ride. Of the 110 Brimstones I saw, 105 were feeding on Wild Teasels, as were all but seven of the 346 Peacocks counted. By contrast, 68 of the 74 Gatekeepers were on Common Ragwort, with none drinking at teasels. Puzzled by this, I later measured the lengths both of the proboscis of these butterflies and the flower-tubes. As can be seen on page 260, Brimstones, like Peacocks, have exceptionally long tongues, enabling them to feed on Wild Teasel, whereas Gatekeepers must make do with the flowers of Common Ragwort, Common Fleabane or Bramble.

Hibernation in the evergreens

Freshly emerged Brimstones remain around their breeding grounds for two or three weeks, but in some regions then migrate to feed and hibernate in flower-rich woods. They rest after a few weeks' feeding, mainly beneath evergreen leaves, such as those of Ivy.

Although nomadic at a regional level, the Brimstone is not a true migrant. Its distribution in the British Isles closely matches that of its foodplants. Thus the butterfly is largely absent from the uplands and west of Wales, and from most of northern England apart from the Lake District; no more than six individuals are reported to have strayed into Scotland. Like many mobile butterflies, the Brimstone has spread north in recent years, but unlike other species, this is due to the planting of buckthorns in amenity schemes, gardens and hedgerows rather than to climate warming. Its distribution can also be patchy at a local scale. It is common, for example, in much of Dorset, yet few sightings are made south of the Purbeck Hills, where both species of buckthorn are rare.

In general, male butterflies emerge from the chrysalis a week or so earlier than the females, allowing them to establish territories prior to successful pairings. However, although male Brimstones develop sperm in autumn, mating is delayed until spring. Christer Wiklund found that while the emergence of both sexes was synchronised in summer, the males awake from hibernation first, and begin patrolling wood edges and hedgerows for 1-2 weeks before the females appear. Once one is encountered, there is a spectacular courtship, with male and female spiralling high into the air. The female seems reluctant to mate, but finally descends into a bush where she sits with wings half-open, quivering as the male pairs with her. It is possible that she releases an aphrodisiac at this stage, for it has been noted that a pair of Brimstones may attract a number of other males, even though they are hidden from sight among the leaves.

Searching out buckthorns

Once mated, females allow their eggs to mature for 1-2 weeks before flying along hedges, and in woods and scrub, searching for either of the two buckthorn species. Although Alder Buckthorn can be abundant on acid soils, both foodplants grow extremely sparsely in many areas, yet it is remarkable how seldom a suitable specimen is missed. Nonetheless, Brimstones do not lay indiscriminately. A study at Monks Wood showed that just two of the 12 buckthorn bushes that grew in its 150ha held 94% of the Brimstone eggs, and that any plant that was not growing both in the sunshine and sheltered from the wind had few or none. The butterflies laid their eggs at all heights on the favoured bushes, but particularly on crowns protruding into sheltered sunshine. Elsewhere, however, they prefer stumpy bushes in more open situations, choosing young buckthorns between 60cm and 1m high, largely ignoring mature shrubs.

Brimstone eggs are easy to find once you learn where to look. Females choose the sunny sides of bushes and lay on unfurling leaflets. Although they lay just one egg at a time, many are eventually placed on the undersides of the tender tips, leaving them protruding like a cluster of tiny, pale bottles. Brimstones are sometimes ready to lay before the buckthorns come into leaf. If forced, they will place the eggs beside unopened buds on the twigs.

The eggs hatch after one to two weeks, and the small caterpillar starts feeding on the upper surface of the leaf, biting holes in the tender tissue below. This makes irregular perforations which expand and distort as the leaf grows, and are very easy to spot. Slightly older caterpillars eat entire leaves, and are also easy to find despite a near-perfect camouflage. Each rests on top of a leaf, aligned head outwards along the midrib, raising the front half of its body slightly above the leaf surface at the slightest disturbance.

Many caterpillars die before completing their development, especially during the final period of growth. Warblers seem to be the main predators, although some caterpillars are killed by common wasps, and tachinid flies parasitise many others. The survivors leave their foodplants to pupate. Only once have I found a chrysalis on Buckthorn, low down beneath a leaf, although I have searched scores that had held full-grown caterpillars. The chrysalis hatches after about a fortnight.

Widespread but patchy

Brimstone butterflies are common throughout the southern two-thirds of England, particularly on calcareous soils, where Buckthorn grows, and on moist acid soils where Alder Buckthorn is the foodplant. Although attracted to flowers in summer, adult Brimstones congregate near buckthorns in spring. It is worth growing two or three small bushes in a sunny, sheltered corner of any southern garden, not only to attract this lovely insect in spring, but also for the pleasure of seeing the eggs and caterpillars. Elsewhere in England, there are strong concentrations of Brimstone in the Lake District, and the butterfly is common in south-east Wales. It is widely distributed across Ireland, but is often uncommon due to the shortage of buckthorns. Only in the Burren are Brimstones plentiful, flying among scrub on the limestone pavement.

It is difficult to say whether the number of Brimstones has changed much in recent years, although it is likely that the species was more common when hedges were cut by hand, and before the widespread grubbing-up of hedgerows. Woods, too, were probably more suitable when coppicing was common, for this produced sunny, sheltered conditions and a succession of vigorous shrub regrowth. The species was certainly well known to early butterfly collectors, and it has been claimed that the word 'butterfly' is a diminutive of its old English name, the 'butter-coloured fly'.

Whether or not this is true, the Brimstone has a unique place in the early history of British butterflies through being the first species known to be the subject of fraud. A Maitland Emmet has given a fascinating account of the 'Piltdown Butterfly'. James Petiver, one of the fathers of British entomology, illustrated this in 1702, writing that it 'exactly resembles our English Brimstone... were it not for those black spots and apparent blue moons on the lower Wings. This is the only one I have yet seen.' The specimen was even given its own species name by Linnaeus. Alas, it proved to be a normal Brimstone, with the spots painted on its wings by a dealer.

▲**Male**
Underwings camouflaged as yellowing leaves; colour variation rarely occurs in either sex.

▲**Female**
Upperwings much paler than those of male; easily mistaken for a Large White in flight.

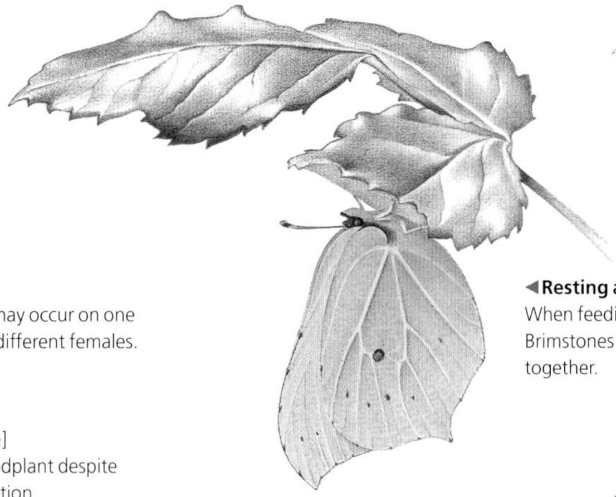

◀**Egg** [× 15]
Laid singly; several may occur on one leaf, but are laid by different females.

◀**Resting adult**
When feeding or at rest, adult Brimstones always hold their wings together.

▼**Caterpillar** [× 1½]
Conspicuous on foodplant despite camouflaged coloration.

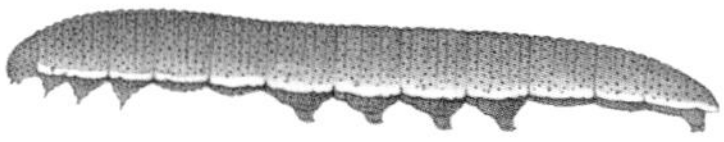

▶**Caterpillar on foodplant**
Fully grown caterpillar rests with body aligned with midrib of foodplant leaf.

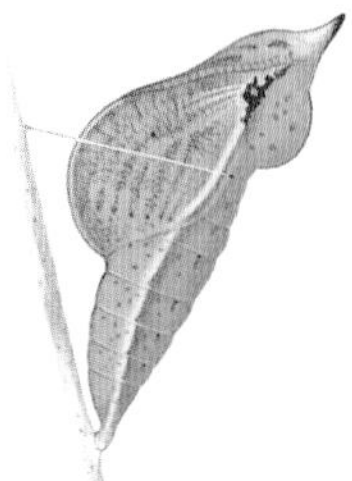

◀**Chrysalis** [× 1½]
Leaf-like colour and shape acts as camouflage; formed away from foodplant.

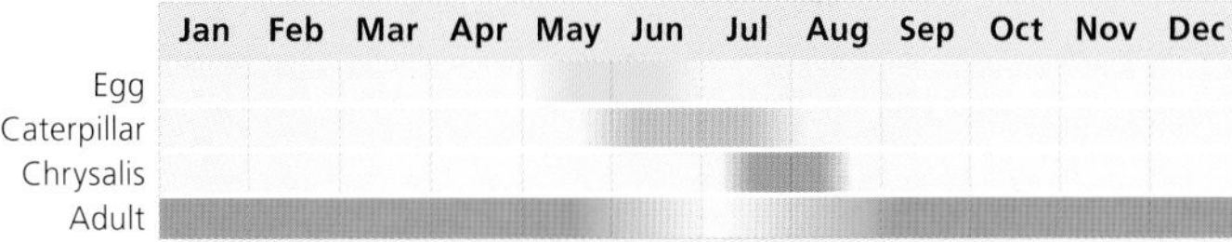

Black-veined White

Aporia crataegi

This striking and magnificent white will be familiar to any naturalist who visits the Continent. It is on the wing as far north as Scandinavia, but is commonest in central southern Europe. The adult is easy to spot, for its powerful wings with their fine tracery of veins are considerably wider than those of the Large White. It is impressive too in flight, as it soars effortlessly over a treetop or glides swiftly past with wings held in a V.

In Britain, the Black-veined White is extinct. Although known to the earliest British entomologists, it was common only in a few regions, and then only for certain periods. Kent was the last stronghold, and until recently collectors could be met who, as schoolchildren, had peeled off the grey, cobwebbed hibernation nests, full of caterpillars, from apple and plum branches in orchards along the Stour. The species then mysteriously disappeared in the early 1920s, and all attempts at re-establishment have failed.

Former distribution
Extinct in the British Isles since the 1920s. The extent of its former range is shown here.

Shedding scales

The Black-veined White is a moderately mobile butterfly that would appear in a locality, breed in more or less isolated colonies for several years, and then die out again. There were also a few more permanent populations, but by and large these were extremely localised. In all areas, numbers would fluctuate enormously from one year to the next, and this is still the case on the Continent.

I have watched Black-veined Whites most often in the Dordogne and in eastern France, in the wooded foothills of the Alps, the Rhône Valley, and the beautiful Hautes Alpes. The butterfly is most easily found by looking along the scrubby edges of woods, in moist, damp valley bottoms, and in any flowering hay meadows nearby. Both sexes are avid feeders on flowers, and are easy to approach.

Black-veined Whites roost on treetops in wooded regions, although the old Kentish colonies would settle in cornfields, where they could be found by the dozen in the best of years. In the morning there is a period of basking before they begin to flutter and then soar in the growing warmth of the day. Pairing is also generally in the morning, and it is common to see them together, the female grasping the male tightly between her powerful wings. She rubs these backwards and forwards against her partner's wings, until many of the scales are missing, leaving a clear black network of veins against a translucent background. I know of no explanation for this curious behaviour.

Early development

British Black-veined Whites were generally seen in late June, reached a peak in early July and often survived into August. The eggs were laid in batches of 50-200, usually on the undersurface of a leaf. Blackthorn and Hawthorn were the two commonest foodplants, although apple, cherry, pear and plum were all eaten in the orchards of Kent. This still happens in some fruit-growing regions on the Continent, where it is regarded as a pest in its periodic outbreak years.

The bright yellow eggs hatch after two to three weeks, and the mass of little caterpillars soon sets about the most tender available leaf tips, living in a group under a fine layer of silk. As they work down the stems they weave a more substantial web, which encompasses the entire group. According to

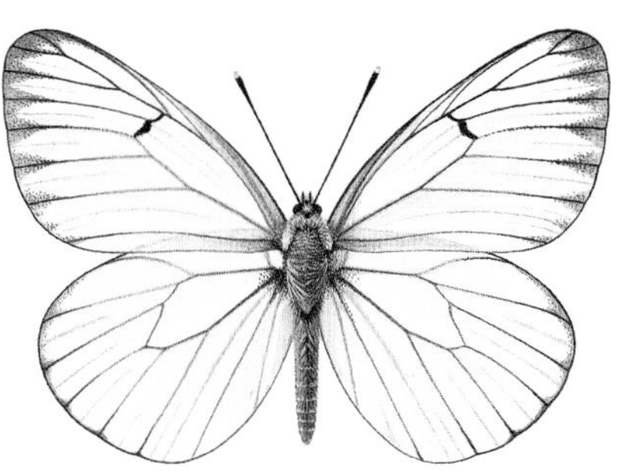

▲ **Male upperside**
Grey triangular patches at end of wing-veins are sometimes absent.

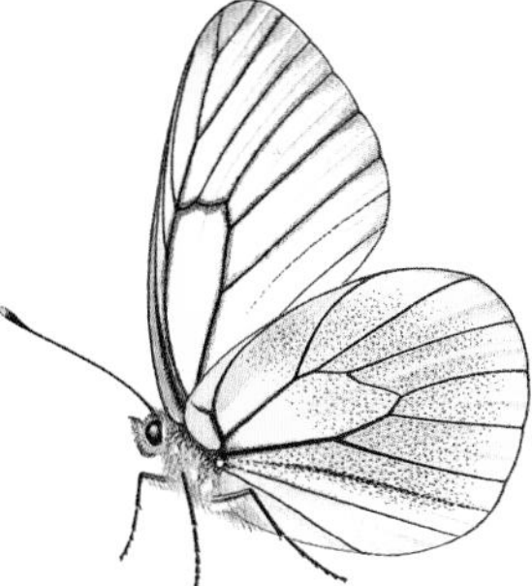

▲ **Male underside**
Undersides of both sexes are similar, although male forewings are whiter.

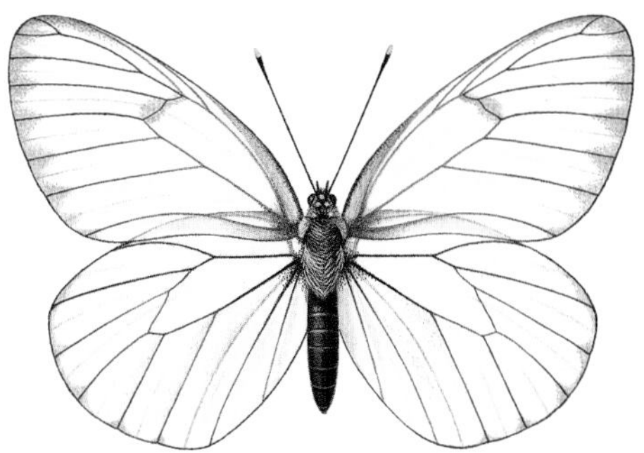

▲ **Female**
Scales lost during mating give the forewings a translucent appearance.

▲ **Egg** [× 15]
Laid in batches of 50-200 on the underside of leaves, hatching after two weeks.

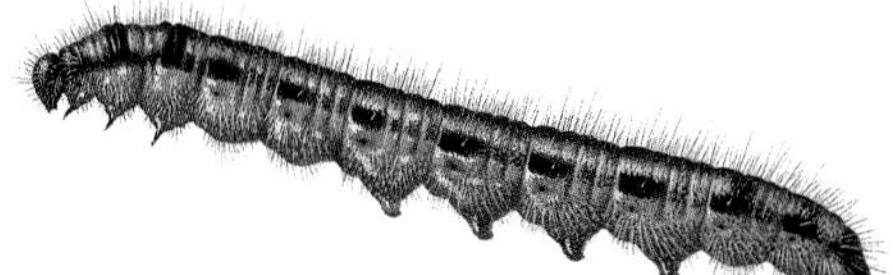

▲ **Caterpillar** [× 1½]
Gregarious at first, living within webs, but later solitary.

▲ **Chrysalis** [× 1½]
As with other whites, the colour varies considerably.

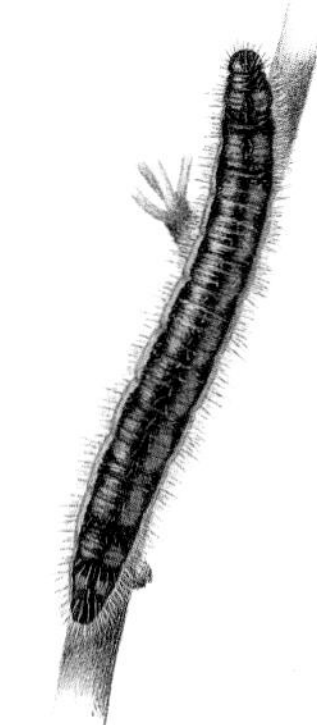

▲ **Caterpillar on twig**
Mature caterpillar rests with its body pressed against the bark of a tree-trunk or twig.

▲ **Mating**
Mating Black-veined Whites on Blackthorn, with male grasped tightly between the wings of the female.

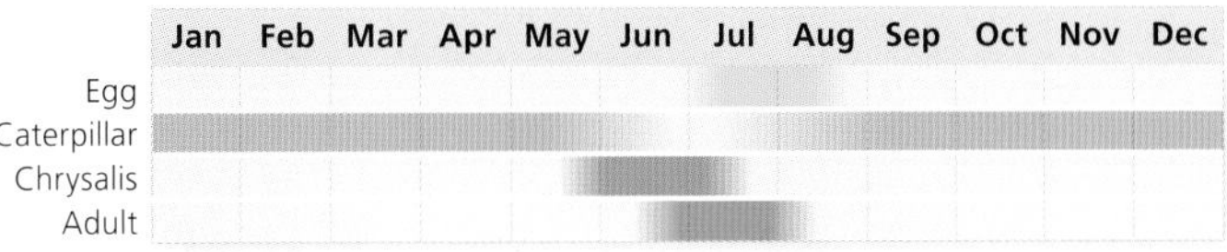

F W Frohawk, who lived in the heart of Black-veined White country, in Kent, at the turn of the last century, they would spend long periods hidden, with batches of one or two dozen caterpillars marching out of the nest in relays to eat the same large leaf, sitting side by side in rows before returning in a pack to digest this in the safety of the nest. In autumn, a much tougher, denser web is spun over the twigs, making, as Frohawk noted, 'a very secure and snug abode'. The small caterpillars hibernate deep within, huddled together in small batches. They re-emerge and are at their most conspicuous in spring, as was well known to the earliest British entomologists. Writing in 1766, Moses Harris described them 'feeding very greedily on the Buds and young tender Leaves. Now is the best time to take them, they being easily seen on Account of their Size, as they lay on their Web altogether'.

The caterpillars remain on their web until quite large, after which they usually split up into small groups. Each rests clasped along the twigs with the dense fringe of hairs pressed firmly against the bark, looking rather like the caterpillar of the Lackey moth. They are still quite easy to find at this stage and, as with many conspicuous and gregarious caterpillars, their hairs can provoke a small rash if touched. This is a form of protection against predators, and is reinforced by a foul smell that lingers malodorously over the entire nest. Finally, they drop off the shrub to pupate inconspicuously in vegetation on the ground.

The Black-veined White in Britain

The story of the Black-veined White in Britain has been one of extreme fluctuations, with periods when it was locally common punctuated by decades when it was rare. It was largely restricted to England, with the most northerly record coming from Yorkshire. It was absent from Ireland, and penetrated only the borders and south-east of Wales, where it was quite common from time to time.

Comprehensive records begin early in the 19th century. There were probably three major strongholds – east Kent, the New Forest, and a wide swathe of wooded, hilly land extending from the Cotswolds southwards into Somerset and westwards through the Forest of Dean and Glamorgan. The occasional colony was also found in most other southern counties, but these were temporary and few and far between, suggesting that this was nothing like so mobile a species as is popularly supposed. There is, for example, just one reliable record for Dorset, in 1815, despite the presence of huge populations in the New Forest 50 years later. It was also rare in other counties beyond the strongholds, such as Surrey and the whole of East Anglia, which had about half-a-dozen known colonies between them.

Colonies disappeared one by one during the first 40 years of the 19th century, until the butterfly was largely restricted to its three strongholds. There was then an extraordinary resurgence, when Edward Newman remembered it 'settled almost by hundreds on the blossoms of the great moon daisy'. But the surge was followed by an equally dramatic demise. The Black-veined White occurred 'in thousands' in several parts of the New Forest in 1860-70, but the colonies had collapsed three years later, and the last Hampshire specimen was taken in 1883. The Kent populations went through the same cycles. The butterfly had been recorded throughout the county in the first half of the century, and occurred in 'phenomenal numbers' in the mid-1850s. At Herne Bay, H Ramsay Cox 'used, by way of amusement, to see how many we could catch at one stroke of the net; we often took four or five at one time'. At Wye it was the commonest butterfly on the wing in midsummer, and yet had disappeared altogether within four years.

So extensive was the decline that the butterfly was not seen at all in Kent in 1875-82. This caused great concern, and hundreds of Continental specimens were released, to the fury of some English entomologists. These attempts at reintroduction failed, and the small resurgence of Black-veined Whites in the north-east corner of Kent is believed to stem from an overlooked pocket of native insects. Whatever their origin, these experienced a temporary recovery along the Stour Valley, and were common between 1902 and 1906. Then another decline set in, and the last reliable Kentish record was from Herne Bay in 1922. The last authentic British colony died out a year later at Craycombe, Worcestershire, and the butterfly has only occasionally been seen as an adult since then, and never, so far as I know, as a caterpillar.

The puzzle of population change

Black-veined White populations show the same kind of fluctuations in Europe as they did in Britain, at least in the north and west. For example, they crashed in Scandinavia in the 1950s, only to increase greatly after 1965. There have been many attempts to re-establish the butterfly in Britain, and although these seldom lasted more than a year or two, deliberate releases are probably responsible for all of the sightings that have been made since 1923. Vast numbers were released in Winston Churchill's grounds at Chartwell, in Kent, only to disappear without trace, and hundreds were released on Holmwood Common, in Surrey, in the 1970s.

Why these attempts failed, and why the butterfly disappeared in the first place, is largely unknown. Its habitat does not appear to have changed, and the only plausible lead is that declines often occurred in years that had been preceded by a wet September. This, however, has not always been the case. We clearly must await studies of the butterfly's population dynamics in continental Europe before we can understand its mysterious collapse.

Large White

Pieris brassicae

This is the larger and more pernicious of the two cabbage white butterflies that infest kitchen gardens and farms, from the Channel Isles to Shetland. Its caterpillars are vastly destructive. They roam over *Brassica* crops in bands, reducing each plant to a skeleton of ribs and leaving it enveloped in the acrid smell of mustard oil. Little wonder, therefore, that this is the least loved of our native butterflies, and that many people who are prepared to encourage the Peacock and Small Tortoiseshell by growing nettles in their gardens will think nothing of killing the Large White's caterpillars or of squashing its clusters of yellow eggs.

It would be unfortunate, however, if this were to blind us to the Large White's more attractive features. It is a handsome insect by any standards, and the eggs that so disgust gardeners are as delightful as those of the Orange-tip when viewed close up. From a naturalist's point of view, this is also one of the most interesting of all European butterflies, owing to its curious parasites, the protective use it makes of mustard oils, and its remarkable migratory flights.

Distribution
A very common vagrant to be seen anywhere except on high mountain-tops. Large swarms sometimes arrive from continental Europe.

Residents and migrants

The Large White has two or three generations a year, although adults can be seen at any time from February to November. The first brood emerges mainly in late April and May, and remains on the wing well into June. As can be seen from the illustrations, springtime adults differ slightly from later ones in having grey rather than black tips to their wings, and were once considered to be a different species. Butterflies from the second emergence are usually three to ten times more numerous than those from the first, and are on the wing from July to September. A third brood often follows in autumn.

Like many pests, it seems likely that the Large White was scarce before Neolithic man first tilled the land and started domesticating the Wild Cabbage. In the centuries that followed, various forms of cultivated *Brassica* became the mainstays of ancestral diets, and every settlement in Europe contributed towards creating one vast continental breeding ground for this butterfly. The populations that developed were awesome: numbers have diminished in recent years, but even today some can be measured in millions rather than thousands of adults. These develop principally in southern Scandinavia, the Baltic Islands and northern Europe, and sweep southwards every year to breed in central Europe, although many other migrations occur.

Our own populations undoubtedly reside in the British Isles, with hibernating chrysalises being reported from as far north as Orkney. But they, too, are regularly reinforced by immigrants, and our home-grown individuals often reach the Continent.

A powerful flier

British Large Whites have a tendency to fly north in spring, and there is scarcely a cabbage-patch in the land that is not colonised in some years. There is a strong sense of purpose to these migratory flights, very different from the erratic flutterings that one sees once adults have settled in a district. The wings are beaten in short, powerful flits as the butterflies press onwards, flying 1-2m above ground level. Large Whites fly at up to 16km per hour given a following breeze, but it is not known how far each can travel without resting. They can certainly cross hundreds of kilometres of ocean, but may be assisted by settling on ships. There is one description of a great swarm that alighted on the sea. Many were resting with their wings erect, while others lay flat on the calm water, but all flew off easily when disturbed.

The main bands of immigrants that reach Britain in high summer are offshoots of the regular southerly flights of central Europe. Some are of extraordinary size. In his classic book on insect migration, C B Williams recounts the earliest record, by Richard Turpyn, of a swarm on the northern coast of France:

'1508, the 23rd year of Henry the 7, the 9 of July, being relyke Sonday, there was sene at Calleys [Calais] an innumerable swarme of whit buttarflyes cominge out of the north este and flyinge south-eastwards, so thicke as flakes of snowe, that men beinge a shutynge in St. Petars fields without the town of Calleys could not see the towne at foure of the clock in the aftarnone, they flew so highe and so thicke.'

Later observers often used the same analogy. Barrett, the Victorian entomologist, wrote that there were 'many cases of... vast flights at sea, sometimes so as to form clouds like a snow-storm, or to cover a vessel and its sails when alighted'.

The number of butterflies in these swarms is usually impossible to gauge. One serious estimate was of 400 million adults on a front almost 5km wide, while a more accurate figure was obtained from a freak disaster in August 1911 on an island in Sutton Broad, in Norfolk. Here, in an area of just under 1ha, 6 million butterflies were caught in the sticky leaves of insectivorous sundews.

There are more conventional attacks, too, on these swarms. Williams describes how flycatchers, sparrows and other birds homed in on one flight through Harpenden, and how the ground became littered with white wings.

The immigrants disperse once they have settled in a region, and may be seen in ones and twos in any flowery habitat. Favourite sites include meadows, downs, hedgerows and wasteland, but they especially gather in gardens, where the cocktail of sulphurous scents that wafts up from mixed rows of *Brassicas* is a far stronger lure than any open field of cabbages. The female first detects these through her antennae, but having homed in and settled on a *Brassica*, she taps the leaves of successive plants with her feet, tasting each to select those with the strongest concentrations of sinigrin, which is the mustard oil she prefers. She then bends her abdomen and pumps out from 40 to 100 eggs, at a rate of about four a minute. At the same time, she deposits a chemical marker on the eggs to deter other females from laying on the plant, for a day or two at least.

Tiny parasitic wasps

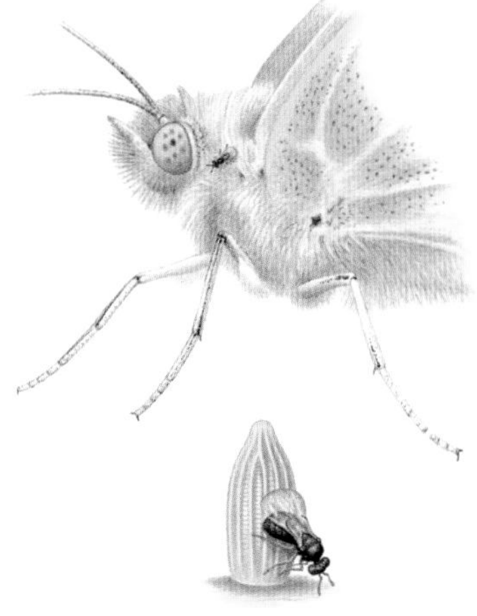

Trichogramma wasp hitching a ride on the head of a Large White.

Trichogramma wasp ovipositing on a Large White egg.

Chemicals and parasitoids

The eggs of the Large White will be all too familiar to gardeners. They stand in small, erect groups on either side of a leaf, pale yellow at first but gradually ripening to a rich orange. Eggs are found on a whole range of *Brassicas*, including cabbages, kale and Brussels sprouts. The leaves of garden Nasturtium are also often used, as occasionally is Wild Mignonette.

Each batch contains a small dose of mustard oil, which is presumed to deter enemies. There is, however, a tiny parasitic wasp that specialises on the eggs of the Large White. It is *Trichogramma brassicae*, a wasp so small that up to 20 adults can be reared in a single butterfly egg. It has a remarkable way of both finding its host and following it on the mass migrations across Europe. The story starts with the male Large White: while mating, he selfishly coats the female butterfly with an anti-aphrodisiac, in the form of the chemical benzyl cyanide, which deters other males from approaching her. Unfortunately for the butterfly, the female *Trichogramma* is sensitive to the odour of benzyl cyanide, and is attracted to pairing butterflies or to freshly mated females. Having found a female, the tiny wasp hops onto the butterfly's head and attaches itself to the fur behind her eye. It then waits for the butterfly to fly away and hitches a ride, accompanying her even, it is presumed, on the great continental migrations. As soon as the female arrives at a

▲Male, first brood
Springtime males have slightly greyer wingtips than second-brood males.

▲Male underside
Both sexes have a similar underside in both first and second broods.

▲Female, first brood
Springtime females are generally lighter in colour than their offspring.

◀Basking adult
Adults sit with their wings half-open to regulate their temperature.

▲Female, second brood
Females of the second, summer brood are often heavily marked with black and grey.

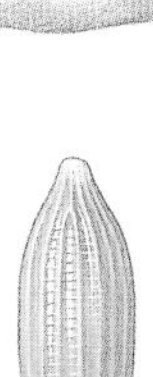

◀Egg [x 15]
Initially pale yellow, becoming orange; hatches after one to two weeks.

▼Egg batch
Clusters of eggs are usually laid on the undersurfaces of *Brassica* leaves.

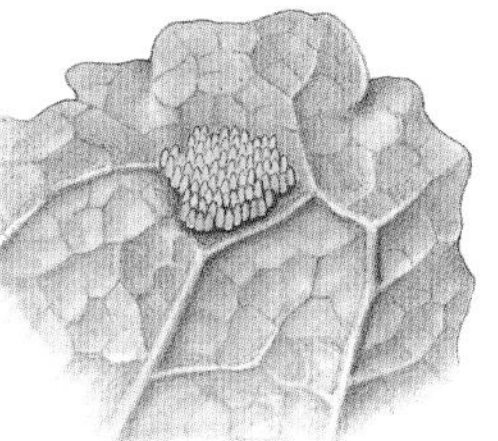

▼Caterpillar [× 1½]
The caterpillars live communally until their fourth moult, feeding or resting at the same time.

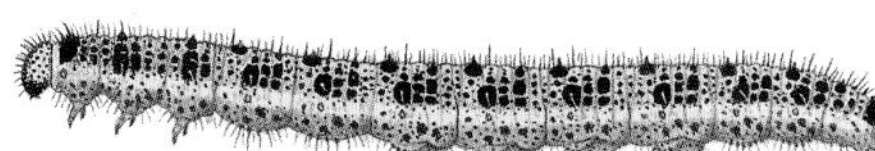

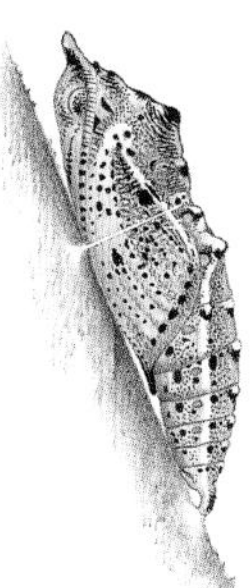

◀Chrysalis [× 1½]
Formed some distance from the foodplant. Markings vary, depending on background.

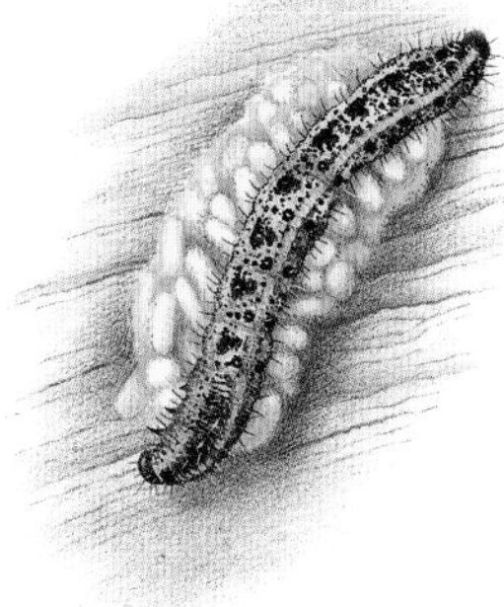

◀Parasites
Tiny *Cotesia* wasps kill many caterpillars. Their larvae live in the caterpillar's body, and then they pupate beside the corpse.

	Jan	Feb	Mar	Apr	May	Jun	Jul	Aug	Sep	Oct	Nov	Dec
Egg												
Caterpillar												
Chrysalis												
Adult												

cabbage patch and starts to lay eggs, the wasp hops off and, with her syringe of an ovipositor, begins to inject her own eggs into those of the butterfly.

No definite record of this inconspicuous parasitoid exists for the British Isles, but it occurs commonly across the rest of western Europe and must inevitably arrive here on the annual swarms of migrating Large Whites. For while it is possible that the incoming females, like Clouded Yellows, are virgins – and hence unattractive to the wasp – many *Trichogramma* hitch rides on male butterflies too. Experts consider that our absence of firm records stems, in part, from the microscopic size of this wasp, and largely from confusion over its identity, for most taxonomists confuse it with a very similar-looking sister species, *Trichogramma evanescens*, an exceedingly common egg parasite throughout the British Isles.

Poisonous caterpillars

Those Large White eggs that survive hatch after one to two weeks, and the little caterpillars remain in a band until their fourth and final skin change. At first, they concentrate on the outer leaves, cutting holes between the veins until only a skeleton of ribs remains. All rest and feed together in synchrony, being stimulated to eat by the oily fumes that escape from the damaged leaves. They spin grubby grey webs of silk over the plant, and between feeds the caterpillars bask on these in the multicoloured groups so detested by gardeners.

One might think that these clusters would be vulnerable to any passing bird, mouse or shrew, but the caterpillars have a most effective defence. While feeding on the leaves, poisonous oils accumulate in their bodies in sufficient concentrations to deter most vertebrate predators. For, as has long been known to the military, mustard oil (or gas) is a burning irritant in low doses and a lethal nerve poison when concentrated.

The enemy within

One group of enemies that is not deterred by these chemicals are further species of parasitic wasps and flies. Both often lay eggs in the bodies of the caterpillars, and their grubs feed as parasites on the caterpillars' tissues. One parasitoid, in particular, concentrates on the Large White: a wasp called *Cotesia glomeratus* which, although a giant compared to the *Trichogramma* egg parasites, is small enough for up to 80 grubs to emerge from one full-grown caterpillar. The female wasp injects eggs into a layer of fat that lies just beneath the caterpillar's skin, and the maggots grow within this, avoiding the vital organs. Then, just as the caterpillar has spun a web on which to pupate, they kill it by piercing through the skin, and form rows of yellow silk cocoons along each side of the flaccid body (illustrated on page 65). Vast numbers of caterpillars are killed in some years, accounting for more than four-fifths of the population.

Both doomed and healthy caterpillars wander some distance to pupate. They often settle beneath the eaves of a building, under a fence or on tree-trunks, where the cocoon-lined corpses or pretty speckled chrysalids are quite easy to find. The latter have various colour patterns, the exact tone being partly determined by the intensity of light surrounding the caterpillar, as with the Swallowtail. This gives the chrysalis some camouflage against its background. It also contains sufficient mustard oils to burn the mouth of any bird that is foolish enough to peck it.

Enough caterpillars survive their enemies for the Large White to remain one of the most common and pestilential butterflies in Europe. It is seldom, however, that the really large swarms of yesteryear develop, although the vast numbers of 2013 may have come close. This reduction is often attributed to another mass-killer of caterpillars, a granulosis virus which reached the British Large White populations from the Continent in 1955. Thus, while it is still common for caterpillars to strip an entire kitchen garden of *Brassicas*, there has been nothing in recent years to approach one outbreak in 1884 when, as John Feltwell relates, 'A train in the Russian town of Kiev was held up by thousands of larvae wandering over the line. The larvae were crushed "like pâté" in front of the locomotive.'

Small White

Pieris rapae

The Small White is a plain, medium-sized butterfly that will be familiar to every naturalist and gardener. It appears, in most respects, to be a less spectacular version of the Large White. Adults have the same basic wing pattern, but are generally smaller and duller. Its migratory swarms are rarely so vast, and although it too can be a serious pest of *Brassica* crops, the damage in Europe is seldom as devastating as that caused by major infestations of Large Whites.

These two species of cabbage white can usually be distinguished by size, although this is by no means an infallible guide. Quite small specimens of the Large White sometimes emerge, especially in years when the caterpillars exhaust their foodplants. However, the male Small White usually has a black spot in the centre of each forewing, a feature missing on the male Large White. Furthermore, in both sexes the dark tips on the upperwings are confined to the wings' extremities and do not extend down the outer edges, as is the case with both the Large White and the Green-veined White. These dark tips are much fainter in the first, spring brood of Small Whites, and are almost non-existent on some males.

Migrations and dispersal

The Small White is a butterfly that lives in loose, open populations – a wanderer through the countryside that reaches every sheltered habitat in its search for nectar and egg-sites. It hibernates, like the Large White, as a chrysalis. After a mild winter, the first adults are seen as early as February, although April is more usual. Numbers then build up to a peak by mid-May, and gradually fall off during June. A second brood emerges towards the end of the month, and continues throughout July. There may also be a third emergence in late summer.

Small White numbers fluctuate considerably between years. The second brood is always the more abundant, sometimes by a factor of several hundred-fold, especially during a warm July. How much this is because of local breeding success or to massive reinforcements from the Continent is unknown. Large migratory flights undoubtedly occur from time to time, occasionally in the company of Large Whites. Some Small White swarms are immense. A

Distribution
A very common nomad, absent only from Orkney, Shetland and high mountain-tops.

famous example is quoted by the Revd F O Morris, from an account in the *Canterbury Journal* of 5th July 1846:

> 'Such was the density and extent of the cloud, that it completely obscured the sun from the people on board the continental steamers on their passage, for many hundreds of yards, while the insects strewed the deep in all directions. The flight reached England about twelve o'clock at noon, and dispersed themselves inland and along the shore, darkening the air as they went... gardens suffered from the ravages of their larvae, even at a distance of ten miles from Dover.'

There has, however, been considerable debate over the regularity of the Small White's migrations. Some people maintain that these occur every year, and that British populations instinctively fly northwards in spring, with their offspring returning south in late summer. Robin Baker, an authority on insect migration, estimates that they can fly over 160km in a lifetime. He cites the spread of this butterfly in Australia, after it was foolishly introduced to Melbourne in 1939, as evidence of its mobility. Within three years, and no more than 25 generations, it had reached the west coast, a distance of 3,000km; it has been a pest throughout the continent ever since. The butterfly was also exceedingly quick to colonise North America after an introduction in the 19th century, and regular migratory patterns have developed there as the species spread to exploit this new region. In addition, definite migrations are regularly recorded passing north, and later south, through the Pyrenees. On the whole, it seems likely that this is a regular rather than a casual migrant, although there seems little doubt that some individuals travel no more than a kilometre or two during their lives.

What is not in doubt is that the Small White is sufficiently mobile to reach every suitable breeding site in the country. This includes town gardens, where it is a familiar visitor to both flowers and vegetables. Small Whites have a penchant for white or pale blossoms, and visit a different range of flowers from the vanessid butterflies. I have, for example, seen scores on my hedge of pale lavender, while the Small Tortoiseshells, Peacocks and Commas gorge on the nectar of buddleia a metre or so away. Small Whites also often roost on white blooms, where they can be difficult to spot.

Like the Green-veined White, both the males and females of this species generally mate more than once. On average, a female pairs three times during a life that will then last about 25% longer – with correspondingly more eggs laid – than an individual that pairs only once. The male transfers two types of sperm during mating: live (eupyrene) sperm to fertilise the eggs, and a 'nuptial gift' of infertile (apyrene) sperm that contains nutrients that are absorbed by the female's body. The male can be parsimonious, however, with his nuptial gifts: he generally retains a portion for the next opportunity to mate, which may occur as soon as an hour after he has fertilised the previous female.

Egg-laying and early development

Female Small Whites become ready for egg-laying two or three days after mating. They then flutter around kitchen gardens in a tireless search for Cabbage and its relatives. The pale, tubular egg is laid singly on the underside of a leaf, and young plants growing in warm nooks often receive large quantities. Every variety of cultivated *Brassica* is infested, and the female also lays on wild members of the cabbage family, such as Garlic Mustard and Charlock. They are partial, too, to garden Nasturtium.

Individual females continue to lay until they die, although at lower rates than when they are young. Egg-laying occurs only in warm weather, but the occasional rainy day is no deterrent, for they simply store ripe eggs inside their bodies and lay twice the number when conditions improve. Female Small Whites have a strong preference for placing their eggs in warm, sheltered situations, and only seldom are large fields of *Brassica* affected beyond the first few rows in from a hedge. A small, sunny kitchen garden, on the other hand, is tailor-made for their requirements, and can attract large numbers of butterflies.

It may take less than a week for the egg to hatch, by which time it will have gone through a succession of colour changes, from almost white to bright yellow to grey. The solitary caterpillar then eats a small hole in its leaf, and bores inwards towards the heart of the plant. There it remains hidden for a week or two, eating a series of ever-larger holes in the tender tissue. It finally lives in the open, resting lengthways along the midrib of a leaf, where the slightly furry green body is exquisitely camouflaged.

Predators and parasites

There has been many a study of the predators and parasites of these caterpillars, much of it prompted by the damage Small Whites can inflict on *Brassica* crops. That by Jack Dempster is especially interesting. He discovered that between half and nearly two-thirds of the caterpillars were eaten by other invertebrates, mainly in the first few days after hatching. Harvestmen and beetles were the chief predators; both are ground-dwellers which scale the crops to hunt by night. He found that there is a strong case for maintaining an untidy allotment or garden, for the weedier the bed, the greater the number of harvestmen and beetles that live there, and consequently a higher proportion of caterpillars will be killed before they can damage the crop. Equally counter-productive is the farmer who sprays insecticides on his crops. These certainly destroy many caterpillars, but they also kill the caterpillars' natural enemies, which take

▲Male, first brood
The black spots and wing margins of first-brood males can be very faint.

▲Female, first brood
Markings bolder than on the males; all females have two spots on the forewings.

◀Male, second brood
Summer males are larger, with blacker markings.

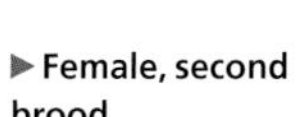

▶Female, second brood
As in the first brood, females are darker than males.

◀Egg [x 15]
Laid singly on a *Brassica* leaf.

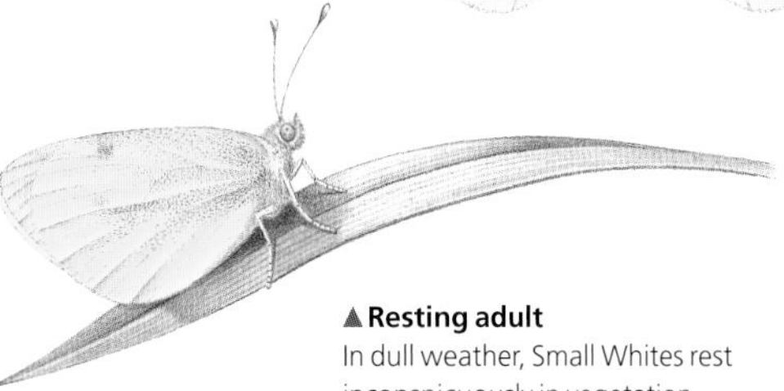

▲Resting adult
In dull weather, Small Whites rest inconspicuously in vegetation.

▼Caterpillar [× 1½]
Slightly furry; green colour camouflages the caterpillar.

Side view

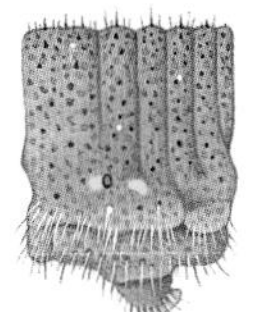

Single segment

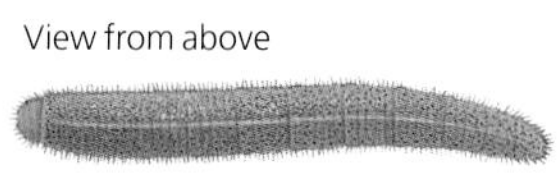

View from above

◀Basking adult
Adult male, basking in the sun on a Cabbage leaf.

▼Chrysalis [× 1½]
Attached to a stem by a silk girdle; two main colour forms.

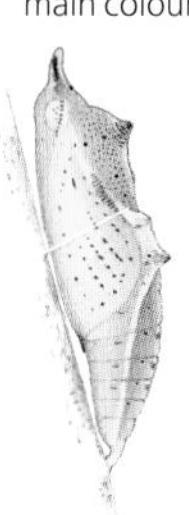

Green form

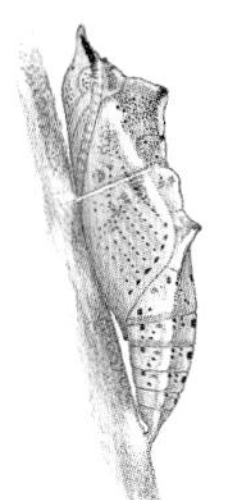

Brown form

	Jan	Feb	Mar	Apr	May	Jun	Jul	Aug	Sep	Oct	Nov	Dec
Egg												
Caterpillar												
Chrysalis												
Adult												

much longer to recover than this very mobile butterfly.

Many Small Whites are also eaten by birds. Sparrows take eggs and young hatchlings, and tits and warblers kill large numbers of older caterpillars. Thrushes inflict a final blow as the caterpillars search for somewhere to pupate. The chrysalis, too, is often attacked. Like the caterpillar, it contains few of the poisons that give such effective protection to the Large White.

Viruses are another killer, especially in cold, wet summers and when caterpillar densities are high. One further enemy is the pernicious *Cotesia* wasp. The Small White is mainly afflicted by *Cotesia rubecula*, a different species to that which attacks Large Whites, but the effect can be equally devastating. It was found in one study that caterpillar and parasite numbers oscillate together, yet out of synchrony: high numbers of caterpillars led to large emergences of wasps, but these in turn killed so many caterpillars that the butterfly's numbers temporarily slumped.

None of these enemies is sufficient to pose a real threat to the Small White, which remains one of our commonest butterflies. It is one of the few species that shows little sign of any long-term decline in recent years, and can be seen in almost every habitat in England, Wales and Ireland. Large populations also abound through the southern half of Scotland, but it is scarce further north. Thus, apart from migrants, the Highlands are largely free of this insect, as are the outer isles such as the Hebrides, Orkney and Shetland.

Green-veined White

Pieris napi

Although often overlooked as a 'cabbage white', this petite and attractive pierid is a much more delicate creature that well repays closer examination. It is an inhabitant of moist sheltered places, and possesses a fluttering weak flight that is perfectly attuned to its peaceful surroundings. The underwings are especially beautiful, with the vein edges picked out by a dusting of dark scales, giving the illusion of a green stripe along either side.

There is much variation in the markings of individual Green-veined Whites, and these also differ between the sexes and at different times of year. Springtime adults of both sexes tend to have darker veins than those of the summer brood, but there is usually less black on their upperwings. This is especially apparent in males, which can be almost pure white in the first generation and also extremely small. They are frequently mistaken for the rare Wood White when seen fluttering slowly down a woodland ride.

Regional variation

There is also some variation in their appearance throughout the British Isles. The ground-colour is a brighter yellow and the veining considerably darker in parts of Ireland and Scotland. These handsome insects are sometimes described as distinct subspecies, and have much in common with the Green-veined Whites of Scandinavia. It has been suggested that they are the modern descendants of Scandinavian stock, which colonised the British Isles during the ebbs and flows of tundra as the last great Ice Age receded. This idea has little support from modern genetic studies, and I suspect that these colour forms reflect local selection for a race better fitted to survive in northern environments. Elsewhere, this is a widespread and successful butterfly which has been described as countless subspecies throughout Europe, Asia and North America.

The existence of so much regional variation suggests that the Green-veined White is a fairly sedentary butterfly, which is confined to isolated breeding groups throughout its range. Adults are certainly quite colonial in the north of the British Isles, but wander more freely in the south, at least on a local scale.

Distribution
A common species in damp pasture and woods, absent only from Shetland and high mountain-tops.

Fending off males

The female rejects unwelcome suitors by sitting with her wings open and abdomen raised.

Butterflies of the damp

Although many flower-rich breeding sites have been destroyed in recent years, the Green-veined White remains a common butterfly of damp grassland and woodland rides throughout the British Isles, and one that is especially abundant in Ireland and the west of Britain. Colonies are more localised in the extreme north, and are not found in Shetland or at altitude in the Highlands. Nor do they often occur in dry, open habitats such as chalk downs. However, there may be small populations among patches of scrub and in copses, particularly on northern slopes, or on pockets of deeper soil, or indeed in any warm sheltered place where the humidity is high.

Green-veined Whites can be abundant where they do occur. Populations of thousands, if not tens of thousands, of adults are common in a good year, but they also fluctuate greatly between the generations. Although numbers increase significantly after a warm May or June, they often crash after a hot dry summer, reflecting the need for a humid habitat. The drought of 1976 was devastating, yet it took no more than two fairly wet years for numbers to recover.

Two types of chrysalis

There are consistent differences, too, in the abundance of the two main broods. Throughout most of the Green-veined White's range, adults emerge from hibernating chrysalises in April and May, and are on the wing almost to the end of June. Egg-laying occurs during this period and, provided the weather is warm, the caterpillars develop quite quickly to form thin-shelled chrysalises which produce a second, much more numerous brood of adults in July and August. They, in turn, lay eggs, but in this case the caterpillars develop into chrysalises that have thick, waxy skins suitable for hibernation; these produce the next brood of adults the following spring.

There is not, however, an all-or-nothing switch between thin-skinned and hibernating generations of chrysalises. Scientists in Japan and Scandinavia have shown that every caterpillar can develop into either form, depending on the temperature and the hours of daylight it receives. Caterpillars that live in regions or at times of year when the days are short invariably produce hibernation chrysalises, whereas those that experience 12 or more hours of light a day go on to produce the thin-shelled form which hatches into an adult two or three weeks later.

Lemon-scented love dust

The natural history of this lovely white is well known, thanks to classic studies by Christer Wiklund, Johan Fosberg and colleagues in Sweden. Male Green-veined Whites start emerging a few days earlier than the females, and soon begin to patrol back and forth in the sunshine, fluttering in weak zigzag flights to investigate the edges of shrubs, woodland rides, and any tussock that might house a mate. Females remain perched among leaves, but are fairly conspicuous. On sighting one, the male flutters around and lands nearby, showering her with a 'love dust' so potent that even we can smell its scent of lemon verbena. Very few females succumb without a chase though, and the two fly off together before she lands and signals acceptance by folding her wings. They promptly pair, and the male then drags her on a short nuptial flight before they settle, locked in tandem.

Promiscuous females

The female is smeared with an anti-aphrodisiac during mating, which deters other males from trying to court her. Its effect, however, is short-lived, and the most attractive females are frequently harassed by suitors. Like most whites, females signify their rejection by opening the wings wide and holding their abdomens upright at 90°, making it impossible for a male to mate. But they are not invariably coy. While some females are genetically programmed to mate but once, others are promiscuous and periodically go foraging for males, mating three to five times during their short lives. This is unnecessary for the fertilisation of the eggs, but the male Green-veined White injects more than just sperms when mating. On average, 15% of his body weight is transferred as a 'nuptial gift', in the form of proteins and other nutrients, into the female's abdomen. This enables females that mate several times to live longer and to lay more and larger eggs during their lifetimes.

Multiple mating is commonest in Green-veined White populations that inhabit warmer sunny environments, which offer prolonged opportunities for egg-laying. A higher proportion of single-mating females occurs in regions with wetter climates, where the disadvantages of multiple mating – for example, of increased exposure to predators – outweigh the advantages of longevity.

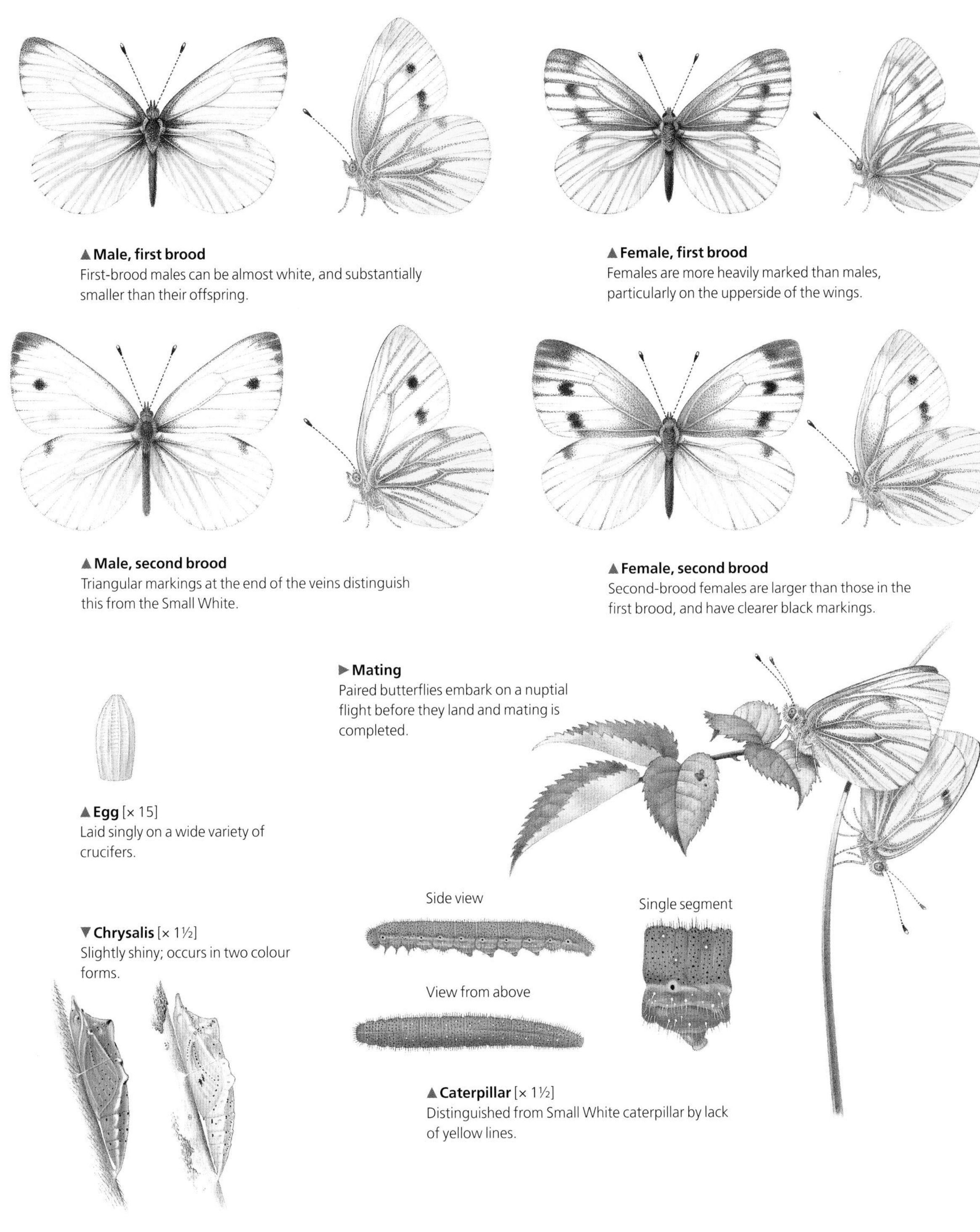

▲ **Male, first brood**
First-brood males can be almost white, and substantially smaller than their offspring.

▲ **Female, first brood**
Females are more heavily marked than males, particularly on the upperside of the wings.

▲ **Male, second brood**
Triangular markings at the end of the veins distinguish this from the Small White.

▲ **Female, second brood**
Second-brood females are larger than those in the first brood, and have clearer black markings.

▶ **Mating**
Paired butterflies embark on a nuptial flight before they land and mating is completed.

▲ **Egg** [× 15]
Laid singly on a wide variety of crucifers.

▼ **Chrysalis** [× 1½]
Slightly shiny; occurs in two colour forms.

▲ **Caterpillar** [× 1½]
Distinguished from Small White caterpillar by lack of yellow lines.

	Jan	Feb	Mar	Apr	May	Jun	Jul	Aug	Sep	Oct	Nov	Dec
Egg												
Caterpillar												
Chrysalis												
Adult												

Nectar and mud

Both sexes of Green-veined White are avid feeders on flowers, and the males sometimes supplement their diet of nectar by mud-puddling. The males of several species of butterfly drink mud, probably because they need sodium and other essential salts to replace the minerals lost when they mate. Nectar is an excellent source of energy-rich sugars, but has a low mineral content.

The female Green-veined White makes do with nectar and nuptial gifts. She spends much of her life fluttering a few centimetres above the ground, constantly landing to tap leaves with her feet. It is a slow, topsy-turvy kind of flight, almost as if she were injured and making desperate attempts to get airborne. In fact, she is tasting every plant, trying to detect the mustard oils of the crucifers on which she will lay her eggs. The female's potential range of foodplants is considerable. She will use almost any species of cress, although in Britain most eggs are found on Water-cress, Cuckooflower, Garlic Mustard and Hedge Mustard. These are common plants in woods and unfertilised grasslands, but like many butterflies the female Green-veined White is extremely selective about the sort of crucifer chosen for her offspring. Large, mature growths are invariably rejected in favour of seedlings or small one-year-old rosette-shaped plants, particularly those growing in moist recesses.

Searching for eggs

The pale, spindle-shaped eggs are very simple to find once one knows this butterfly's favourite spots. They are perhaps easiest to see in boggy grassland, but it is quite possible to find them in woods. Here, I generally search towards the bottom of ditches, in recently disturbed rides, or within the body of a wood, concentrating on quite shady humid areas. The eggs may be on remarkably small plants. Although they are laid singly, I have frequently found three or four together on the undersurface of a Water-cress or Cuckooflower seedling that consisted of no more than the four initial leaves, sprouting in old hoof-prints in a boggy meadow, or along the crumbling banks of ditches and streams.

The eggs hatch after a week or two, and the caterpillar feeds only on the tender leaves, growing rapidly to pupate after about four weeks. It looks like a Small White at this stage, except that there is no yellow line running down either side of the body, merely a circle of yellow around each of the spiracles. This is easily seen with the naked eye, and is shown in the artwork of the enlarged segments of the two species. Their chrysalises are harder to distinguish. Both exist in two main colour forms, regardless of the season. Those of the Green-veined White are perhaps the better camouflaged, for I have yet to find either form in the wild.

Bath White

Pontia daplidice

This lovely dappled white has a curious pattern of immigration. Although excessively rare in most years, it was known and named by the earliest entomologists, and had a brief resurgence in the 1940s when, in one year alone, over 700 individuals were reported. It has been virtually absent ever since, yet is common enough in southern Europe. I well remember the thrill of my first sighting – in the Colosseum in Rome – where scores of Bath Whites were fluttering around the scrubby vegetation surrounding the monuments.

This vagrant's true home is the Mediterranean and North Africa. As with so many whites, the adults migrate north each spring, breed for two or three generations, and their offspring return south in autumn. But migrants seldom penetrate very far north, and the species is generally killed by British winters, despite some circumstantial evidence that a few hibernated successfully after the extraordinary influx of 1945.

Like the Orange-tip, whose female this somewhat resembles, the Bath White lays its eggs on crucifers, although in this case the leaves rather than the flowerheads are the caterpillars' food. Another difference – in Sweden at least – is that the Bath White has a penchant for laying on seedlings and low-growing rosettes, and deliberately avoids large mature crucifers. This has little to do with the nutritional value of the leaves, but reflects the fact that the caterpillar needs to lie in the warmest spots available when living near the northern limit of its range.

Records from the past

The first British record of a Bath White came from Gamlingay, near Cambridge. It was possibly a temporary resident, for specimens appear to have been caught in more than one year by William Vernon in the late 17th century. One of these survives as a remarkably well-preserved specimen in the Hope Collection at Oxford; its image can be seen in the first plate of E B Ford's incomparable New Naturalist, *Butterflies*. 'Vernon's Half Mourner', as the butterfly was known, was extraordinarily rare thereafter, and hardly featured among the patchy records from the 18th century. It did, however, appear in an embroidery by a young lady of Bath, and

Distribution
Most sighting occur in southern England, close to the coast. Numbers are erratic, with large numbers being seen in some years.

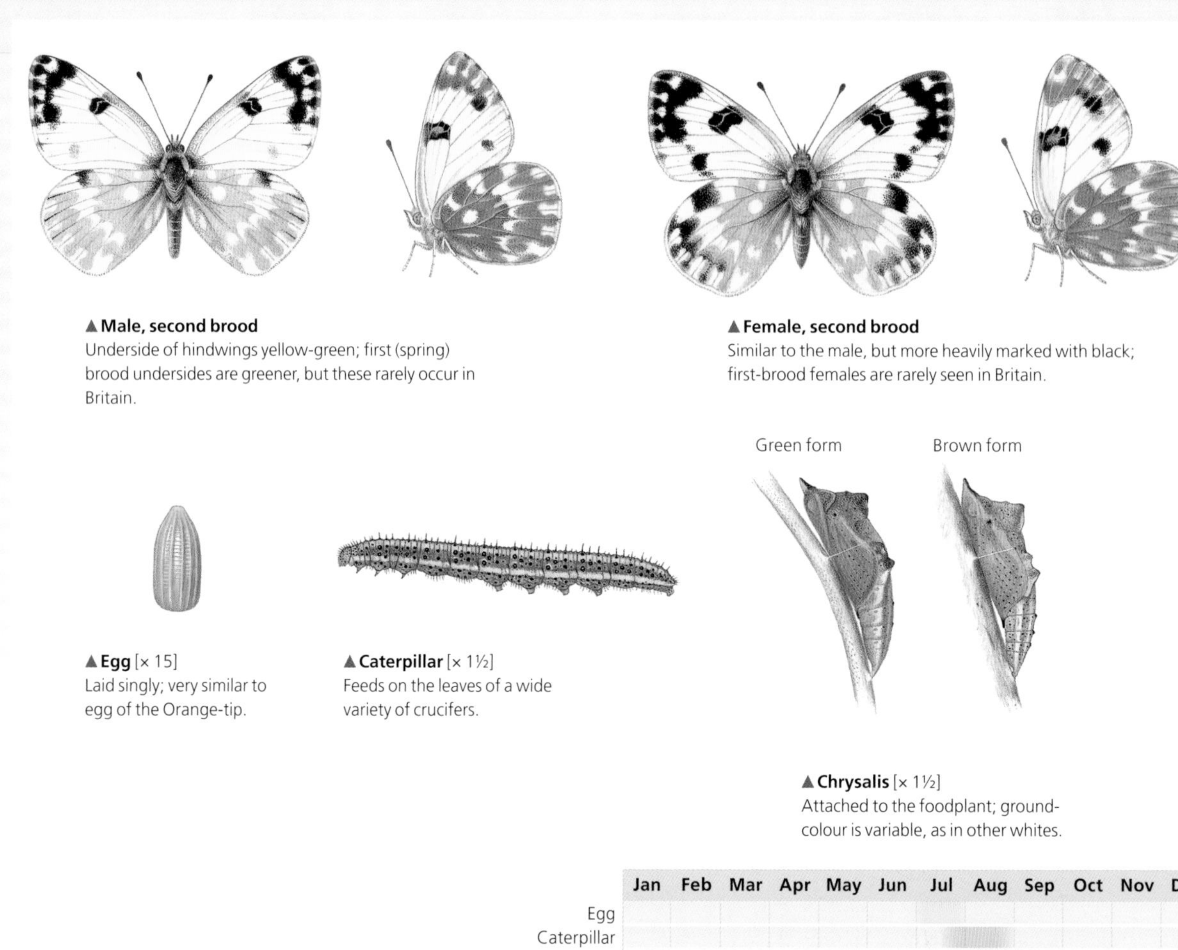

▲ **Male, second brood**
Underside of hindwings yellow-green; first (spring) brood undersides are greener, but these rarely occur in Britain.

▲ **Female, second brood**
Similar to the male, but more heavily marked with black; first-brood females are rarely seen in Britain.

▲ **Egg** [× 15]
Laid singly; very similar to egg of the Orange-tip.

▲ **Caterpillar** [× 1½]
Feeds on the leaves of a wide variety of crucifers.

▲ **Chrysalis** [× 1½]
Attached to the foodplant; ground-colour is variable, as in other whites.

Vernon's name was supplanted from 1795 onwards.

Records continued to be few and far between during the next 150 years. The butterfly was often absent, and fewer than 400 specimens were reported between 1826 and 1944. Moreover, half of these stem from a less than satisfactory record by W W Collins, who in 1906 reported seeing a 'swarm' of over 200 Bath Whites on the Dorset cliffs, west of Durdle Door. Unfortunately, Collins's account was written 30 years later, when just a single specimen survived to verify his memory, although the Revd F L Blathwayte recollects seeing four of them.

A year without parallel

The year 1945 was the most extraordinary known for many of our rarer immigrants. It started, for the Bath White, on 14th July, when in less than an hour C S H Blathwayt netted 38 specimens in a Cornish field, just exceeding Bernard Kettlewell, who took 37 in Cornwall that day. Together, they had collected more than the total number seen in Britain over the previous 50 years! Moreover, Blathwayt estimated that perhaps 200 were present in his field at Looe.

This was clearly part of a freak migration that embraced the whole of southern England, and even reached Ireland. Between July and late October, over 700 were reported. Most were probably immigrants, but numerous observations were made of egg-laying on Sea Radish, Wild Mignonette and, especially, on Hedge Mustard, and many caterpillars, chrysalises and emerging adults were found. Kettlewell reports seeing the females 'quivering over low weeds'.

The phenomenon was, alas, short-lived. Although sightings were more frequent than usual in 1946-51, there has since been a very lean period indeed. Around 20-25 Bath Whites have been reported in the past 58 years, similar to the poorest stretches of the 19th century. It is, however, one of the fascinations of butterfly-watching that there remain a few erratic rarities that can yet be found on British soil. Mid-July to late August is when the majority are seen, almost all from the southern English counties.

Orange-tip

Anthocharis cardamines

The Orange-tip is the prettiest of the springtime butterflies, and is fortunately still common across most of the British Isles. The male is especially conspicuous, not simply because of his orange wingtips, but also because he is a patroller *par excellence*, and spends much of the day wandering through the countryside searching every shrub and tussock for a mate. The female is more elusive. She hides among bushes for many hours each day, and has grey rather than orange tips to her wings. But she is equally attractive in her own way and has the same exquisite undersides as the male. Not for nothing was this butterfly once called the 'Lady of the Woods' or 'Wood Lady', a superior description to its prosaic modern name. Continental names are also more imaginative, including Aurorafalter (Sunrise Butterfly) in Germany and L'Auroré (Rising Sun) in France.

Warning signals and camouflage

Male Orange-tips are fine examples of both warning coloration and camouflage. When flying, they are conspicuous and hence quickly spotted by predatory birds. However, they are also highly distasteful, because their bodies contain large amounts of bitter mustard oils, accumulated from their foodplants during the caterpillar stage. It pays for the males to advertise this fact, and the bright tips of the upper forewings act as a warning signal every time they fly. Once a bird has tasted an Orange-tip it is reluctant to repeat the experience.

The resting butterfly sits still, with its wings closed, and is less likely to be noticed. Only the undersides of the hindwings are visible, and camouflage takes over as the most important form of defence. The exposed wing surfaces are delicately mossed with green, or rather with the illusion of green, for the lovely lichen-like mottling is created from an intricate mixture of black and yellow scales. They offer near-perfect camouflage when the butterfly rests on flowers of Cow Parsley or Garlic Mustard, the latter being one of its principal foodplants. This is the only protection for females, which fly so seldom that orange wingtips are unnecessary.

The Orange-tip hibernates as a chrysalis, and the first males emerge as early as March during a warm spring, although mid-April is more usual. The main flight is in May to early June. In exceptionally early seasons, there may be a small emergence of second-brood adults which fly, and

Distribution

A common nomad breeding throughout the lowlands of England, Wales and Ireland. Much more localised, but expanding, in Scotland. Irish specimens belong to a distinct sub-species, *hibernica*, distinguished by its slightly smaller size and darker markings across the upperwing fringes.

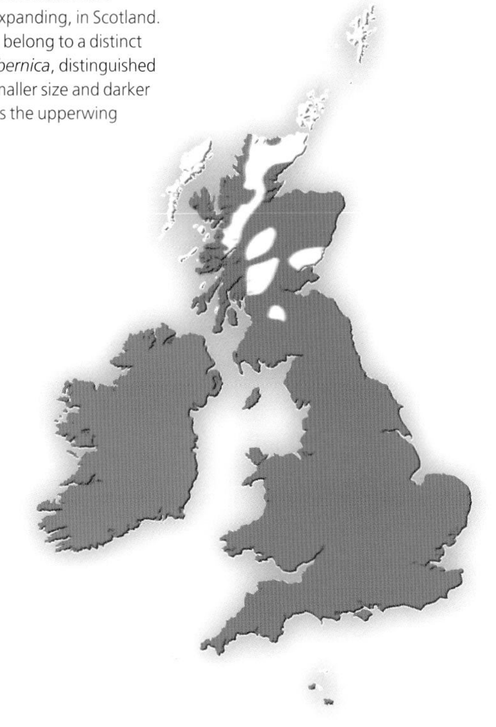

seem out of place, in high summer.

The natural history of this species is well known, thanks to studies by Stephen Courtney, Jack Dempster and Christer Wiklund. The adults in most regions live in loose, open populations, with both males and females wandering in no particular direction through the countryside, merely following hedgerows and wood edges as they search for mates or foodplants. This takes them into a wide variety of habitats in the south, including almost every garden. But they seldom stay long unless flowering crucifers are present. Honesty and sweet rockets will detain them a while, but I grow Cuckooflower (Lady's Smock), both for its lovely pale flowers and because the many eggs laid on it are more likely to survive. In the vegetable plot I grow Garlic Mustard.

Orange-tips behave differently further north, where they are more localised and live in more compact colonies. Nevertheless, this has not prevented a considerable expansion in range across Scotland in recent years, in response to the warming climate. On one site studied in Durham, between 175 and 300 males emerged each year, probably accompanied by equal numbers of females. These spent their entire lives patrolling back and forth along a short stretch of riverbank. This appears to be a typical size for a northern colony. Low densities are normal throughout most of the south, although there are many pockets where higher numbers can be seen.

Mating and egg-laying

For as long as the sun shines, the male Orange-tip flutters along hedges, shrubs and bushes in search of a mate. His initial approach is not discriminating, and he investigates any white object, including the Green-veined Whites that emerge in abundance on most Orange-tip sites. They are soon rejected, for they lack the female Orange-tip's scent. Little will stop a male once he has recognised a young female of his own species, especially if she is a virgin, whereas older females are examined for no more than three seconds. But once a fresh female is detected, he will force his way through the densest foliage, whereupon she, if receptive, signals her readiness by raising her abdomen at right angles for about four seconds, and mating begins. Curiously, the same posture is adopted by older females in an attempt to deter males that are pestering them.

The female later emerges to feed and lay eggs. She has a swift, no-nonsense approach compared with the investigative flutterings of most butterflies. Flying mainly along hedge-banks and the margins of fields, rides and glades, she brushes against or alights on the taller plants, quickly taking off again if they prove not to be flowering crucifers. Suitable foodplants are recognised first by sight from the air, and then chemically through sensitive cells on her feet. Although swift when egg-laying, she is extremely selective about which plants are chosen. Prominent, isolated, unshaded flowering crucifers growing within 1m or so of a hedgerow, bank or wood edge are greatly preferred, as are those with large flowerheads, for on Cuckooflower at least, small plants produce too few pods to feed a single caterpillar. Perhaps because of this, the young caterpillar is cannibalistic. Females generally avoid laying on flowerheads that already contain an egg, which they detect by the presence of deterrent chemicals (pheromones) deposited on the fresh shells.

Developing caterpillar

The pale, spindle-shaped egg is easy to find if you examine the undersides of crucifer flowers in June. It becomes even more conspicuous two to three days later when it turns pink and then deep orange. The vast majority are found on Cuckooflower on heavy soils and on Garlic Mustard on dry sites. For the reasons given above, eggs tend to be evenly distributed between suitable plants, with most containing a single example. They hatch after a week to a fortnight, and the small caterpillar burrows into the flower, which by now is well expanded. It feeds on the developing seed and soon sits as a black, long-haired maggot, exposed on the seed-pods. It has been claimed that these hairs are really glands, which produce droplets of sweet liquid to attract ants that protect the caterpillars from enemies. If true, this must be very rare in England, for I have examined hundreds of young Orange-tip caterpillars in the wild, and have yet to find one being attended by ants, or to hear of anyone who has.

The growing caterpillar soon feeds only on the seed-pod of its plant, becoming beautifully camouflaged as it grows. It lies along the top of the pod, eating inwards from the tip. Cannibalism accounts for about 10% of deaths in young caterpillars, yet despite this thinning out, a further 15% may starve as they grow larger. In addition, they have to contend with a host of natural enemies, despite the elaborate camouflage of the later stages. Overall, no more than a fifth to a third of eggs survive to form chrysalises. The other dangers early in life come from invertebrate predators that climb the foodplants at night. Once larger, birds are the main hazard, and on some sites over a third of the survivors then succumb to *Phryxe vulgaris*, a parasitic fly (see page 185).

The surviving full-grown caterpillars leave their foodplants to pupate. The chrysalis is exceptionally beautiful and, like those of most whites, exists in two colour forms. It hibernates in this stage, and we can estimate that about four-fifths of them are killed, probably by small mammals and birds. Even in the first weeks after pupation, it is extremely hard to find in the wild, but occurs in bushes and tall vegetation

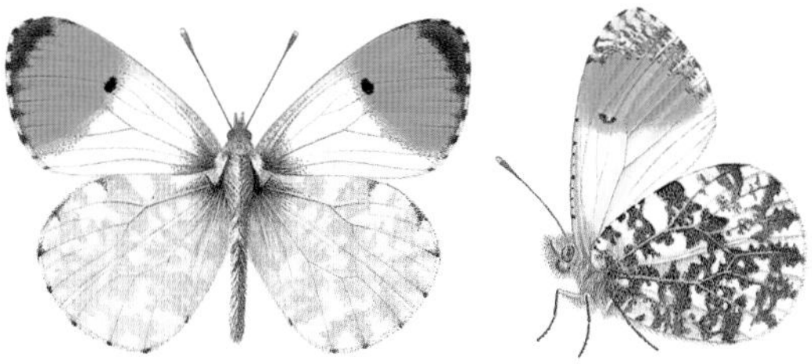

▲Male
The distinctive male emerges about a week before the female; underside of hindwings mossed with green.

▲Female
Similar to other whites, but undersides of hindwings mossed with green.

◄Resting adult
At rest, forewings are concealed by the camouflaged hindwings.

▲Egg [× 15]
Greenish-white when first laid, then becoming orange.

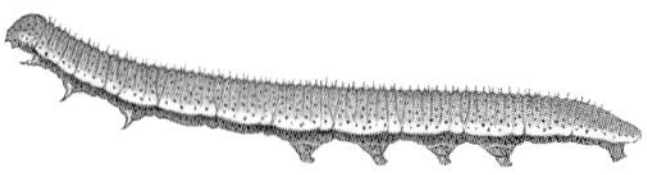

►Egg-laying
Eggs are laid singly under flower buds.

▲Caterpillar [× 1½]
Caterpillars may be cannibalistic if they meet on the same foodplant.

▼Chrysalis [× 1½]
Two colour forms occur but the brown form is by far the most common.

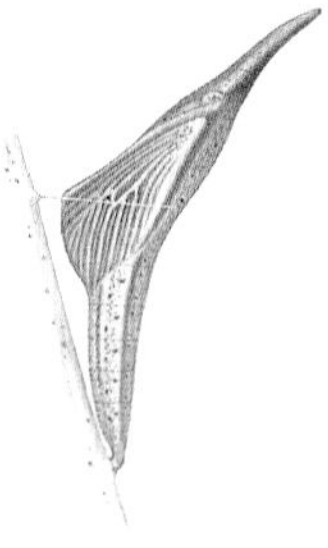

Brown form

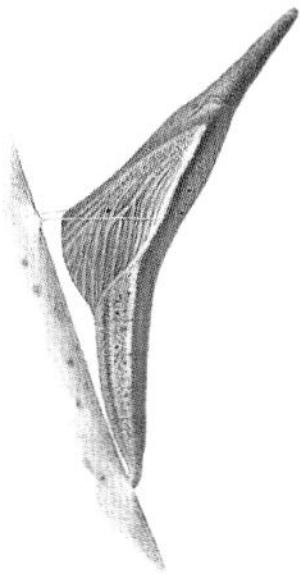

Green form

▲Caterpillar on seed-pod
Mature caterpillar feeds on the seed-pods of crucifers.

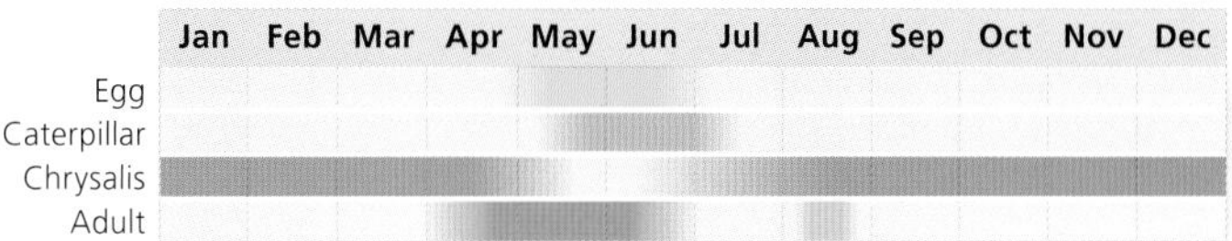

near the foodplants, where the virgin females later sit, waiting for males, the next spring.

Regaining northern breeding grounds

In spite of its many natural enemies, the number of Orange-tips seen in any year depends mainly on the weather when the eggs were laid during the previous year, and in the longer term on the availability of suitable foodplants. Grassland management can have a major effect. It is common to find that a meadow full of Cuckooflower and Orange-tip eggs has been grazed or mown long before the caterpillars are ready to pupate. Similarly, the policy of cutting many road verges in June – to increase visibility for drivers – must destroy countless thousands of caterpillars every year.

Such losses probably do no more than depress Orange-tip numbers in any region, preventing it from reaching the abundance often seen in abandoned meadows and woodland glades. Far more serious has been the widespread agricultural improvement of meadowland, which has eliminated Cuckooflower from the vast majority of damp meadows in the British lowlands.

Despite this decline in its foodplant, the Orange-tip remains a common butterfly throughout southern England, Wales and Ireland, emerging in record high numbers in 2011 on sites that have been monitored annually since 1976. It is still found in almost every wood and sheltered lane, and along a great many hedgerows and ditches. It is more localised further north where, however, it has experienced a remarkable spread throughout the lowlands of northern England and Scotland over the past 60 years. To a large extent, the butterfly was regaining ground it had lost after the early 19th century, and despite localised losses further south, its distribution today matches that of any earlier period in its history.

Green Hairstreak

Callophrys rubi

No British butterfly has a wider range of foodplants than this hairstreak, and few European species can match a distribution that extends from the coast of Lapland to arid Mediterranean *maquis*. Yet the butterfly is seldom common: in Britain, it abounds only on lowland English heathland and in warm western valleys from Cornwall to Inverness. It survives elsewhere in a handful of localities in most counties, and while not yet rare, it is a localised species that is declining everywhere.

Green Hairstreaks emerge once a year. The first adults generally appear in mid- to late April, although the date is variable depending on the warmth of spring, and reach a peak from mid-May to mid-June, with stragglers surviving well into July. The butterflies usually live in small, self-contained colonies. A typical population produces no more than five to ten males at the peak of the flight period, representing a probable emergence of 30 to 50 butterflies over the whole season. However, the species can be much more abundant. I have watched Green Hairstreaks teem by the dozen in Cornwall, hopping and jinking a few centimetres above Western Gorse, their vivacity contrasting with the sleepy atmosphere produced by the warm, still air, heavy with the almond scent of furze blooms. This kind of sighting is by no means unique: the species can be equally abundant on sheltered moors in western Scotland.

Perching posts

The Green Hairstreak is a charming butterfly to watch. Males are highly territorial, and sit for long periods spaced around the edges of their breeding sites, each perched on a prominent shrub. They always sit with closed wings, and adjust their body temperature either by leaning sideways to catch the sun or by positioning themselves head-on to avoid it, in the same way as the Grayling. Males usually perch 1-2m up, and can be hard to spot among the fresh green leaves. To find them, search in particular along the lower edges of a breeding site, tapping any prominent shrub that is in full sunshine. It is surprising how often a male will flip out from under your nose. Stand back and wait when this occurs, for after a few violent circuits he will return to his vigil, often alighting on the same leaf.

I once marked every male in a small colony in Devon, and was surprised to find how constant these perching posts

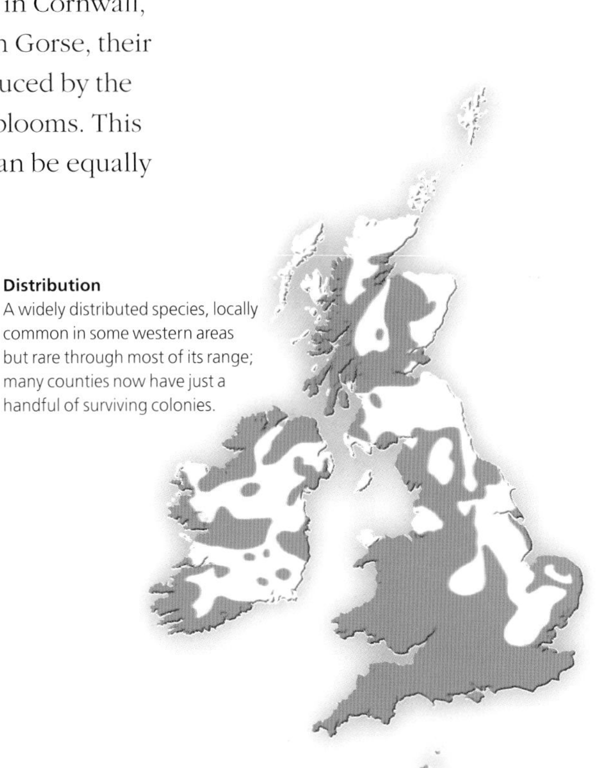

Distribution
A widely distributed species, locally common in some western areas but rare through most of its range; many counties now have just a handful of surviving colonies.

were over the season, yet how often the different individuals switched between them. Each would occupy his perch for perhaps an hour before launching into the air to investigate a passing insect. If his quarry was another male, a ferocious battle would ensue, with the rivals spiralling and looping around each other in tight circles until one was eventually vanquished. Females were pursued with greater persistence, during which time another male often stole the occupant's perch. This game of musical chairs carried on between fixed spots across the site, with males regularly changing places as perches became vacant on favoured shrubs.

Iridescent wings

The male Green Hairstreak is not only easy to photograph and watch while perching, but is also remarkably tame. It is easy, in cool weather, to persuade one to crawl onto your fingers, where he will happily probe for salts should your hand be slightly sweaty. He can be examined in detail if you orientate his head towards you and blow gently as you lift him, which makes him tighten his grip. Notice the black oval eye, ringed by gleaming bands of white and black, and the exceptionally hairy fringes and mouthparts, or palps, that gave this creature its generic name of *Callophrys,* which is Greek for 'beautiful eyebrow'.

The wings are lovely, too. They have a velvety look when fresh, a tail that is reduced to a mere stump, and a hair-line that ranges from a continuous white streak, through the typical series of dashes, to being absent altogether on a few specimens. Their colour is a most unusual feature: this is the only British butterfly to have wings that are truly green. The Orange-tip and Bath White both have pretty olive mottling on their undersides, but theirs is an illusion created by an intricate mixture of black and yellow scales. The Green Hairstreak's iridescent colour results instead from the refraction of light through a microscopic lattice of closely-packed cubes within each scale, producing a diffraction grating with peak reflectance at 0.55μ, which at any angle and in the dimmest light appears clear green.

Egg-laying

The female Green Hairstreak is more elusive than the male, but can be found slowly fluttering around shrubs and ground plants, frequently alighting to crawl and probe fresh leaves or buds to determine whether these are suitable for egg-laying. She uses a wide range of plants. The main British hosts are Bilberry flowers and fruit on acid moors, soft gorse shoots on heaths and neutral soils, and Common Rock-rose on chalk and limestone downs. Other foodplants include Broom, Common Bird's-foot-trefoil, various vetches, and the flowers of Dyer's Greenweed, Buckthorn, Dogwood and Bramble. The last is especially used in woods, and was probably a more important food in the past when British woodlands were open, sunny and more suitable for breeding. Indeed, Bramble was the only foodplant known to early entomologists, which is why the butterfly has the specific name of *rubi.*

Despite the wide range of foodplants, females are fussy over their choice of plants for egg-laying. No thorough study has been made, but I have noticed that they lay only among the tenderest young tissues, which probably contain the most nitrogen, an element essential for the caterpillar's growth. The female has an unusually flat ovipositor, which enables her to inject eggs into the tightest crevices of plants, such as deep between the soft growing points of gorse leaves. The egg itself is thin-shelled and flexible, and is moulded by the space into which it is squeezed. This is in marked contrast to the robust shells of the hibernating eggs of our four other hairstreaks, and thinner than the waxy, white eggs of most other lycaenids. This feature accounts for the green, shiny look of Green Hairstreak eggs.

The growing caterpillar

The eggs hatch after a week or two, and the small caterpillar burrows into nutritious, soft plant tissue, often feeding on flowers and growing pods or fruit. It is notable among caterpillars for being able to feed without harm on Dyer's Greenweed flowers, which contain strong concentrations of a toxic quinolizidine alkaloid. Indeed, the butterfly shows a strong preference for this unusual foodplant on sites where it has the choice. It does not, however, sequester this poison in its body to deter birds or other predators, but instead has some system for detoxification. As the caterpillar grows older it lives more openly on its foodplant, but even then is extremely difficult to find because of the excellent camouflage of its green and yellow body. On gorse, one trick is to look for a tip that is grey and withered, for here the caterpillar's camouflage fails. However, most of the wild caterpillars I have seen are those that I have observed from the egg stage. They are unexciting creatures that live hunched over the growing tip of their foodplants, apparently motionless, with their heads buried deep in its tissues.

By August, the caterpillar is fully grown. At this stage it deserts the foodplant to search the ground for a pupation site, and it remains here as a chrysalis until the following spring. The only one I have found in the wild was deep inside an ant nest. This is probably where all Green Hairstreaks hibernate, because captive chrysalises are highly attractive to ants. Not only do the ants lick the secretions that ooze over the hairy, brown cuticle, but they may also be attracted, or at least appeased, by the cluckings and churrings made by the chrysalis's sound organ. Most lycaenid butterflies seem to communicate with ants in this way, but the Green Hairstreak

▲Male
Each forewing has a pale sex-brand. The only British butterfly with green wings.

▲Female
The forewings lack the sex-brand, otherwise markings are indistinguishable from those of male.

◀Male
Form *punctata*, with white streak on forewings.

◀Female
Form *caecus*, lacking white streaks.

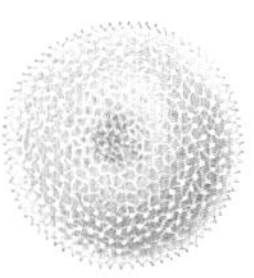

◀Egg [× 22]
Thin-shelled; laid in crevices of a wide variety of foodplants.

▼Perching adult
Male on gorse flower, ready to dart out at any passing insect.

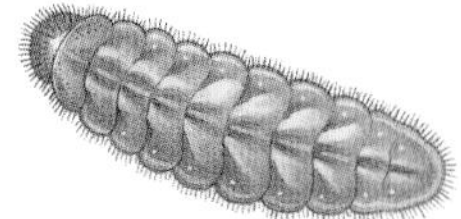

▲Caterpillar [× 2¼]
Feeds on a wide variety of foodplants, depending on habitat.

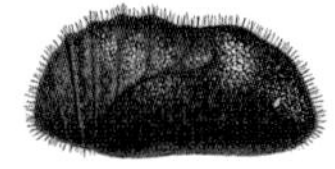

▲Chrysalis [× 2¼]
Formed on or below ground, tended by ants, and may overwinter in an ant nest. Produces audible squeaks.

▲Caterpillar on gorse [× 1½]
Perfect camouflage protects the caterpillar while it feeds.

	Jan	Feb	Mar	Apr	May	Jun	Jul	Aug	Sep	Oct	Nov	Dec
Egg												
Caterpillar												
Chrysalis												
Adult												

is remarkable for the loud volume of its stridulations, which are clearly audible to the human ear as a series of squeaks. Indeed, it was in this species that the extraordinary phenomenon of 'singing' chrysalises was first noted, more than 200 years ago.

Scattered colonies

Colonies of Green Hairstreak are found in a wide range of habitats, including sunny woods, wet moors and dry chalk downland. The two features common to most sites are that they are warm and sheltered, and that shrubs are always present. No-one has yet explained why this is such a local butterfly in most of its range, although its need for an abundance of succulent foodplants and high densities of ants, possibly of particular species, may be part of the story.

At present, the largest British colonies of Green Hairstreak are found in the west, from the huge populations that sometimes develop on Bilberries in Scotland to those of the Welsh, Devon, and Cornish coasts, where Gorse and Western Gorse are the main foodplants. In the west of Britain, Green Hairstreaks can still be expected among dunes and in any warm, sheltered valleys that contain its foodplants, as well as on most lowland heathland throughout the country where Gorse is common. It is also to be found across much of Ireland on sheltered patches of rough ground containing its foodplants.

Elsewhere in the British Isles it is usual to find a small population of Green Hairstreaks on any southern chalk or limestone down where Common Rock-rose is abundant, provided the site contains reasonable areas of scrub. On flatter land, however, there have been innumerable extinctions. These result from modern forestry (which leaves most woods too cool and shady for this sun-loving insect), the reclamation or agricultural improvement of old grassland, heaths and moors, and the general tidying-up of the countryside. Together, these changes have reduced the Green Hairstreak to a very few colonies per county in central and eastern England, occurring mostly in woods or along railway embankments and cuttings. Despite these losses, this remains the most widely distributed of the five British hairstreaks, and easily the commonest of the group in Ireland, Scotland and northern England.

Brown Hairstreak

Thecla betulae

This is the largest, brightest, and perhaps most attractive of our five native hairstreaks. Unfortunately, few people see the adult in the wild for, although the butterfly is scarce rather than rare, it spends almost its entire adult life perched out of sight on a treetop. The golden-coloured females are somewhat easier to spot, because they briefly descend in August and September to lay eggs on twigs of Blackthorn and its relatives, along hedgerows, wood edges and around sheltered patches of scrub.

Distribution
Very local in wooded regions in the south. Common only in the Burren, south-west Wales, north Devon and the west Weald.

I had the pleasure of studying the ecology of this butterfly in all the stages of its life cycle during a six-year period in the 1970s. My main site – a hotch-potch of clearings, wood edges and overgrown hedges on the Wealden clays of west Surrey – contained one of the largest known colonies in Britain. Nevertheless, only about 40 adults emerged in the worst year, and no more than 300 survived in the best, when they laid just over 4,000 eggs. By comparison, in some years at least a million Purple Hairstreak eggs were laid on the oaks growing in the same area.

Brown Hairstreak eggs are laid at low densities over wide areas of countryside, although it is not unusual to find two, three, or even four eggs on a particularly suitable young twig, almost always Blackthorn. The colony I studied was supported by nearly 6,000 Blackthorn bushes growing along 42km of hedges and wood edges, compressed into an overall area of about 30ha. This, however, was an unusually compact example. Most Brown Hairstreaks breed over considerably larger areas encompassing hundreds of hectares. Yet, even on large sites, the same areas are used for breeding year after year, with few eggs laid beyond the traditional boundaries.

Master trees

This widely dispersed distribution presents the Brown Hairstreak with some difficulties when it comes to finding a mate, for no more than one or two adult butterflies may emerge per kilometre of hedgerow during the three- to four-week emergence period. Like the Purple Emperor, the species solves the problem by using 'master trees' – particular treetops where the males congregate and to which virgin females fly as soon as they emerge. Very few Brown Hairstreak master trees have been discovered in Britain. Typical examples are Ashes growing near the lowest point of the basin of countryside that contains a colony: all are large, bushy trees that tower high above the canopy of the adjoining woodland.

In the site I studied, the same Ash tree was used year after year, and I never saw males anywhere else. Using binoculars, it was possible to watch the butterflies fidgeting and wandering over the leaves in sheltered pockets on the canopy, basking in the sunshine or drinking the sticky honeydew that coats so many Ashes in August. Sometimes they took to the air, spiralling in rapid loops before returning to the leafy platform.

Adult males appear to spend their entire lives perched

on the master tree, except occasionally in years when they descend to feed on flowers. They are extremely tame when feeding and can be approached closely and photographed. But these descents are unpredictable, and probably occur only when honeydew is scarce on the master tree.

Treetop courtship

The courtship and mating of the Brown Hairstreak has been observed only once, so elusive are the butterflies on their high treetop canopies. No-one knows quite how long the females remain at the master tree, but my own observations suggest it may be for the six to ten days that it takes for their eggs to mature. They then disperse over the extensive breeding areas. There is evidence that only a limited number of females remains within the fixed areas of the colony, and that once 15 or so have gathered on the master tree, the surplus emigrates beyond the colony boundaries. This appears to limit individual colonies to densities seemingly well below that which could be supported by the surrounding Blackthorn bushes. Quite why this should be is, as yet, unknown. It may be relic behaviour that has survived from earlier times, when the Brown Hairstreak lived in a very different landscape of mixed woodland and clearings, and had to colonise the periodic gaps that arose when decrepit trees were blown over in the ancient Wildwood. Nowadays, however, there are few unoccupied breeding sites, and it seems likely that few females that desert their colonies will find new habitat in our modern agricultural landscape.

Females that remain in the colony fly only on the warmest days, and are seldom seen before 10am or later than 4pm. They spend long periods basking in weak sunlight, with their wings opened wide, allowing the dusky upper surfaces to absorb the maximum heat. As the sun gets warmer, less of the upper-wings are exposed until, in the hottest weather, they are kept tightly closed. The shiny undersides and white hairs on the lower half of the body then reflect rather than absorb the light, and prevent the butterfly from overheating.

It is unusual for a Brown Hairstreak to fly in air temperatures lower than about 20°C. When it is warm enough, the females descend in rapid, jinking flights, hugging the wood edges or hedgerows and seldom crossing bare ground. Their golden colour is particularly noticeable in flight, so much so that they were once thought to be a different species from the duller males, and were known as Golden Hairstreaks.

Egg-laying

Unlike the males, female Brown Hairstreaks regularly feed on late-summer flowers such as Common Fleabane, Bramble and thistles. Each female feeds in distinct bouts, punctuated by long periods of egg-laying when she flies along woodland edges and hedgerows, periodically alighting on a projecting leaf. She then taps the upper surface with her front legs, tasting it through the chemical receptors near their tips. If the plant is a Blackthorn or other species of *Prunus,* she next begins a curious, crab-like descent down the twig, edging sideways and backwards into the bush while probing every nook and cranny with the curved tip of her plump, rounded abdomen. This slow, methodical examination may take several minutes, but a suitable spot is usually found, and she squeezes out a single, bun-shaped egg. This is always laid on bark, on young growth usually at the base of a spine, or where one-year-old wood branches from a two-year-old stem. In Britain, the vast majority of eggs are laid on Blackthorn, although I know a few sites where Bullace supplements it. The eggs are generally no more than 1m above the ground.

The Brown Hairstreak has one generation a year, and is one of our last butterflies to take to the air. Adults emerge in late July and August, but egg-laying females are seldom seen before mid-August, and reach a peak of activity in early September. There is little time for the egg to develop, so it enters hibernation when the embryo is a mere ribbon of white tissue suspended in the rich fluids within the shell. Our three other woodland hairstreaks, by contrast, hibernate as fully formed little caterpillars within the egg.

Early development

The eggs remain on their twigs for eight months before hatching in late April and early May. They are gleaming white, with a beautiful raised sculpture over the surface, rather like intricate icing on a cake. Despite being widely spread, the eggs are easy to find from November to April when the dark Blackthorn twigs are bare. It is, in fact, far easier to see the egg than any other stage of this elusive butterfly: I have often found 100 or more in a day on a good site. To locate them, look particularly on the lowest projecting growth that is both exposed to the sun yet sheltered from the wind. Small one- to two-year-old suckers projecting from the body of a shrub also tend to be selected.

On undisturbed sites between half and three-quarters of the eggs hatch, with an unidentified disease accounting for most deaths. Predatory insects also kill small numbers, although, unlike other hairstreaks, they are only rarely attacked by parasitic *Trichogramma* wasps. There are, however, very few undisturbed Brown Hairstreak sites in Britain, and the majority of eggs in most colonies are destroyed by hedge-trimming. A severe trim can remove every egg from a hedge, but on average about one-fifth survive on cut hedges. Uncut woodland edges form vital sanctuaries for the survival of the species.

Like other butterflies that hibernate as eggs, the eggshell is thick and extremely tough. In spring, the tiny caterpillar

▲**Male**
Wing-tails orange, but otherwise predominantly dark. Underside is marked with white stripes.

▲**Female**
Larger than male, with an orange-gold band on forewings. The white stripes on underside are more pronounced.

▼**Egg** [× 22]
Laid singly on Blackthorn twigs, near the junction of one- and two-year-old growth.

►**Male**
Rare aberrant form *unistrigata* with reduced stripes.

►**Female**
Rare aberrant form *uncilinea* with broken stripes.

◄**Chrysalis** [× 2¼]
Formed on the ground, where it is tended by ants.

▲**Caterpillar** [× 2¼]
After first moult, caterpillar lives beneath leaves of foodplant.

▲**Basking adult**
Females spend much of their lives basking in the sun, flying only on warm days.

▲**Eggs** [× 22]
The thick overwintering egg case of a Brown Hairstreak (left) compared with a Common Blue egg.

takes up to a day to nibble a neat, round hole in the top of its shell before squeezing out, leaving the empty case so firmly attached to the bark that this can be found up to a year later. The caterpillar immediately crawls into an unfurling leafbud, but emerges after the first moult about a fortnight later, to live the rest of its life dangling upside down on a silk pad spun on the undersurface of a leaf. It sheds its skin twice more, looking the same but growing larger each time. Wedge-shaped in profile, with yellow stripes on a pale green background, it is extraordinarily well camouflaged beneath the leaf. Despite this, up to four-fifths of caterpillars are found and killed by predators. Harvestmen, spiders and insects are the main culprits during the first month, and the moment they grow too large for these enemies they are instead picked off by Willow Warblers, tits and other insectivorous birds.

Locating the caterpillars is not easy, although it is by no means impossible if you patiently turn the leaves of a bush known to have contained eggs. A surer method is to beat the bushes sharply with a stick, and to sort through the variety of caterpillars that tumbles onto a sheet held beneath. It is a method that I avoid, though, for the branches are left twisted, broken and brown later on in the year when the adults are around, reducing the pleasure of watching them.

Chrysalises and ants

The slug-like caterpillars remain motionless on their pads all day, but slowly wander over the bushes at dusk, browsing the tenderest leaf tips. Each usually returns to the same pad after its meal, although a new one may be spun nearby every week or so. Then, between 40 and 60 days after hatching, the caterpillar becomes a mottled purple colour and crawls to the ground to find a pupation site. Captive caterpillars are highly attractive to ants at this stage, although I have yet to see one tended by ants in the wild.

Ants certainly tend the chrysalis, which calls them with chirruping noises, as some blues do. I have seen few wild chrysalises, but, through following many caterpillars that were placed in natural positions, I am convinced that the majority pupate in cracks in the ground, in tussocks, or within the curl of a dry, dead leaf. Some were soon found by ants and buried in a loose cell of dry earth, and tended incessantly for the four weeks before emergence. It may be that, in the wild, all chrysalises are tended and protected by ants, or even that they enter the nests, like the Purple Hairstreak.

Despite its retinue of ants, this is a dangerous stage in the butterfly's life cycle. Small mammals, in particular, find the speckly brown chrysalis irresistible, and very large numbers are eaten. Up to four-fifths of the whole colony was killed during the pupation period on my Wealden site. The main culprits appear to be mice and shrews. In captivity, both show an extraordinary ability to sniff out chrysalises. Once one is unearthed, a shrew pounces on it in a frenzy of excitement and squealing, tearing and scattering the case into tiny fragments, while gobbling up the sticky contents. A mouse is more restrained. It sits up on its hindquarters holding the chrysalis in both hands, as a squirrel might a nut. It neatly nibbles the chrysalis until not even the hard cuticle remains.

Both shrews and mice could find and eat up to one chrysalis a minute in my captive pens. Large beetles, by comparison, managed about one chrysalis a week. In this case, the shell is slit right around the edge, as if cut by an old-fashioned tin opener, and the jagged halves are prised apart to enable the beetle to feed on the rich contents. Only rarely have I found this damage in the wild, and it appears that ground beetles pose little danger to Brown Hairstreak chrysalises in comparison to small mammals.

Hedgerow habitat

Brown Hairstreaks lay enough eggs to withstand these heavy losses, provided their habitat remains intact. A colony needs several kilometres of bushy hedgerows or woodland edges, and at least a third of these must remain untrimmed in any one year. Suitable breeding sites occurred more frequently in the past, when hedges were cut by hand. The trim was neither uniform nor as deep as it is today, while the practice of layering ensured that every hedge was left uncut for roughly two years out of ten. Most modern hedges are annually trimmed down to uniform rectangles, often more in the name of neatness than from a necessity to keep them stockproof. The widespread grubbing-up of hedgerows has also contributed to the disappearance of this butterfly from East Anglia and many flat regions of Britain.

It is likely that the Brown Hairstreak was considerably more common in centuries past, for despite its reclusive behaviour, this 'fly' was well known to the early English collectors. In *The Aurelian Legacy*, Michael Salmon shows a photograph of a female caught in 1702 by James Petiver, the father of British butterflies, in Croydon, a neighbourhood from which the butterfly has long since disappeared. It is beautifully preserved between two transparent sheets of mica, bound with gummed paper round the edges like an old-fashioned lantern slide. Petiver named it, rather clumsily, 'The Golden brown double Streak', and considered it a different species from the male, 'The brown double Streak'.

Although the Brown Hairstreak has declined steeply in recent decades, it remains common in a few localities, albeit generally much overlooked. There are three regions where the butterfly is so widespread that one can hope to find eggs on any suitable-looking stretch of Blackthorn. One is on the heavily wooded clays of the western Weald, stretching from Horsham to Haslemere. The second is in the sheltered, low-lying valleys sandwiched between Exmoor and Dartmoor, from as far west as Torrington extending in a band across Somerset around the southern borders of Sedgemoor to the limestone base of the Polden Hills. The third is in similar countryside in the south-west quarter of Wales. There are several colonies in addition to these – one in the Blackmoor Vale of Dorset, three in Hampshire, and perhaps several centred on Bernwood Forest, of which the latter appear to be expanding so widely that eggs were recently found on the fringes of Oxford's city centre. It also breeds in sheltered valleys in Gloucestershire, on the Isle of Wight, and in one or two colonies in Worcestershire and Lincolnshire. Sadly, the butterfly has disappeared from the vast majority of eastern and northern sites, including the whole of East Anglia and the Lake District.

There are no authentic records of Brown Hairstreak from Scotland, and for many years there was fierce controversy about its occurrence in Ireland. Fortunately, surveys have confirmed that this most elusive of butterflies maintains a very healthy stronghold among the shrubs and scrub growing over a wide area of the Burren, where I have myself found eggs with ease in previously unrecorded spots. There is every reason to hope, therefore, that this lovely hairstreak will be rediscovered in its other reputed localities, in the southern counties of Cork, Waterford, Wexford and Kerry.

Purple Hairstreak

Favonius quercus

This is the commonest hairstreak in England and Wales, but colonies are often overlooked because the adults spend most of their lives in treetops. Individuals occasionally descend to flowers, but these tend to be faded and tattered, and approaching the end of their lives. Alternatively, pristine adults that have just emerged from chrysalises hidden within ant nests can be found resting on foliage a metre or so above ground level in the early morning, especially after periods of rainy weather, too cool for flight.

The Purple Hairstreak is easy to identify when seen at close quarters. It is the only British hairstreak to have an eye-spot on the underside, next to the short tail, and the only one with blue or purple markings on the upperwings. The uppersides are frequently displayed, for the adults bask with their wings open, unlike their relatives, the Green, Black, and White-letter Hairstreaks, which always sit with folded wings.

Life in the treetops

Like most lycaenids, Purple Hairstreaks live in self-contained colonies, and are rarely seen far from the oaks on which they breed. A single isolated tree can support a colony, but most populations are confined to woods. Numbers fluctuate enormously from one year to the next. I counted the hibernating eggs for seven years running on the same boughs of 14 oaks on a site in Surrey. There were only 26 eggs on these branches between 1973 and 1974, but three years later the number had increased to almost 1,500. This count was from a minute proportion of the oak boughs on the site, and the whole wood must have contained well over a million eggs that year, producing about 50,000 adults. Such population explosions have always been a feature of Purple Hairstreak colonies, and tend to occur after warm springs and summers. Numbers were particularly high, for example, in the early 1980s.

This butterfly has one generation a year and hibernates as a fully formed, unhatched caterpillar inside the egg. The first adults emerge in early July, reaching a peak towards the end of the month, then flying throughout August and often lasting into September. They fly to a treetop on emergence, but until recently little was known of their subsequent activities. Quite the best description of how adult Purple Hairstreaks behave has been provided by David Newland,

Distribution
The commonest British hairstreak, much overlooked. Probably present in every southern oakwood, but scarce in the north and in Ireland.

who observed them for 23 hours over four days from a gazebo overlooking the canopy of Sheringham Park, an oak plantation in Norfolk.

Newland confirmed that the Purple Hairstreaks seldom flew during the main part of the day, least of all in dull weather. Most of their lives were spent perched in sheltered, sunny nooks in the canopy, where they rested, basked or slowly walked in circles, dabbing at the leaves for honeydew. Like Purple Emperors, they seldom, if ever, perched on the highest trees. On the contrary, the adult butterflies congregated on one old oak that was shorter than its neighbours, providing a sheltered yet unshaded canopy. During the daytime, males and females perched, basked or crawled over the leaves, with little interaction between members of each sex. As evening approached they became more active. From about 5pm to 8pm the males established perching posts at higher levels than their daytime positions, yet still avoiding the uppermost branches of the highest trees. From there they launched themselves after any passing female, or engaged in prolonged aerial battles with rivals, spinning and diving at high speed above the treetops. From the ground they look like a handful of silver coins that has been tossed into the sunlight, and are one of the wonders of the British butterfly season; watch them through binoculars to appreciate their speed and dexterity of flight. At Sheringham Park, Newland found females were less conspicuous in the early evening, but those that did fly were chased by males: he describes how one encounter led, on landing, to mating, after a brief face-to-face contact of the two butterflies' antennae.

In the daytime, it is possible to walk beneath a tree bearing several hundred Purple Hairstreaks without being aware of their presence. I search for the butterflies by tapping the lower boughs. It is surprising how often an adult is dislodged, and occasionally scores fly up together in a shimmering cloud of purple and silver. Nor, in mixed woodland, are adults confined to oak, especially when other trees or bushes drip with honeydew. Ash is a particular favourite, and I have regularly dislodged hundreds around the wooded edges of Dartmoor from 3-6m-high Buckthorns, whose flowers they find irresistible. Later in the season, most sightings are of females. These have a more drunken flight, as they weave around oak boughs, searching for places to lay eggs throughout the sunlight hours.

Eggs and caterpillars

Purple Hairstreak eggs are laid on branches that are in full sunshine and, for preference, partly sheltered from the wind. They are laid at all heights over the canopy and most are found on the southern edges of trees. Each is about the size of a pinhead, but is reasonably conspicuous after the oak leaves have fallen, looking like a small, pearl-grey bun stuck on the base of a plump flower-bud, or on the rough parts of adjoining twigs. Although laid singly, two or three may be found together in good years. Hatching occurs in April, just as the oak flower-buds are starting to break. The tiny caterpillar bores into the heart of a bud, and feeds out of sight on the tender tissues. The first skin is shed around ten days later, and at about that time the caterpillar emerges to spin a silk web around the base of its expanding bud, which traps the brown scale-leaves when they peel off their buds, forming a cocoon.

The mottled, woodlouse-shaped caterpillar is perfectly camouflaged in its cocoon, emerging only under the safety of darkness to browse on fresh oak leaves. This affords it considerable protection: about half the caterpillars hatching from 104 eggs I once studied in a Surrey wood survived on oak trees, whereas four-fifths of the colony died during the three- to four-week period between leaving their trees and emerging as adults.

Living with ants

There has always been uncertainty about where the Purple Hairstreak forms its chrysalis. Some claim that it pupates under moss on branches, but I have never found them there, nor know of any first-hand evidence that this is so. Early collectors sometimes found them while sieving soil for moth chrysalises on the ground beneath oaks. Having failed to find any myself, I once carried out a systematic search of every crack, tussock, and patch of soft earth beneath an oak that had held a large number of caterpillars. Working outwards from the trunk, I found nothing in the first 5m, until I reached the outer edge of the oak boughs, where the grass grew in sparse clumps. I then pulled open a tussock that contained a large nest of the red ant *Myrmica scabrinodis*. As the crusty earth walls subsided and the ants poured out, I was amazed to find two Purple Hairstreak chrysalises in the heart of the brood chamber. Further searching soon uncovered chrysalises in four more nests, including those of the ant *M. ruginodis*. Yet not one of 76 antless tussocks that I examined contained a Purple Hairstreak.

This observation does not prove that every Purple Hairstreak forms its chrysalis inside an ant nest, but it suggests that this is the case. Both caterpillar and chrysalis attract ants through secretions, and both are able to 'sing': the former, we believe, by contracting its muscles to force air through the trachaea, the latter by rubbing together a tooth-and-comb structure on adjacent segments of the abdomen (see page 92). It would be intriguing to know whether the fully grown caterpillar finds its own way into the nest, or is dragged there by the ants. The latter explanation is a distinct possibility, for red ants carry home any sweet object that they can drag, and the caterpillar, with its swollen lobes that are easy to grip, is just light enough for determined ants to carry.

If, as is likely, the Purple Hairstreak depends on ants during

▲**Male**
Male's purple sheen appears almost black from certain angles. Both sexes have an eye-spot on the underwing next to the wing-tail.

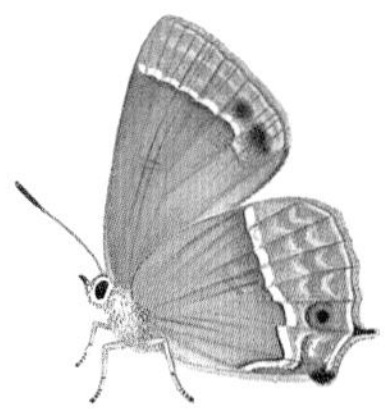

▲**Female**
Reduced purple area is visible from all angles. Underwings are very similar to those of male.

◀**Variation**
Very rarely aberrations occur like this specimen with obscured markings.

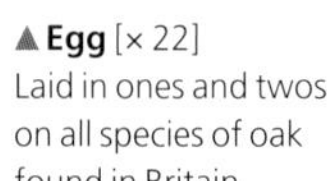

▲**Egg** [× 22]
Laid in ones and twos on all species of oak found in Britain.

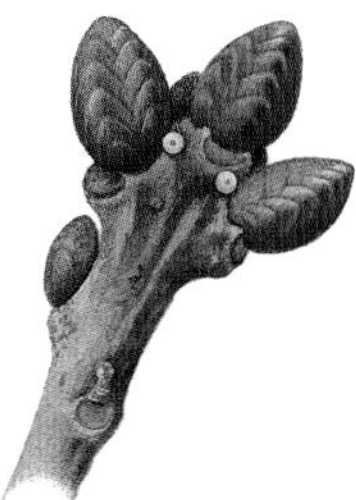

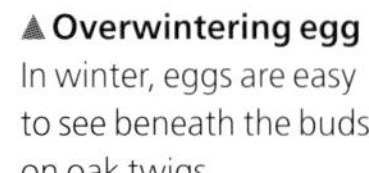

▲**Overwintering egg**
In winter, eggs are easy to see beneath the buds on oak twigs.

▼**Basking adult**
Adults feed and bask, with their wings open, on the tree canopy.

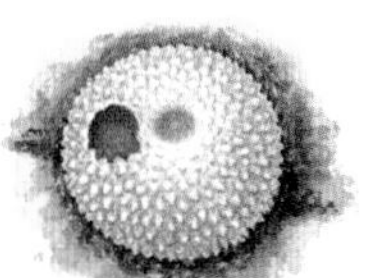

▲**Parasitised egg** [× 15]
Egg showing exit hole of a *Trichogramma* wasp.

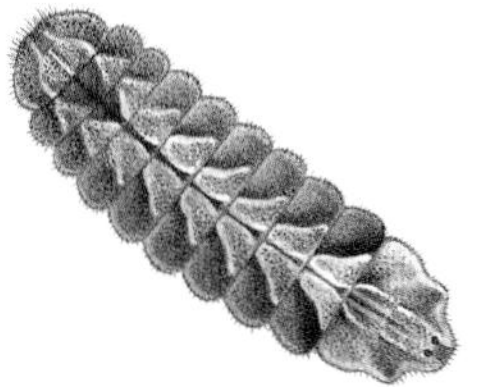

▲**Caterpillar** [× 2¼]
Colour and shape camouflages the developing caterpillar on oak buds.

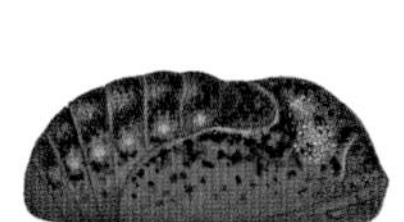

▲**Chrysalis** [× 2¼]
Formed on or below ground, sometimes in ant nests.

	Jan	Feb	Mar	Apr	May	Jun	Jul	Aug	Sep	Oct	Nov	Dec
Egg												
Caterpillar												
Chrysalis												
Adult												

Attracting ants

The chrysalises of most lycaenid butterflies are protected by ants. They attract ants with secretions from microscopic pores, and by squeaking like adult ants.

Chrysalis 'songs' are produced by the toughened edges of two abdominal segments. These bear a set of teeth and grooves, which squeak when they are rubbed together.

its vulnerable chrysalis stage, this might explain why certain woods consistently support far larger colonies than others. The ideal site for this butterfly is probably a wood that contains oaks of mixed ages, so that the canopy undulates, providing a large surface area of warm, sheltered boughs and sunny clearings beneath which ant nests exist in large numbers. Old-fashioned coppices that contained low densities of standard timber trees probably fulfilled all these conditions. However, the butterfly has survived well in modern woods. Even in conifer plantations, it is common practice to leave a scattering of oaks as nurse trees, or to have a fringe around the edge for cosmetic reasons. These can support remarkably high densities of Purple Hairstreaks, and even isolated hedgerow, park or garden trees can support a small colony, at least in the south.

Thus, despite some losses as a result of the mass planting of conifers, the Purple Hairstreak is comparatively unscathed, and still exists throughout its traditional range. It is especially common in southern England and lowland Wales, occurring in almost every wood that contains a reasonable scattering of oaks, and even being found in many places where oaks are scarce. Colonies are few and far between further north, but are probably much overlooked. The Scottish entomologist, George Thomson, considers that colonies can be found in almost all suitable habitats north of the border. But the butterfly appears to be genuinely scarce in Ireland, where it is largely confined to hillside oakwoods between Wicklow and Derry, and around the southern coast.

White-letter Hairstreak

Satyrium w-album

This vivacious little hairstreak is usually seen as a silver speck, tumbling in the July sunshine and circling around a treetop. Its flight period overlaps with both the Black and the Purple Hairstreaks, and it can be mistaken for either relative on the wing. All three species fly in the same erratic way, and all have a sheen to their undersides that glints at certain angles in the sun. The Purple Hairstreak looks very different when settled, but White-letter and Black Hairstreaks are similar. The White-letter Hairstreak actually has the blacker wings. Indeed, it was known as the 'Dark Hairstreak' to early collectors.

In practice, the White-letter Hairstreak's dusky uppersides are never seen in the wild, for the butterfly always settles with its wings closed, regulating its body temperature in the same way as a Grayling, by standing at different angles to the sun (see page 256). The markings on the undersides are generally constant. Individuals may vary in the boldness of the white hairline and in the length of the tail, but major aberrations, such as the famous specimen we illustrate, are extremely rare.

Distribution
Widespread but local species that has declined greatly due to Dutch elm disease. Elusive and often overlooked where it does occur.

A diet of elms and honeydew

Adult White-letter Hairstreaks emerge during the first week of July in a typical year, and are on the wing for at least a month. Each colony breeds on either a small clump of elms or on a single elm tree. Most are found on the sunny edge of a wood, but there are many others that are supported by an isolated grove or a single hedgerow elm. Solitary adults have occasionally been reported from gardens and other unusual habitats in recent years, perhaps because their breeding elms had succumbed to Dutch elm disease. But, as a rule, this is a fairly sedentary species: hedgerow colonies remain strictly around each breeding tree, and even in woods the adults seldom fly further than a few treetops away, often settling on a neighbouring oak or Ash that is sticky with aphid honeydew, their principal food. A typical colony contains a few dozen adults, although in occasional years the butterfly is more abundant. Even then it is usually elusive. There are a few sites where the adults regularly descend to feed on Wild Privet, Creeping Thistle and other flowers, and this may occur widely in certain years. But this is rather a rare event caused, I suspect, when honeydew is in short supply on treetops.

It is easy to underestimate the numbers present in a White-letter Hairstreak colony. I remember once sitting for more than an hour beneath an oak in Monks Wood, in Cambridgeshire, watching through binoculars as the adults spun up above the canopy to alight on the leaves. I guessed that there were perhaps 20 or 30 butterflies altogether, and then climbed an adjoining tree for a closer look. Here I had a good view over about half of the oak's canopy, and I was astonished to see at least 70 White-letter Hairstreaks, hidden high above the ground. Some slowly rotated their hindwings as they perched in the sunshine, while others walked over the leaves, trailing their probosces between their legs. From time to time one would encounter a patch of honeydew, causing it to twitch and spiral while it dabbed at the sticky syrup.

Egg-laying

Pairing probably also occurs on the canopy, although I know of no-one who has seen this take place. The egg-laying female is more conspicuous. She flutters around her breeding elm before alighting to crawl crab-like along the twigs, probing every nook with a plump, bent abdomen before laying a single egg on a rough patch of bark. The egg is shaped like a miniature flying saucer, and is one of the most attractive of all British butterflies. Many are killed in their first three weeks of life by parasitic *Trichogramma* wasps, but they appear to be immune from mid-August onwards. By this time, each contains a perfectly formed little caterpillar, which remains in the egg until the following spring.

White-letter Hairstreak eggs are laid at all heights on elm trees, and are easy to spot on low boughs once the leaves have fallen. They are, indeed, somewhat easier to find than the adult butterfly. Most eggs are laid beneath flower-buds, in forks, and particularly on the wrinkled girdle scar at the junction of the current and previous year's growth, usually on a sunny sheltered twig. English Elm, Small-leaved Elm and any of the hybrids may support a colony, but the favourite foodplant is Wych Elm.

Camouflaged young stages

The eggs hatch just as the elm flower-buds are swelling, usually in mid- to late March. The tiny caterpillar lies first on the dark sepals, burrowing its head into the opening bud to feed on the soft tissues within. It gradually changes colour as it grows, in much the same way as the Black Hairstreak, although in this case the camouflage bears a remarkable resemblance to an expanding elm bud. By its second moult the caterpillar has developed a pretty lilac saddle, and moves from bud to bud, scooping out the contents but leaving the outer scales intact. If the tree is more advanced, it hides and feeds among elm flowers from the outset.

As it grows older, the caterpillar feeds increasingly on leafbuds, and finally on the fully expanded leaves. At this stage it rests on a silk pad spun beneath the leaf, which it closely resembles. Despite this camouflage, full-grown caterpillars are quite easy to spot as dark silhouettes if you stand beneath an elm in early June, looking upwards at any branch that supports fruit.

The chrysalis is also formed beneath an elm leaf or, less often, at a fork on a twig. Brown, speckled and hairy, it looks exactly like a dead elm leaf, yet is fairly easy to find. Like most lycaenids, it possesses a sound organ and has a curious rasping song. I know of no-one who has found the chrysalis being attended by ants, but this is perhaps because among common British species, only wood ants *Formica* spp. climb trees, and few White-letter Hairstreak colonies occur where these abound.

Loss of elms

The White-letter Hairstreak has traditionally been regarded as a local species in England and Wales, but there is no doubt that a great many colonies were overlooked in the past. At one time it occurred in all counties south of Yorkshire, and was especially common throughout the Midlands and in the Welsh Borders. Colonies also occurred in many parts of lowland Wales and the West Country but, for some unexplained reason, have always been scarce in these regions.

The situation changed dramatically in the 1970s when Dutch elm disease caused a devastating loss of elm trees, and a large number of White-letter Hairstreak colonies disappeared. It is unfortunate that this butterfly has such a preference for mature, flowering elms, for it is exactly these specimens that are most vulnerable to the disease. Young trees, suckers, and elm hedges may also be infected, but it is only when they attain flowering height that they begin to succumb. The hairstreak colonies do not always perish: there are several examples of them persisting on the young suckers that regenerate around dead stumps, but such cases seem to be unusual.

On the credit side, the preferred foodplant of the White-letter Hairstreak, Wych Elm, has been less severely affected by disease than other species of elm, especially in the north of the butterfly's range. Moreover, perhaps because of the warming climate, colonies have been expanding at its northern boundaries, for example in the Wirral area of Cheshire. It is also now clear that this hairstreak previously bred on a much higher proportion of elm trees than had been supposed. Thus it is worth searching any flowering elm.

Despite elm disease, the White-letter Hairstreak survives in many localities. Its remaining strongholds are, as ever, in central England, from Yorkshire in the north down to Hampshire, extending as far west as Cheshire, Staffordshire, Herefordshire and Worcestershire. It is still present, and probably much overlooked, in East Anglia, south-east England, Wales and the West Country. There is also a possibility that the occasional Scottish colony survives, although there has been no definite record for at least a century. It does not occur in Ireland.

▲**Male**
The pale sex-brand on the forewing distinguishes the male from the female.

▲**Female**
Ground-colour of female's wings is slightly paler than the male's.

◄**Male**
Aberrant form with reduced white stripe on underside.

►**Female**
Very rare aberrant form *albovirgata*.

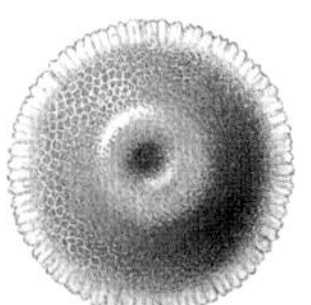

▲**Egg** [× 22]
Laid singly; overwinters on twig of elm.

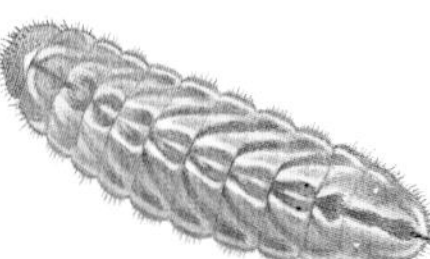

▲**Caterpillar** [× 2¼]
Camouflage conceals the caterpillar among elm buds, and later among expanded leaves.

▼**Chrysalis** [× 2¼]
Pupation usually occurs beneath the leaves of the foodplant. The chrysalis is anchored to the plant by a silk girdle.

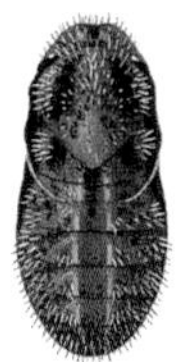

View from above   Side view

▼**Feeding male and female**
Adults feeding at flowers of Creeping Thistle. Nectar is used only occasionally as a supplement to the usual diet of aphid honeydew.

	Jan	Feb	Mar	Apr	May	Jun	Jul	Aug	Sep	Oct	Nov	Dec
Egg												
Caterpillar												
Chrysalis												
Adult												

Black Hairstreak

Satyrium pruni

This little butterfly has an attractive caterpillar and chrysalis, but as an adult it is undeniably plain. Nonetheless, naturalists travel far to see it. This is partly because of the Black Hairstreak's great rarity, and partly because the butterfly is delightful to watch, whether spinning high above the canopy of an Ash, tumbling in the June sunshine over immense banks of Blackthorn, or settled on a leaf, tap-dancing and fidgeting as it drinks the honeydew.

Discovery and deception

The native Black Hairstreak is restricted to one basin of low-lying clays that stretches from Peterborough to Oxford. It was one of our last butterflies to be discovered, with the southern colonies not being found until 1918. As often happens, it took a schoolboy, W F Burrows, to locate these, in Bernwood Forest's Hell Coppice, in Oxfordshire. No-one believed him at first – least of all Oxford's professional entomologists – but an expedition was eventually mounted and Burrows's find was officially verified. Other rarities were soon discovered in the same wood, and 'See you in Hell!' quickly became a catchphrase among the university's butterfly collectors.

There was controversy, too, surrounding the first British discovery. This was in 1828, when an entomological dealer, a Mr Seaman, visited Monks Wood, Cambridgeshire, and caught some small dark lycaenids. These were exhibited to the Entomological Club in the belief that they were White-letter Hairstreaks. However, the great Victorian entomologist Edward Newman spotted the mistake, whereupon Seaman changed the name of the locality to 'Yorkshire' to give himself a monopoly over this valuable new species.

Distribution
A rarity confined to small patches of Blackthorn in about 40 east Midland woods; also introduced in one area of the Surrey Weald.

Seaman's selfishness was ineffective, for in 1829 Professor Babington of Cambridge also took specimens in Monks Wood, and by 1837 the curate of Polebrook, the Revd William Bree, had discovered it in his parish at Barnwell Wold, in Northamptonshire. Bree was a hospitable man, and Barnwell soon thronged with the leading Victorian collectors. Barnwell Wold contained an extraordinary range of rarities, most in large numbers. The Purple Emperor, Chequered Skipper, Wood White, Duke of Burgundy and six fritillaries were there, together with all five hairstreaks and Barnwell's greatest prize – the only substantial colony of Large Blues known at the time. Sadly, these rarities have long since disappeared, leaving the Black Hairstreak as the sole relic of this classic collecting ground.

A diet of honeydew

The Black Hairstreak is an elusive butterfly, with a comparatively short flight period. Adults are usually seen during the last ten days of June and the first week of July, but this varies by several days depending on the warmth of the spring; in the past decade the first sightings were generally made around 10th June. They are not easy to find even at the peak of the emergence, since most colonies are confined to very small parts of a wood, perhaps 100m of sheltered edge or one particular glade. They seldom move far, and scarcely fly at all on some days, preferring to sit out of sight on a treetop. Field Maple and Ash are favourite resting sites, probably because the leaves are often coated with sweet aphid honeydew. Although most colonies contain just a few dozen adults, large numbers of Black Hairstreaks gather together on good sites, leaning with closed wings sideways to the sun. They then tap the leaves with their forelegs to detect honeydew, or slowly crawl from leaf to leaf with their tongues trailing between their legs, lazily drinking these sticky secretions.

Behaviour and identification

At other times there is greater activity, with adults spiralling through the air at high speed in a characteristic jerky flight. The jerkiness is exaggerated by the fact that the underwings reflect the sun, creating a silvery flash on every upbeat. This makes the flying adult extraordinarily difficult to distinguish from a White-letter Hairstreak, or even a Purple Hairstreak. Both of them commonly occur on Black Hairstreak sites, and overlap with the last Black Hairstreak adults in July. The White-letter and Black Hairstreaks also look similar at rest: unlike the Purple Hairstreak, neither opens its wings to bask, so only the undersides are visible. Despite its name, the Black Hairstreak is more golden than the dusky White-letter Hairstreak. It has a series of black spots along the inner edge of the orange band on its hindwings, whereas the White-letter Hairstreak has a black line. This is a far safer distinguishing feature than the white hairline halfway in from the edge opposite the tails, which can look like a sideways-on W in both species.

When it does descend to ground level, the Black Hairstreak compensates for its normal inaccessibility by being extremely tame. Adults will often bask for long periods and can be closely approached: in cooler weather I have often had them crawling over my fingers. They are also oblivious to any distraction when feeding. While honeydew is undoubtedly their main food, they also descend to the flowers of Wild Privet and Dog-rose. Both shrubs are common among the Blackthorns on most Black Hairstreak sites, and for any entomologist who has lived in the east Midlands, their scents filling the air on a warm summer's day will forever conjure up images of this little butterfly.

Cryptic younger stages

The female Black Hairstreak is mated almost immediately on emergence, sometimes before her wings are fully dry. There is then a gap of a few days before egg-laying begins. Up to 30 eggs can be laid in a day, each placed singly on a *Prunus* twig. Blackthorn is by far the commonest foodplant in Britain, but entire colonies have been supported by its relative, the Wild Plum.

The eggs remain on their twigs for nearly nine months before hatching in spring. They are surprisingly difficult to spot, even in winter when the twigs are bare. To find two or three in an hour is very good going, whereas one might find 100 Brown Hairstreak eggs for the same amount of searching. Most are laid on twigs that are from one to four years old. To find them, search sheltered sunny *Prunus* bushes of all ages and at all heights, for contrary to popular belief, the eggs are not confined to the twigs at the tops of ancient bushes.

The Black Hairstreak egg varies in colour from pale yellow to rich brown, and by spring many have a green coating of algae that makes them even harder to spot. The first three weeks of life are particularly dangerous, many eggs being killed by a minute parasitic wasp, *Trichogramma evanescens*. By August, each egg contains a perfectly formed caterpillar, which is immune to further attack. Roughly two-thirds of the eggs survive. They hatch from mid-March to late April while the *Prunus* is still in tight bud. The caterpillar slowly nibbles a neat hole in the top of the eggshell, often taking two days before it can squeeze out, leaving the empty shell so firmly fixed to its twig that it can be found, still cemented in position, up to a year later. The caterpillar immediately crawls onto a plump flower-bud, which it pierces with sharp jaws. However, rather than entering the bud to feed, it can reach the soft tissues in the furthest corners by means of a very long and thin extensible neck, leaving the remainder of its body resting outside on the brown scale-leaves.

The chestnut-coloured caterpillar is superbly camouflaged at this stage, and the resemblance becomes even more remarkable as the *Prunus* buds break and expand. The little caterpillar first develops a white 'saddle' that matches the breaking bud on which it lies. As the leaves expand, it moves to the base of a clump, its appearance gradually changing to match its background. By late May, it sits exposed in sunshine among a fresh clump of leaves, eating the tender tips by day, but almost indistinguishable from them due to its fleshy green body and the purple tips on its serrated back, exactly like the edge of a young *Prunus* leaf. Several stages of this changing pattern, which alters continuously rather than at moults, are illustrated on page 99.

The chrysalis, too, is a masterpiece of disguise. It resembles a bird-dropping and is attached quite openly to the top of a leaf or, more often, to a twig. With practice it is quite easy to find, for the camouflage is by no means perfect, but search in early June at the beginning of the chrysalis stage, for by the time the adults emerge up to four-fifths may have died. Willow Warblers are the chief predators.

Relics of an ancient landscape

Black Hairstreak numbers fluctuate considerably from one year to the next. Larger numbers emerge after a warm May and June, probably because the vulnerable caterpillars and chrysalises develop more quickly in warm weather, leaving little time for them to be eaten by birds. There are also big differences between the average sizes of colonies on different sites. The largest are found where massive banks of Blackthorn grow in exceptionally sheltered, yet sunny, situations. Typical examples are the south-facing edge of a glade, scrub of mixed ages growing on the sunny side of a wood, and where a tall, unkempt hedgerow runs parallel to a bushy woodland edge. Smaller colonies breed in slightly more exposed or shaded situations, such as in nooks along woodland edges or in the gaps beneath the canopy in mature, open woodland.

Sunny banks of sheltered Blackthorn are not particularly rare in the British landscape, so it is curious that just 80 colonies of Black Hairstreak had been found in the 150 years since its original discovery in 1828. All were in the east Midlands forest belt, a string of ancient woodlands, relics of the once-continuous forest that extended from Rockingham to Narborough in the north, via the Huntingdon Fen edges southwards through Yardley Chase, Salcey, Whittlewood, Waddon Chase, Grendon Underwood, and finally to Bernwood and Wytham near Oxford. At one time or another, it has occurred in more than half of the larger woods in this region.

The clue to this restricted distribution lies in the history of these ancient forests rather than in any intrinsic suitability they possess for Black Hairstreaks. True, the Blackthorns are magnificent in this region, growing tall before blowing over when they become too massive for the shallow suckering root systems, but many Black Hairstreaks feed on small young plants. It is the extraordinary reluctance of the adults to fly far that explains why they became restricted to this part of England.

Only in these Royal game forests has there been an unbroken history of forest management that has proved to be sympathetic to the Black Hairstreak. Such management includes very gradual scrub clearance, with coppice cycles lasting 20-40 years before being cut again, together with the encouragement of Blackthorn regrowth because of the excellent cover it provides for species of game. Almost everywhere else in the countryside it was commonplace to have much shorter coppice cycles, with any particular stand of shrubs being harvested up to 20 times in a century. This very sedentary butterfly could not colonise new shrubs quickly enough to keep pace with short coppice cycles, leaving the east Midlands as the only place possible for its survival.

Successful conservation

The traditional management of east Midlands woods ended around the end of the 19th century, being replaced by more intensive forestry in some areas and by abandonment in others. This led to the extinction of many Black Hairstreak colonies, and by 1980 only about 30 survived, in several cases thanks to enlightened conservation measures. Fortunately, this is quite an easy butterfly to save, for it needs little space and only occasional management to maintain its breeding sites. Great advances have been made in the last few decades, especially in the southern half of the Black Hairstreak's range. Tall banks of sheltered but unshaded Blackthorn that would otherwise have been cleared have been allowed to persist alongside normal forestry operations in Salcey and Bernwood forests. In addition, sheltered Blackthorns have been positively encouraged on several nature reserves, including those managed by the Berkshire, Buckinghamshire and Oxfordshire Wildlife Trust. As a result, the Black Hairstreak has spread considerably during the past 30 years: indeed, Stuart Hodges and the Upper Thames branch of Butterfly Conservation recently identified 50 colonies in their region alone, far more than was known in the heyday of traditional woodland management. One or two new colonies are still being found in most years by this active group, mainly on previously unexplored sites, and a record high of 1,095 adults sightings was counted in this region in 2010. Sadly, numbers seem to have crashed in the cold wet summer of 2012 (although this was also a very difficult year for recording), before a modest recovery in 2013.

The largest of these new colonies is also the most curious. Nearly 20 years ago, when the M40 motorway was being constructed through intensive farmland bordering Bernwood Forest, Katherine Bickmore and I were invited by the Department of Transport to design the Black Hairstreak's idea of heaven over several hectares of former arable fields, to compensate for the new road clipping the wood edge. We designed a sinuous maze, with an infrastructure of shallow winding banks made from the enriched top-soil that was stripped from the rest of the field. The banks were planted with 22,000 Blackthorn bushes grown from local cuttings, interspersed with Wild Privet, Dog-rose, maple, oak, elm, sallow (for Bernwood's Purple Emperors), Buckthorn and other shrubs. The intervening ground, now reduced to a bedrock of sticky Oxford Clay studded with fossil oyster shells, was to be grassland, so was treated with hay, laden with seeds, cut from a neighbouring nature reserve.

Within a year, Brown Hairstreaks were breeding on the new *Prunus* hedges, and in four years the grassland supported Green-winged Orchids, Cowslips and other characteristic flowers. In contrast, we predicted that the Black Hairstreak might take 20 years to colonise. It arrived five

▲Male
Forewings have inconspicuous sex-brands. Black spots distinguish both sexes from the White-letter Hairstreak.

▲Female
Orange markings are more extensive than those of the male. Patterning similar to male's, but tails on hindwings longer.

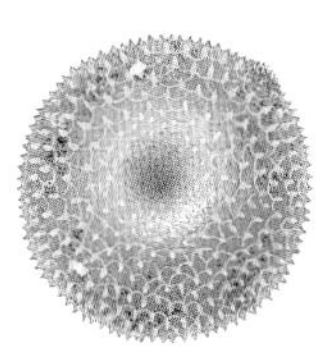

◀Egg [× 22]
Laid nine months before hatching; often becomes mottled with algae.

▲Mating
The female mates soon after emergence, often on Blackthorn twigs.

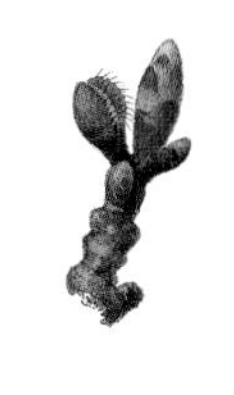

◀Young caterpillar [× 2¼]
On hatching, the caterpillar feeds on flower-buds.

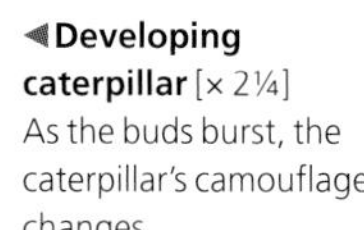
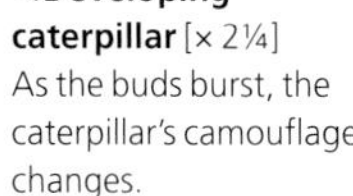

◀Developing caterpillar [× 2¼]
As the buds burst, the caterpillar's camouflage changes.

◀▶Mature caterpillar [× 2¼]
As the leaves expand, the caterpillar's colour changes to match its background. When fully grown, it is almost wholly green.

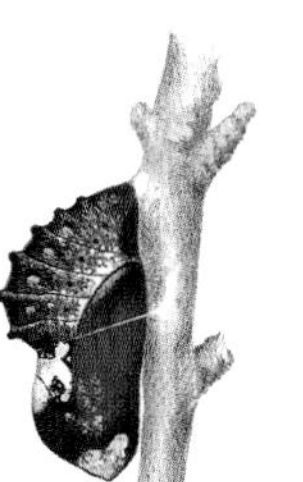

◀Chrysalis [× 2¼]
Camouflaged as a bird-dropping; attached to a twig or leaf.

View from above

Side view

	Jan	Feb	Mar	Apr	May	Jun	Jul	Aug	Sep	Oct	Nov	Dec
Egg												
Caterpillar												
Chrysalis												
Adult												

years ahead of schedule, and made up for the wait by rapidly growing into the largest colony I have known. On its peak day in 2007, I made 77 sightings in an hour and, unforgettably, saw six together on a single spray of Wild Privet flowers, twisting and jostling for nectar, oblivious of the unceasing roar of traffic from the motorway just a few metres away.

The Surrey colonies

Equally encouraging has been the recent spread of this butterfly outside its historical range. There had long been rumours of a Surrey colony, but these were dispelled when the original source was traced as a misidentified White-letter Hairstreak. However, as coppicing was abandoned in the Weald, the abundant Blackthorn of the western clays often grew into massive, sheltered banks, apparently ideal for this little butterfly. Their suitability was put to the test in 1952, when A E Collier, a master at Oundle School, moved south to Cranleigh, in Surrey. Oundle is in the heart of Black Hairstreak country, and Collier took a few caterpillars with him. These produced about a dozen adults, which were released in a suitable-looking wood near Cranleigh, and were promptly forgotten. Two years later, while walking in this wood, Collier was delighted to see a few Black Hairstreaks 100m or so from the original release point. These prospered for several years, until the whole wood was converted into a cornfield.

No more Black Hairstreaks were seen for 15 years. Then, in 1975, while watching Purple Emperors in a glade about 1.5km away, I was amazed to notice Black Hairstreaks jinking in abundance over the banks of thorn. Evidently, a small group of colonists had reached another wood in the neighbourhood before reaching this site, and they were slowly spreading. In the next five years about five separate colonies developed in this area, which soon held more Black Hairstreaks than any other known wood in Britain. Unfortunately, much of this area has now been cleared, but the Surrey hairstreaks kept one jump ahead of the developers. In the 1980s they reached a scrubby, overgrown railway line, abandoned 20 years earlier, where they bred for several years. Although this colony too appeared to die out, recent sightings suggest that the Black Hairstreak persists in Surrey, more than 50 years after Collier's original introduction.

It is particularly encouraging to find a rarity like the Black Hairstreak thriving in the Home Counties, but it does show how very slowly this little insect spreads. Presented with a vast area of predominantly suitable habitat, it has moved no more than an average of 100m a year. This suggests that there may well be other places in Britain where Black Hairstreaks can thrive, if only they could get there; for example, an apparently successful introduction was recently made to Lincolnshire.

Prospect of further introductions

It seems likely that further introductions will be made as it is realised that this charming butterfly is locked in the east Midlands as a relic of a long-defunct form of woodland management. Indeed, there is good evidence that many colonies within its traditional range originate from introductions. The great entomologist, Lord Walter Rothschild, paid H A Leeds to catch 'large numbers' of Black Hairstreaks between 1900 and 1917, for release in east Midlands woods. These were 'doing well some years later', although, alas, no record remains of the exact woods that were used.

Even the famous colony at Monks Wood is an introduction. Early in the 20th century, Monks Wood stock was used to replenish a nearby wood at Warboys that had lost its colony. However, Monks Wood itself was largely felled during the First World War, and all the Black Hairstreak's breeding sites were cleared. When the butterfly had not been seen for five years, adults were caught in the Warboys colony and released in Monks Wood. Here, nearly a century later, they can still be seen in excellent numbers.

Large Copper

Lycaena dispar

A century and a half has passed since our native Large Coppers disappeared. Yet the creature continues to fascinate, partly through the specimens in museums, and partly because a Dutch subspecies of the butterfly could, until recently, be seen flying at Woodwalton Fen, where a fragment of ancient habitat has been preserved among the agricultural prairie-lands of north Cambridgeshire.

This colony was delightful to watch, but there was also some sadness involved. For the Large Copper belongs to a landscape and a culture that had virtually vanished from England by the middle of the 19th century. It once bred in small clearings in the fenlands, where 'the Bog Myrtle used to grow in profusion'. Although abundant on a few sites, colonies seem always to have been few and far between. The first was found on Dozen's Bank, in Lincolnshire, by John Green, the Secretary of the Spalding Gentlemen's Society. He named his discovery the 'Orange Argus of Elloe', and painted an exquisite watercolour of a male in 1749. This record, however, remained unknown until the entry was discovered in the Society's minute book more than two centuries later, in 1982, and it was not until Lewin's *The Papilios of Great Britain* was published in 1795 that the entomological world became aware of British Large Coppers, which had been found on a 'moorish piece of land' in Huntingdonshire.

Former distribution
A fenland species extinct by the mid-19th century. A colony of the Dutch subspecies *batavus* was, until recently, maintained at Woodwalton Fen.

A British subspecies

The discovery of this handsome butterfly caused enormous excitement among the collecting fraternity. It was soon noted that British specimens were more beautiful and larger than any known from the Continent, and were rightly considered to be a distinct subspecies, named *dispar*.

The next 30 years saw the heyday of the Large Copper, so far as English naturalists were concerned. Colonies were found in the Norfolk Broads, in Cambridgeshire and in Suffolk, but most of all in the old county of Huntingdonshire, in about 2,000ha of unbroken fenland surrounding Ugg and Whittlesea Meres. Although wild and inhospitable, this was nevertheless a worked landscape of peat and reed cuttings, with sheep and cattle grazing on the fen clearings in summer. The ancient communities maintained a patchwork of fen meadows, marshes, reedbeds and waterways in which Water Dock, the caterpillar's only foodplant, was widespread and common. The butterfly itself was found 'in spaces covered with sedge and coarse grass' that were frequently submerged during winter.

A disappearing habitat

But already the fenlands were changing. The Earl of Bedford began the process when, in 1634, he employed Cornelius Vermuyden to introduce Dutch methods of drainage. By the 19th century, the fens of Lincolnshire and Norfolk had all but disappeared. Huntingdonshire and Cambridgeshire withstood the onslaught longer, and it was not until 1851 that Whittlesea Mere was drained and the last great obstacle to efficient agriculture was removed. Today, the landscape is utterly transformed, and consists of endless expanses of flat, ploughed peatlands, criss-crossed by neat, parallel drains and virtually bereft of wildlife. It is a loss that we nowadays deplore, but it was viewed very differently at the time. Even the Revd F O Morris, author of the most beautiful of the Victorian books on butterflies, took the change with Christian fortitude:

'Science, with one of her many triumphs, has here truly achieved a mighty and a valuable victory, and the land that was once productive of fever and of ague, now scarce yields to any in broad England in the weight of its golden harvest... The entomologist is the only person who has cause to lament the change, and he, loyal and patriotic subject as he is, must not repine at even the disappearance of the Large Copper Butterfly, in the face of such vast and magnificent advantages. Still he may be pardoned for casting "one longing lingering look behind," and I cannot but with some regret recall... the time when almost any number of this dazzling fly was easily procurable.'

In fact, the extinction of the Large Copper slightly preceded the draining of Whittlesea, with the last five specimens being caught at Holme Fen in 1847 or 1848. Another colony lasted a further three or four years at Bottisham, in Cambridgeshire, and the final British record is from the Norfolk Broads in 1864. These losses were probably caused by the decline in traditional fenland management, which had been giving way to more intensive farming for many years. Butterfly dealers were another hazard, and may have tipped the balance on a few sites. Certainly, vast numbers of specimens were taken. One collector gave this account in the early 1840s:

'It soon got known among the fen folk that [Large Coppers] were worth two shillings each in London, and two men came from Cambridge and secured a large number, which they took to London in boxes, and sold at sixpence each. I went down three years after, and got some of the larvae. They appeared to be very local, and most numerous where their food plant – the water-dock – was most abundant. The larvae were collected by all persons, young and old. I bought two dozen off an old woman for ninepence, from which I bred some fine specimens and sold them at one shilling each.'

Many were later to write with regret that they had not secured more, so valuable did the old English specimens become once the butterfly was extinct. J W Tutt wryly charted their appreciation, culminating in the sale of 14 specimens for £71 15s. in 1900, and predicted that 'these prices are as nothing to what may be expected in the not very distant future, when "coppers" may produce figures approaching the prices for Great Auk's eggs'. Perhaps because of their value, many specimens have survived in immaculate condition and can be seen in museums around the country. The zoological department at Cambridge has a particularly fine series.

Continental subspecies

Dutch colonies of Large Copper were discovered in Friesland early last century, in recently abandoned peat cuttings. They belong to a subspecies called *batavus*, but as can be seen from our illustration, are similar in size and markings to the old English form. They clearly had a common ancestry in north-west Europe before being divided, perhaps 6,000 years ago, when the continental land-link was broken by the rise in sea level.

Both are very different from the third subspecies, *rutilus*, which is found in marshlands throughout central and eastern Europe. This feeds on a wider variety of plants, has two rather than one generations of adults a year, and is significantly smaller, especially in the second brood. It also has duller underwings, but is a beautiful insect nonetheless. The *rutilus* form of Large Copper remains locally common where European wetlands survive, especially in east Europe, where it has spread and increased in recent years.

Colonies and courtship

The natural history of the Dutch Large Coppers has been thoroughly studied by Fritz Bink in the Netherlands and by Eric Duffey, Andrew Pullin and Mark Webb in England. It appears to have similar requirements to the old English subspecies, living in small, self-contained populations that range from a few dozen adults to several hundred, with a low level of interchange between neighbouring colonies. The butterflies emerge in early July, reach a peak in the second half of the month, and often survive well into August. They are very dependent on sunshine, jinking across fens at high speed while the sun is out, and hiding in reedbeds in gloomy weather. The males are as pugnacious as those of other coppers, although they do not establish such obvious territories. Instead, they sit on flowers or reeds, and dart up to intercept any insect that flies by. If this proves to be a virgin female, the pair soon lands on the reeds. Courtship is then a brief, fluttering affair, with the male first flapping his wings over the female before the two butterflies couple.

The females mate shortly after emergence, but three to five days pass before the first eggs are laid on the glossy broad leaves of Water Dock. They are extremely fussy when choosing plants. They almost invariably avoid large, leafy specimens growing on the edges of dykes, and the few eggs that are laid on these seldom survive the winter. Instead, they hunt for medium-sized plants growing in open marshland that is regenerating after a disturbance, particularly

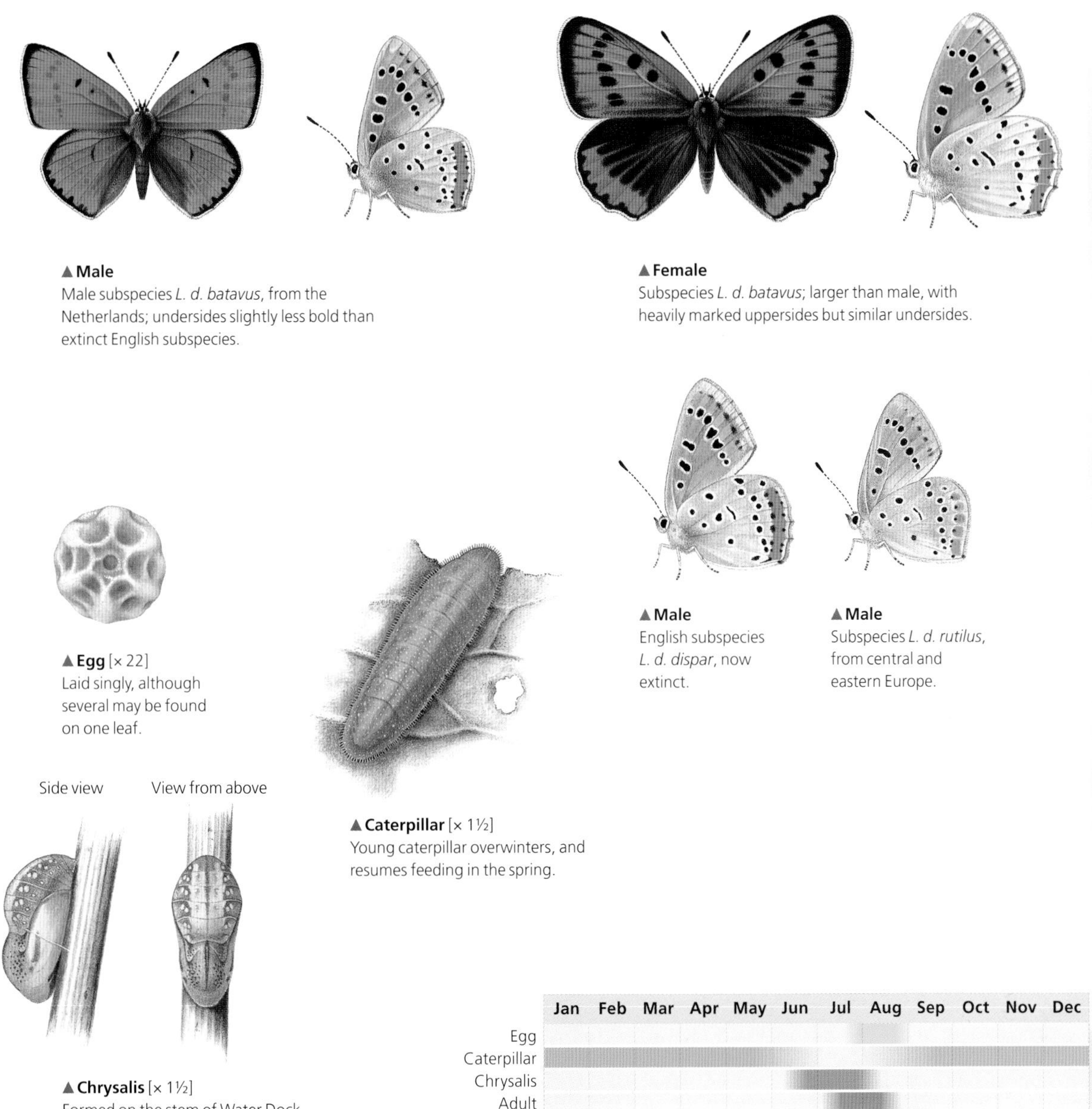

▲**Male**
Male subspecies *L. d. batavus*, from the Netherlands; undersides slightly less bold than extinct English subspecies.

▲**Female**
Subspecies *L. d. batavus*; larger than male, with heavily marked uppersides but similar undersides.

▲**Male**
English subspecies *L. d. dispar*, now extinct.

▲**Male**
Subspecies *L. d. rutilus*, from central and eastern Europe.

▲**Egg** [× 22]
Laid singly, although several may be found on one leaf.

▲**Caterpillar** [× 1½]
Young caterpillar overwinters, and resumes feeding in the spring.

▲**Chrysalis** [× 1½]
Formed on the stem of Water Dock.

favouring specimens along the edges of taller vegetation.

It has been suggested that the females choose warm and sheltered plants for egg-laying, and that waterside specimens are simply too cold. This may well be so, for the butterfly occurs at the extreme of its northern range in Britain and the Netherlands, and small-scale clearings among the reeds would undoubtedly be several degrees warmer than the edges of dykes. But they may also be seeking exceptionally nutritious plants. Bink showed that growth was hastened, and that larger butterflies were produced, when the caterpillars fed on young, vigorous dock leaves that were rich in nitrogen.

Whatever the explanation for this selectiveness, the egg is easy to find once one knows the sort of plants that are chosen. Indeed, they are so conspicuous that Eric Duffey was able to use students, myself included, to count every one of the thousands of eggs laid each year at Woodwalton Fen in the 1970s. Although laid singly, small clusters of two, three or four eggs were often found together, usually beside the prominent leaf ribs. They have a curious robust sculpturing that is highly attractive when seen through a hand-lens.

The eggs hatch in mid-August, leaving conspicuous empty shells on the dock leaves. The little caterpillar starts feeding on the undersurface of its leaf, excavating shallow grooves like those produced by the Small Copper, and which are equally easy to find. It then hibernates after one or two skin moults, still small and hidden on a silk pad spun on a twisted dead leaf. These take some finding, but are not impossible to spot when water levels are low. Many are later submerged, and can survive for at least three months underwater in this hibernating stage. It is much more vulnerable when older, and isolated colonies are believed to have been lost through exceptional flooding in May.

The caterpillars re-emerge in spring, eating holes in the dock leaves but avoiding the midribs. This leaves a distinctive lattice pattern, looking not unlike a Swiss cheese plant. If these leaves are examined in May, the fat, slug-like caterpillar can be found on the underside, usually resting on the midrib. It is sometimes attended by ants, for although it lacks a honey-gland, there are secretory pores all over the body. These, unfortunately, do not save many caterpillars from being killed in their later stages, mainly by birds or by parasitic ichneumon wasps and tachinid flies. The survivors form plump, rounded chrysalises on the grasses and reeds near the docks.

Woodwalton Fen

As with most declining butterflies, a clue to the Large Copper's decline may lie in the very fussy egg-laying requirements of the females. Whatever the primeval habitat of this species may have been, it has long depended on man to make piecemeal but regular clearings or cuttings in the fens. This, alas, ceased almost entirely in Britain, even in those fragments that were not claimed for agriculture. Woodwalton Fen is a prime example. Peat-cutting continued through the 19th century, and the fen in 1896 was still 'full of flowers'. Ten years later, the practice had ceased, and by 1910, when the Hon N C Rothschild bought the land as a nature reserve, 'the marsh meadow-land was dominated by a thick growth of reed'. This saved it from drainage and ploughing, but by 1926 much of the fen had become a birch and sallow carr woodland. This was partly the result of 30 years of little interference, and partly because the water level fell as a result of increasingly intensive agriculture on the surrounding land.

At this stage, Charles Rothschild took command. About 9ha were cleared and planted with Water Docks, which had long been eliminated by carr, creating the famous 'Copper Field' into which Capt E Bagnell Purefoy and James Schofield released 25 male and 13 female Dutch Large Coppers in 1927. They were an immediate success, and in 1928 it was estimated that over 1,000 adults emerged. The colony continued to prosper, but the management of the site was difficult, especially during the Second World War. From the earliest days, the butterfly was 'helped' by caging caterpillars to protect them from predators, and numbers were regularly topped up with captive-bred specimens. In many ways this was a shame, but almost every naturalist was prepared to accept this artificial situation for the joy of watching the adults in July. Despite cagings and top-ups, the colony died out in 1969, and did so again on several later occasions after successive re-introductions had been made. The programme was suspended in the 1990s.

This was not the first re-introduction of Large Coppers, nor the most successful. As early as 1913, before the Dutch subspecies had been discovered, Purefoy cleared and planted a small bog at Greenshields, Tipperary, to which caterpillars and adults of the *rutilus* subspecies were introduced. The colony prospered for 43 years, without any caging or supplements, until the bog became overgrown. It was cleared again in 1942, when Dutch adults were introduced. This colony survived without interference until the bog again became overgrown in 1955.

Future hopes

Several other places have seen introductions. The most successful was at Wicken Fen, near Cambridge, where Dutch Large Coppers flourished for about ten years, disappearing only when the wettest area was ploughed to plant potatoes during the war. No other attempt has met with long-term success, although in some cases the butterfly briefly thrived.

Something is clearly lacking in the quality of habitat generated on British restoration sites, for Pullin and Webb found considerably higher overwintering survival of caterpillars in the Netherlands. British breeding areas may also have been restored on too small a scale to support populations of Large Copper in the long term. Conservationists have not, however, abandoned this very attractive butterfly. In the Weerrriben and other areas of the north Netherlands, where *batavus* was approaching extinction, the restoration of adequate water-tables, coupled with extensive cattle-grazing or rotational cutting of numerous small reedbeds, has led to a welcome recovery. There is hope, therefore, that the Great Fen project, which will reconnect two of the Large Copper's most famous Victorian breeding grounds – Woodwalton and Holne Fens – by flooding and restoring 3,000ha of intervening farmland to fenland, may generate suitable habitat on a sufficient scale. Furthermore, the lessons learnt from recent fen management in the Netherlands should increase the prospect of a successful introduction to the Norfolk Broads.

Small Copper

Lycaena phlaeas

This exquisite insect is one of the liveliest and brightest of lowland butterflies in the British Isles. It is, above all, a creature of dry, dusty places, yet the occasional adult can be seen almost anywhere other than on the intensive arable prairies of East Anglia, and even there it has not been entirely eliminated.

A variable butterfly

The adult Small Copper is unmistakable, and almost too familiar to merit description. Nonetheless, it is worth noting that this is a somewhat variable species, given to aberrations throughout its range, and existing as distinct colour forms in some places. In the north of Scotland, there is more gold and less black on the upperwings than further south, and in Ireland the underwings are distinctly greyer. By far the commonest and most attractive variety has a series of blue spots along the bottom edge of each hindwing. This is *caeruleo-punctata*, a form that is particularly frequent in northern Scotland, where up to half the adults in a colony may sport this livery. It is also common enough in the south – I saw several blue-spotted specimens in my garden during the warm summer of 1989. Much rarer is an albino form, called *alba*. Here, the normal copper markings are replaced by white, making a curiously attractive insect in the wild. The inheritance of this unusual characteristic appears to be controlled by a single gene. Being recessive, this perhaps finds expression in small, inbred populations founded by single females bearing the gene. One small colony near my home in Dorset produced a fair proportion of albino adults for many years.

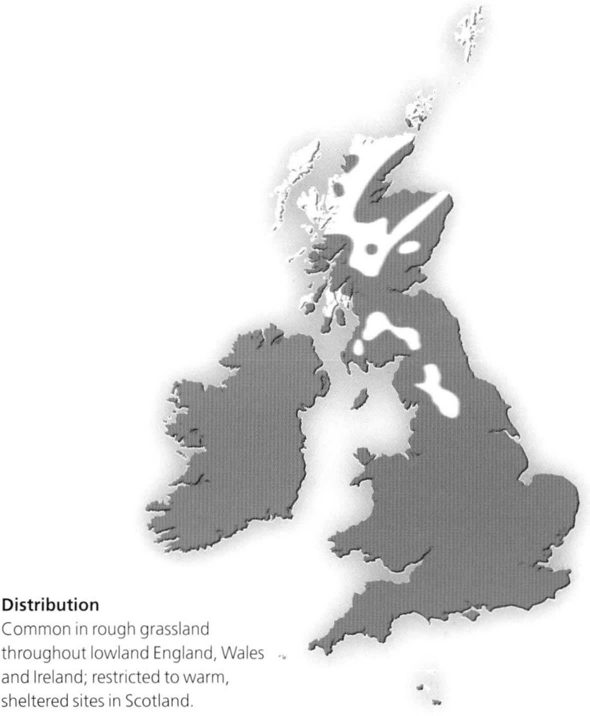

Distribution
Common in rough grassland throughout lowland England, Wales and Ireland; restricted to warm, sheltered sites in Scotland.

The Small Copper is variable, too, in the time of appearance. A first brood is generally seen in May, but can be a good month later in the north of Scotland. This is typically followed by a second emergence in July or August, and often by a third in September and October following a warm summer, when numbers are generally high. There is some evidence that it squeezes in a fourth brood in exceptionally hot years, but further monitoring is needed to confirm this. By and large, the butterflies in the second emergence are more numerous than the first, but even so colonies tend to be small. A comparatively large colony, in Cambridgeshire, was measured by Jack Dempster, and contained no more than 400 adults in high summer. However, most populations contain far fewer individuals, and the butterfly is usually seen in ones and twos.

Small Coppers generally live in close-knit colonies, often supported by small areas of land such as a well-drained bank, a road verge, or a patch of urban wasteland. Some individuals are prone to wander, and are frequent visitors to country gardens where they sup at flowers and, in the case of males, establish territories which they vigorously defend. One is even recorded to have reached the *Royal Sovereign* light vessel, more than 12km off the Sussex coast.

Vivacious adults

The adult Small Copper appears almost hyperactive, so fidgety are the movements and so rapid the flight. It cannot sit still even when feeding, but rotates its body on the flowerhead while constantly probing for fresh nectar. Dusk sees a halt to this frantic activity, when the butterfly roosts head-down on tall grass stems. I have occasionally found them roosting in aggregations, gathered on grass stems like blues, but this seems to be unusual; more often than not, a solitary sleeper is found.

Each male establishes a small territory in a warm, exposed position on a flowerhead or on the ground, often favouring discarded rubbish, even silver paper or a matchbox. It is sometimes said that he marks this territory with scent, but although this is an attractive theory, I have yet to find evidence of it. In fact, the male appears to have little need of a chemical deterrent, for he is a pugnacious scrapper that waits on his perching pad with wings ajar, ready to hurl himself upwards at any passing insect. He usually returns after a brief skirmish to regain his post, although females are pursued with zeal.

The Small Copper's courtship is so swift that it is hard for the eye to follow. The Danish entomologist, H J Henrikson, gives a good reconstruction: there is first a rapid zigzag flight, a metre or two above ground level, with the two insects flying in close formation. They then crash-land and rest before resuming the aerial courtship. Finally, the female, followed by the male, descends into the ground vegetation, where mating occurs.

Eggs and their development

The female Small Copper is more sedate when egg-laying, and flutters slowly above ground level, often settling to crawl over and tap the vegetation with her antennae in an attempt to find suitable plants. Like most butterflies, she is exceedingly choosy. The eggs are more or less restricted to Common Sorrel and Sheep's Sorrel, but only certain growths of these plants are chosen. Springtime butterflies lay in quite tall but sparse vegetation. They prefer sorrels that are 10-30cm high, and crawl down to deposit their curious, golf-ball-like eggs under tender leaves or, more often, on the upper surface at the joint between leaf-stalk and blade. Later broods almost invariably choose small, sprouting sorrels that are growing in warm, exposed positions, often selecting plants with no more than two or three little leaves.

I suspect that this seasonal distinction is less curious than it seems. Like other lycaenids, the females probably favour young sorrel growth because this contains a high concentration of nitrogen and other nutrients. In summer, the same conditions would be found only on smaller regenerating leaves. This may also explain the occasional record of eggs being laid on tender broad-leaved docks, when few sorrels on the site have lush leaves.

Whatever the explanation, there is no doubt that small, fresh sorrels can be peppered with Small Copper eggs after a good summer. F W Frohawk describes finding an extreme example on 17th October 1933, with 'over 300 eggs and many young larvae upon very small plants of Sorrel growing on a dry bank in a space of 15 or 20 yards'.

This is generally one of the easier butterfly eggs to find, and it is well worth collecting one or two. The sculpture of the eggshell is fascinating when seen through a lens, and the caterpillar is easy to rear and attractive – a fleshy, green grub, occasionally adorned with pink stripes, whose sluggish gait gives no hint of the vivacity of the adult.

Small Copper eggs hatch after one to two weeks, and the little caterpillar eats a small groove on the underside of its sorrel leaf, leaving the transparent upper surface intact. It fits snugly in this channel, which it elongates for a day or two before starting another. Soon the whole leaf looks as if it has been vandalised from beneath by a miniature chisel, and these silvery grooves are one of the easiest ways of detecting the butterfly. Caterpillars eat larger portions as they grow older, perforating the sorrel leaves, again in a characteristic way, although the creature beneath can be hard to find. Finally, after about a month, an attractive, dumpy chrysalis is formed. I know of no-one who has found this in the wild, but I suspect from its structure that it may sometimes form an association with ants.

Small Copper colonies

Colonies of the Small Copper can be found throughout most of the British Isles, and are absent only from the high ground of northern England and Scotland, from the extreme north-west of Scotland, and from Orkney, Shetland, the Outer Hebrides and remoter isles. The butterfly is particularly common in the south, although many colonies have been eliminated from the most intensively farmed areas; one analysis in Wales suggested a 90% decline. Nevertheless, this is a butterfly to look for on all warm, well-drained soils, for example on heaths (where Sheep's Sorrel is the main food), on chalk and limestone downs (where Common Sorrel is eaten), and above all on rough patches of wasteland, warm banks, embankments, dunes and in old quarries. Road verges also often support appreciable numbers, as do dry wood edges and the sides of ditches.

The Small Copper is not numerous every year, but its numbers can swell markedly during warm summers – unless there is a drought, in which case the sorrels shrivel and become unsuitable for breeding. Fortunately, this is a sufficiently mobile butterfly and can overcome most temporary setbacks, and it remains a delightful member of our fauna, easily spotted by amateur naturalists almost everywhere.

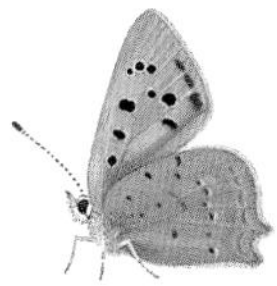

▲**Male**
The male is smaller than the female, and has more pointed forewings.

▲**Female**
Larger than male; when egg-laying flutters slowly over vegetation and is easily identified.

▲**Colour variant**
Blue-spotted aberrant form *caeruleo-punctata*.

▲**Colour variant**
Rare aberrant albino form *alba*.

▶**Perching male**
The alert male chases away any butterfly or other insect that enters his territory.

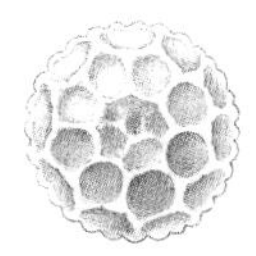

▲**Egg** [× 22]
Laid singly; several eggs may be found on the same leaf.

▲**Caterpillar** [× 2¼]
Usually plain green; some are heavily lined with purple.

▶**Feeding damage**
Young caterpillars eat grooves in the leaves of sorrels. These are easily spotted, as are the eggs, which are usually laid near the midrib.

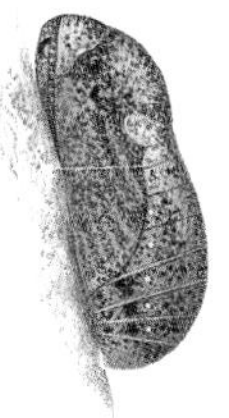

▲**Chrysalis** [× 2¼]
Formed among leaves; adults emerge after about 3-4 weeks.

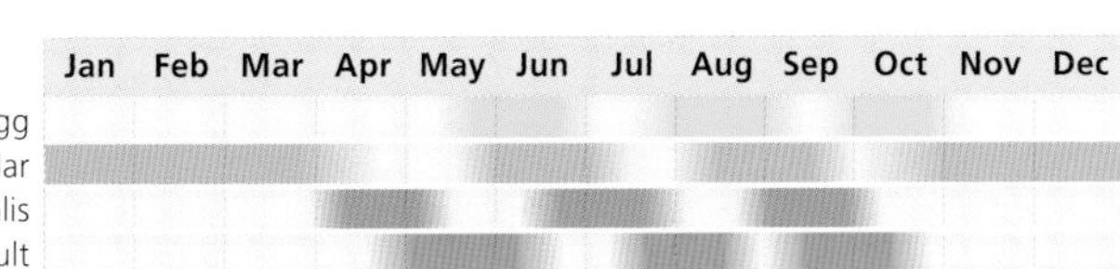

Long-tailed Blue

Lampides boeticus

The Long-tailed Blue is one of the most widely distributed and common of the 6,000 species of lycaenid butterfly known throughout the world. It occurs over the whole of Africa, in southern Asia and Australia, and through the entire southern half of Europe. But in Britain it is a rare immigrant, causing understandable excitement when it does arrive.

The Long-tailed Blue is a lively insect, with a rapid, jerky flight more reminiscent of a hairstreak than a blue. It also resembles hairstreaks in possessing long, delicate wing-tails, with an eye-spot where each joins the wing, and it too has a curious way of slowly rotating them in unsynchronised circles. This makes the tails wave, distracting the attention of predators from the more vulnerable parts of the body. One often sees the butterfly with a beak-mark where a bird has attacked the wrong end.

Distribution
A rare immigrant from southern Europe.

Visitor from the Mediterranean

Unfortunately, this charming butterfly cannot withstand our winters, and in Europe is probably a permanent resident only in Mediterranean regions. There it can be extremely common; I have often watched Long-tailed Blues buzzing in the searing heat around the flowerheads of brooms and Bladder-senna, the main leguminous plants on which they lay. The egg is a small white disc, and the tiny caterpillar feeds first among the flowers before boring into the broom and Bladder-senna pods. It grows quickly on this unappetising medium, and the plump green caterpillars can be found gorging the developing seeds and soft tissues, by splitting pods that have a hole bored in their sides.

Happily, the Long-tailed Blue has become an increasingly frequent visitor. The first two British adults were caught near Brighton, and another at Christchurch, in 1859, with just 30 adults spotted in the following 80 years. Others were undoubtedly overlooked, for this is one of our less conspicuous immigrants. The earliest sighting of a Long-tailed Blue was in June, but most are seen in August and especially September, with a few in October and November. Nearly all were in the southern coastal counties, from Devon to Kent, with the remainder reaching the Midlands and north Wales. They are particularly attracted to gardens growing everlasting-pea. Records for Long-tailed Blues showed an abrupt change in the *annus mirabilis* of 1945, when an unprecedented number of rare migrants arrived. This wave of visitors included 38 Long-tailed Blues – about a quarter of all British records up to that date. Sightings then settled down to their normal pattern during the next 35 years, but in 1990 there was an immigration greater even than that of 1945. The butterfly was found in Kent and Surrey, but the highest numbers were seen in central London, where two colonies briefly established themselves. One was in Kensal Green Cemetery, where 'about a dozen' Long-tailed Blues were counted; the other was at Islington, as Brian

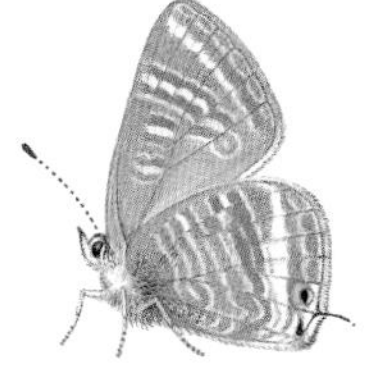

▲Male
Flight rapid and jerky, like a hairstreak's. Wavy patterning and broad white band on underside distinctive.

▲Female
Amount of purple on the uppersides varies; the undersides are constant in both sexes.

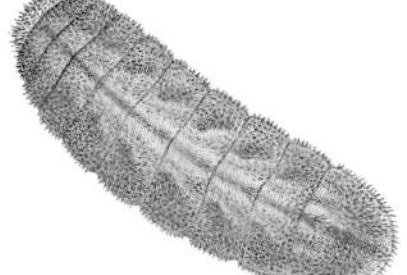

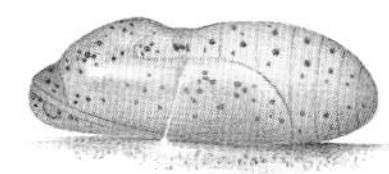

◀Chrysalis [× 2¼]
Formed inside a dry, curled leaf, loosely attached by silken threads.

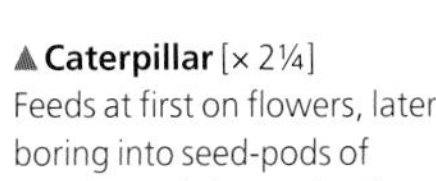

▲Caterpillar [× 2¼]
Feeds at first on flowers, later boring into seed-pods of members of the pea family.

	Jan	Feb	Mar	Apr	May	Jun	Jul	Aug	Sep	Oct	Nov	Dec
Adult												

Wurzell reports in the *Bulletin* of the Amateur Entomologists' Society:

'For about four weeks, butterflies were observed on most warm days at Gillespie Park ... the majority, up to 10-12 adults on the wing at one time, tended to quarter the dry railway land above. Their lively interest in Lucerne flowers was presumed to be for nectar, but I detected several eggshells of unmistakable Lycaenidae structure on the tube-shaped sepals surrounding the bases of bladder senna flowers (*Colutea arborescens*)... Two slug-like larvae were found nestling deep within the keels of flowers whose perforated sides had prompted closer examination.'

Breeding and migration

In captivity, the Long-tailed Blue will feed on a wide range of legumes, and was known as the 'Pea-pod Argus' to Victorian naturalists. Early evidence of breeding in the wild in Britain includes 26 eggs found in 1952 by a Mr Chevalier on everlasting-pea in Dorking, while in 1945 eight adults emerged from Bladder-senna pods that had been brought indoors for a flower arrangement. Caterpillars are also occasionally imported with lentils, and are responsible for adults that emerged from two batches of mange-tout peas imported from Kenya in 1998 and 1999, and another found in an Essex greengrocer's shop in 1998.

During the present century, Long-tailed Blues were reported with increasing regularity up to 2013, when a migration occurred in early August that was to dwarf all previous ones: in Sussex alone – a historical hotspot for the species – a minimum of 65 different adults was seen (and often photographed) by the year's end, compared with a total of 22 adults reported since its first country record of 1859. Adult Long-tailed Blues were seen in other southern counties, too, in 2013, but the strongholds were in Kent, where ephemeral colonies established at Minnis Bay near Margate and on Kingsdown Leas, east of Dover, and in Sussex, where breeding was confirmed at 15 locations. Many eggs and young larvae were found in both counties, mainly on everlasting-pea, and these produced an autumn emergence roughly four times more numerous than that of its immigrant parents. I am indebted to Colin Pratt and Neil Hulme for a meticulous account of Sussex's *boeticus* year. Hulme also noted that these adults abruptly disappeared in mid-October, and that whilst eggs had been found with ease in August and September, none was seen from the autumn generation, despite mating pairs being observed. He plausibly suggests that they embarked on a return migration to winter breeding grounds, as has recently been established in the Painted Lady (pages 178-181).

The Long-tailed Blue is certainly a powerful flier, given to annual migrations throughout the world and regularly crossing large seas. These have been little studied in Europe, but a southerly migration is known to occur through the Pyrenees each autumn. Further east, in Asia, the Long-tailed Blue migrates in fantastic numbers, regularly ascending the foothills of the Himalayas in each dry season and later returning to the lowlands for the wet season.

Geranium Bronze

Cacyreus marshalli

Like the Bloxworth Blue, the claims of this attractive small lycaenid for inclusion in our book are borderline: it makes the cut as much for its expected occurrence in future years as for the few records of breeding reported in the last decade.

Apart from the natural colonisation of southern Spain by the Monarch, this is the only exotic species of butterfly to have colonised Europe in historical times. It is a native of South Africa, where it breeds on wild pelargonium plants, and is continuously brooded, with no hibernation stage. The first record for Europe of a free-flying adult was on the island of Majorca in 1987, although its caterpillars had been found on plants imported to Britain about ten years earlier. It probably reached Majorca on potted plants imported by the horticultural trade, but soon established itself as a resident. Today, the Geranium Bronze can be seen flying in any month of the year in the Balearics, especially in – but not restricted to – villages and towns, where it finds a constant supply of foodplants in hanging baskets, containers and gardens, as well as in municipal gardens and flower displays.

Mixed emotions

The Geranium Bronze rapidly spread around the Mediterranean, presumably via gardeners and the horticultural trade. Today, it is a common urban butterfly in lowland regions of Portugal, Spain, southern France and Italy, as well as in the Canary Islands and Morocco. It raises mixed emotions: to horticulturalists it is unequivocally 'a plague... due to its aggressiveness it is necessary to combat it on time to impede its progress, which will finish [by] destroying our geraniums', as an Iberian website proclaims. To many entomologists, knowing that it attacks only cultivated pelargoniums and not native species of *Geranium*, it is a not unattractive addition to the European fauna.

At first sight, the adult Geranium Bronze looks rather like a Long-tailed Blue, but closer examination shows more contrasting, zigzagging grey and brown undersides, described by the same horticultural website as 'brown with stains in the base of the wings'. It lays the typical white disc-like eggs of a lycaenid on the flower-buds, and sometimes leaves, of cultivated pelargoniums, often several on the same plant. The young caterpillar buries into a flower-bud and then usually bores into a stem, where it lives for two or three skin moults, feeding on the soft interior tissues. In due course, these hollowed stems begin to rot, but not before the final instar caterpillar has left to feed openly on pelargonium leaves, where its hairy green body, edged in rose-madder, is beautifully camouflaged. The chrysalis has similar markings.

Arrival with the horticultural trade

The first account of an adult Geranium Bronze in the British Isles was provided by Crispin Holloway, who reported that one had been seen in their garden at Lewes, East Sussex, by his father John, on 21st September 1997. After a while, they saw the butterfly egg-laying on potted geraniums. Two days later a second adult was seen, which continued to flutter around and lay more eggs. Then a third appeared, thought to be a male, and at one stage four Geranium Bronzes were

▲**Male upperside**
Markings are similar in both sexes.

▲**Male underside**
Both sexes have distinctive zigzag pattern on underside.

▲**Female upperside**
The female is usually larger than the male.

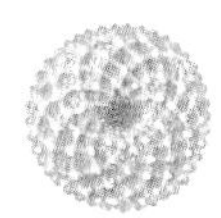

▲**Egg** [× 22]
The white egg with its tracery of ridges is typical of a lycaenid.

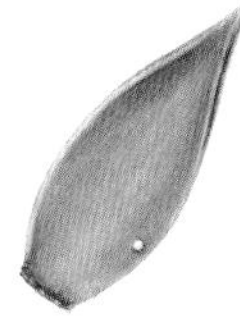

◀**Egg on pelargonium** [× 1½]
The single egg is usually laid on a flower-bud.

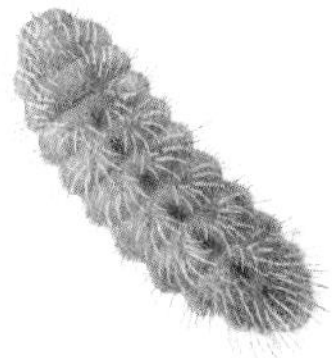

◀**Caterpillar** [× 2¼]
The mature caterpillar feeds openly on the plant, where its green and pink colouring provides protection.

▲**Chrysalis** [× 2¼]
The hairy chrysalis is sometimes attached to a withered leaf near the foodplant.

flying in their garden. The last to be seen was on 1st November, roosting in the garden with a Small Copper. Although not expected to survive the winter, a further adult appeared the following May, presumably having survived the winter in a neighbour's greenhouse. It is believed that these butterflies were introduced as eggs or young caterpillars on the neighbour's pelargoniums, bought from a local nursery that imported them from Spain or Portugal earlier in the year, although when the nursery received enquiries about their origin the story changed, perhaps to avoid responsibility for introducing an exotic insect.

There have been several reports of Geranium Bronzes from gardeners since then, some more convincing than others, including a few definite records, the most recent – with photographs – reported from an allotment in Liverpool and a garden in Sussex. However, the butterfly appears unable to withstand British winters, except in a conservatory or glasshouse, and is unlikely to establish itself here as a resident under current climates. Yet with its rapid spread across southern Europe, coupled with the mass movement of pelargoniums in the horticultural trade, it is likely that there will be further, probably temporary, establishments of this hairstreak-like butterfly in the future.

The Bloxworth or Short-tailed Blue

Everes argiades

The Bloxworth or Short-tailed Blue was first reported in Britain by the Revd Octavius Pickard-Cambridge, the Victorian expert on spiders, whose son Arthur took a female on 18th August 1885 while butterfly collecting on Bloxworth Heath, Dorset. Two days later they caught a male on the same heath, and the next day another schoolboy, Philip Tudor, took a third near Bournemouth, about 15km away.

1885 was one of the great years for immigrant Lepidoptera, and this clearly included a small band of Bloxworth Blues, as they were promptly christened. Entomologists were few and far between in east Dorset a century ago, and the fact that three of these butterflies were found within three days suggests that a good many more may have been around. Unlike the Monarch, the Bloxworth Blue is not the sort of rarity one spots from afar. On the contrary, it can easily be mistaken for a Small, Silver-studded or even a Common Blue when on the wing.

These, in fact, were not the first British Bloxworth Blues to be caught. Two specimens had already been netted 11 years earlier in a small quarry near Frome, and one, more dubiously, was taken at Blackpool in 1860. They remained unidentified, however, until Pickard-Cambridge published the beautifully illustrated account of his findings in the *Proceedings of the Dorset Field Club* in 1886.

The next 55 years were very sparse indeed, with only three more records of what was increasingly being called the 'Short-tailed Blue', after the minuscule wisps on the hindwings. Then, in 1945, four specimens were seen. After that there were two records of Bloxworth Blue in east Dorset in 1952, and two from Sussex in 1958 and 1977. In the past decade sightings have occurred more regularly, but by no means in every year, some of which are believed to have resulted from accidental escapes.

Distribution
A very rare immigrant from continental Europe. Most sightings have been in Dorset and the South West.

▲Male
Tiny wing-tails and twin orange spots on a silvery ground-colour on undersides distinguish this from other blues.

▲Female
Purple coloration is variable, and almost absent on some specimens.

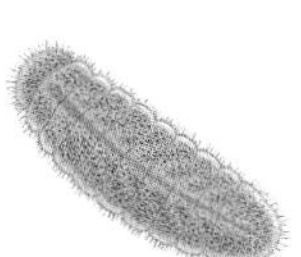

▲Caterpillar [× 2¼]
Feeds on leguminous plants; caterpillars are sometimes cannibalistic.

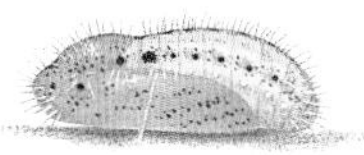

▲Chrysalis [× 2¼]
Formed on the caterpillar foodplant; adult emerges after two weeks.

	Jan	Feb	Mar	Apr	May	Jun	Jul	Aug	Sep	Oct	Nov	Dec
Adult												

Immigrants and offspring

It is interesting that 11 of the first 17 Bloxworth Blues recorded in the British Isles were from a comparatively small area of central southern England, comprising east Dorset, west Hampshire and south Somerset. This led R F Bretherton, an expert on insect migrations, to suggest that our specimens may originate from the heaths of Brittany, where the butterfly is still locally abundant. Breeding has never been verified in Britain, but it is likely that some sightings were of the offspring of earlier immigrants. The butterfly has two broods a year in northern Europe, the first emerging in May and June, and the second in July to September when all British records were made. Although immigrant butterflies often arrive in bands, it would be extraordinary if two had stayed together as far inland as Frome or Bloxworth. It is more likely that both pairs were the offspring of earlier immigrants. There would certainly have been an abundance of the caterpillar's foodplants present – Red Clover, Lucerne, various trefoils, Tufted Vetch and, reputedly, gorse.

For most naturalists there is little chance of seeing this butterfly in Britain, although it is common enough in central and southern Europe, where I have watched it many times. It does not appear to be a strong flier, and is seldom mentioned among the true migrants of Europe. No-one, to my knowledge, has studied its flight patterns, and it remains possible that it is more a local wanderer than a true migrant, akin to the Holly Blue.

The male Bloxworth Blue is particularly attractive in flight, producing alternate flashes of deep blue and silver as the upper- and underwings catch the sun. It has a jerky stuttering flight and, like most butterflies, is on the wing during the morning in central southern Europe, before resting during the heat of the day. There then follows a second short period of activity in the late afternoon. On the Continent, it can often be seen along the edges of woods, in wasteland and flying around the edges of scrub. Females frequently flutter over moist hay meadows a week or two after the first cut, laying profusely on sprouting clumps of clover that regenerate from rootstocks in the warm, damp soil.

This butterfly is an easy species to rear in captivity, and worth attempting, for the chrysalis is attractive, with long white hairs, and black spots and lines on a pale green background. The caterpillar is pleasant but unexceptional: fleshy, slug-like and a uniform pale green.

Small Blue

Cupido minimus

This dainty little butterfly has a wider distribution than any blue other than the Common Blue, with colonies found as far apart as John O'Groats in north-east Scotland and Kerry in south-west Ireland. But despite being widely scattered, it is rare in almost every region it inhabits. Even in its strongholds of Gloucestershire, Salisbury Plain and south Dorset, it is no more than locally common.

Why the Small Blue should be so much scarcer than its foodplant, flowering Kidney Vetch, is unknown, but a need for unusually sheltered conditions may be one answer. Colonies are seldom found on open downs or exposed cliffs. Instead, most butterflies breed in sunny nooks, where the soil is thin and unstable and the plant cover sparse and warm. Dune slacks, old quarries and steep embankments are all favourite habitats, and when visiting these places in late June it is always worth searching Kidney Vetch flowers for the tiny, pale blue eggs.

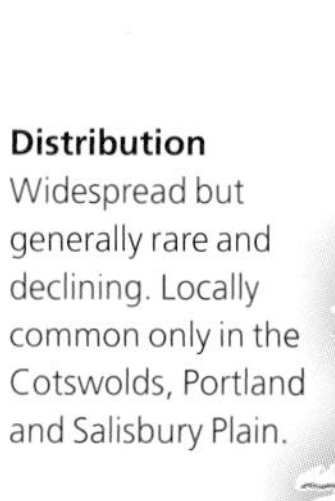

Distribution
Widespread but generally rare and declining. Locally common only in the Cotswolds, Portland and Salisbury Plain.

Small colonies

The natural history of this, our smallest butterfly, has been well studied by Ashley Morton. Adult Small Blues are highly colonial, and are often confined to no more than 200 square metres of land, supported by perhaps two dozen Kidney Vetch plants. Typical colonies contain fewer than 30 adults each, and breed in the same isolated patches for generation after generation. This is not to say that larger colonies do not occur – I have watched Small Blues teeming by the hundred in the Cotswolds and on the Isle of Portland, and sites with over a thousand individuals are known elsewhere – but these are few and far between, and the vast majority contain just a few dozen adults.

A typical emergence begins in mid-May, with peak numbers flying about three weeks later. A few linger on into July, and almost overlap with a small second generation that emerges in ones or twos throughout high summer on most southern sites. In Scotland there is strictly one generation a year, which emerges about a fortnight later than the English colonies.

Adult behaviour

This is an easy butterfly for beginners to overlook, but colonies are simple enough to locate once its distinctive behaviour is familiar. Males gather in a sheltered nook, usually a sunny depression at the base of a slope, where there is a scattering of shrubs or tussocky grass. Here they perch for most of the day, 30-120cm above the ground on prominent leaves, with each male spaced 1-2m apart. Their smoky black wings are held half-open towards the sun, so that the silvery blue dusting of scales is clearly visible.

Virgin females fly to these perching sites and are rapidly courted by the males, which spin upwards to investigate any small butterfly passing overhead. But there is no elaborate

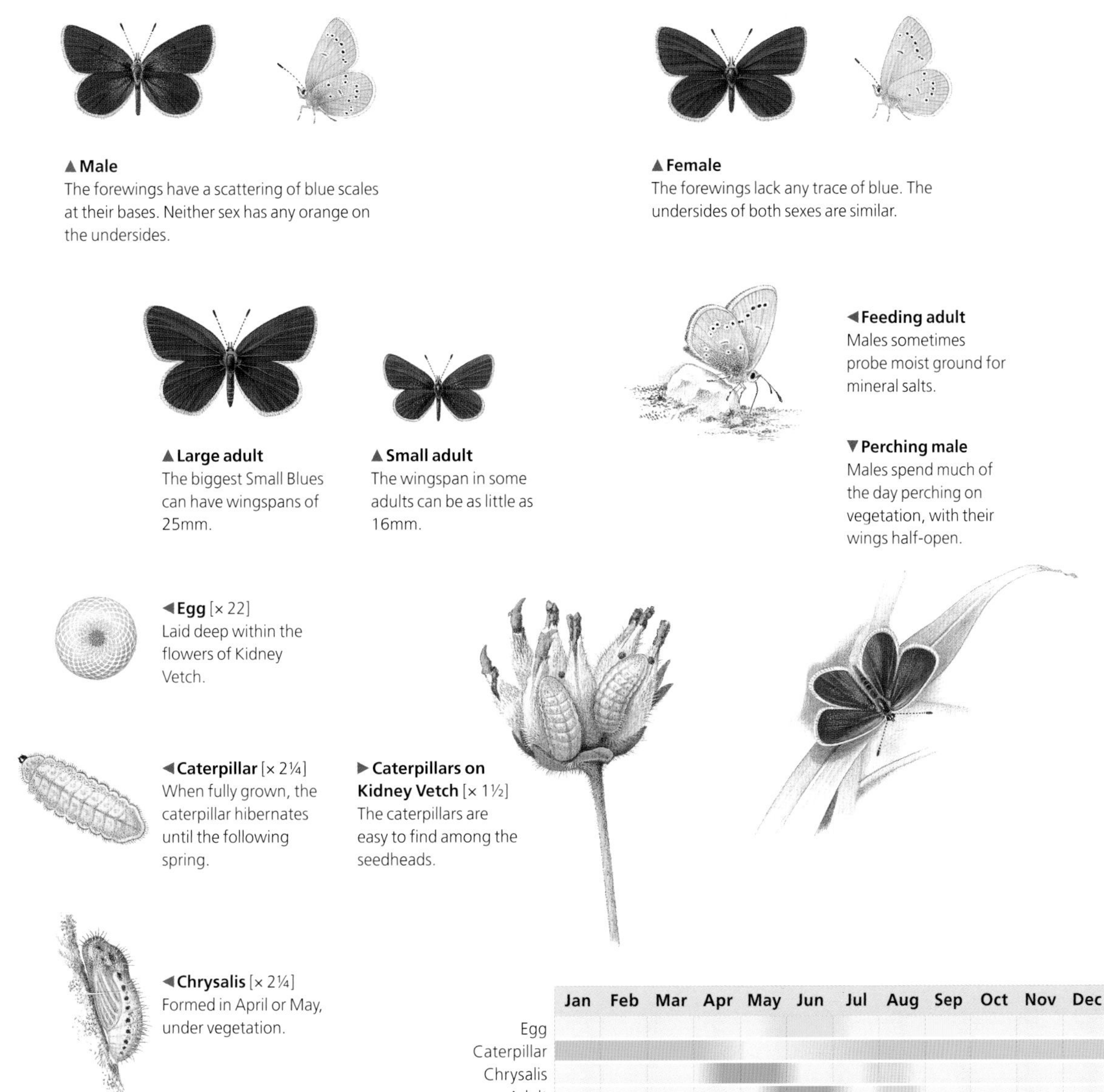

▲**Male**
The forewings have a scattering of blue scales at their bases. Neither sex has any orange on the undersides.

▲**Female**
The forewings lack any trace of blue. The undersides of both sexes are similar.

▲**Large adult**
The biggest Small Blues can have wingspans of 25mm.

▲**Small adult**
The wingspan in some adults can be as little as 16mm.

◀**Feeding adult**
Males sometimes probe moist ground for mineral salts.

▼**Perching male**
Males spend much of the day perching on vegetation, with their wings half-open.

◀**Egg** [× 22]
Laid deep within the flowers of Kidney Vetch.

◀**Caterpillar** [× 2¼]
When fully grown, the caterpillar hibernates until the following spring.

▶**Caterpillars on Kidney Vetch** [× 1½]
The caterpillars are easy to find among the seedheads.

◀**Chrysalis** [× 2¼]
Formed in April or May, under vegetation.

courtship and, once mated, the female avoids the perch areas altogether, concentrating instead on patches of Kidney Vetch, where she feeds, rests, basks and lays eggs between the yellow florets.

Adult Small Blues drink nectar almost exclusively from Common Bird's-foot-trefoil, Horseshoe Vetch or Kidney Vetch, creating a charming sight against the yellow blooms. Males also have a penchant for wet, salty patches, and in the Alps and the Cevennes I have seen them by the hundred jostling together along the stony borders of streams, probing the moist soil with their tongues. On British sites they gather to feast instead on the unsavoury substitute of dog dirt. Adults also congregate in loose groups when roosting, each perched head-down on tall vegetation. Curiously enough, these roosting sites are in distinct spots that are visited every evening, and may be some distance from the feeding and egg-laying sites, or from the males' perches.

Both sexes look weak and fluttery in flight, and much bluer than would be expected from examining the settled butterfly. With a little practice, this is an easy species to distinguish on the wing, although beginners often confuse it with the more silvery Brown Argus. When the butterflies are at rest there should be no confusion: the little, round wings have distinctive undersides of clear silver-blue, peppered with black dots, and no trace of the orange found on most blues. The Small Blue is rather like a miniature Holly Blue from below, and the largest specimens approach Holly Blue size.

Egg-laying and development

Few female Small blues fly more than 40m from their emergence spots during their brief adult lives, although some long-distance flights evidently do occur, since new breeding sites created up to 10km from the nearest population are generally colonised within five years. A typical female spends much of her life fluttering around Kidney Vetch flowers, probing the yellow florets for a suitable egg-laying spot. She concentrates on prominent, long-stemmed plants growing in warm, sheltered depressions, and inserts a single egg between the tightest florets of a young flowerhead that is still largely in bud. She then rubs her abdomen over the buds, probably to deposit a scent that will deter other females from laying there. As with the chemicals released by whites, its effect soon wears off, and it is not uncommon to find three or four eggs of various vintages on the same flowerhead of a prominent vetch.

Although it is tiny, the egg is the easiest stage in which to find a colony of the Small Blue. It is necessary simply to prise the 'fingers' of each flowerhead gently apart, and eggs will be found along their downy sides; do this with care, for the flowers and seedheads have brittle bases and are easily detached. The eggs hatch after one to three weeks, and the young caterpillar burrows deep inside a floret to feed on developing anthers and seed. It is cannibalistic at this stage, and if two or more caterpillars enter the same floret, only one survives. As they grow older, the grey-pink caterpillars become easy to find once more, for they live openly on the flower clusters. Each lies head-down while it bites a series of holes into the base of flowers in order to feed on the seed. This damage is noticeable long after the caterpillars have left to pupate; their droppings are also a tell-tale sign.

After the third moult, when Small Blue caterpillars enter the final stage of their growth, they possess all of the ant-attracting organs illustrated on pages 119 and 143, although the tentacle organ does not function. They can also produce rasping 'songs' as loud and persistent as those of the Adonis Blue. Yet in captivity the caterpillars are less attractive to ants than those of Common – let alone Adonis – Blues, and are seldom milked by them in Britain. This may partly be because few of our native ants climb the 20cm or so to reach Kidney Vetch flowerheads. In central Europe, I have examined scores of these vetches, and never once found a Small Blue caterpillar without ants present.

By late July the caterpillars are fully fed and desert their flowerheads, which by now are starting to disintegrate and shed seed. Each caterpillar settles on the ground, reputedly in a crevice, under soil and especially under moss, where it remains dormant for nine months. Then, in late April or early May, it seeks a pupation site, again under vegetation on the ground. It is probably earthed up by ants at this stage, for the chrysalis also attracts them.

A declining species of skeletal soils

The most productive places to seek this charming little blue are nooks containing Kidney Vetch flowers, not necessarily in profusion but situated in a sun-baked, sheltered terrain of rough and broken ground, where an annual supply of new seedlings is assured. Typical sites include warm limestone pavement and abandoned limestone and chalk pits. Colonies also occur on unstable ground, including steep, thin-soiled banks and calcareous sand dunes composed of fragmented seashells. Railway lines and road verges are also excellent places: I know of five isolated colonies on the cuttings of dual carriageways in southern England, where the butterflies hop from flower to flower up to the kerb, oblivious of the constant stream of traffic. There are also a few regions where the butterfly breeds on open downland. Chief among them are steep, unfertilised valleys in the Cotswolds, and the south-facing slopes of the Mendips and Salisbury Plain.

The Small Blue's current strongholds of the Cotswolds and Salisbury Plain support an estimated 145 and 185 colonies respectively, including some very large populations. A further 100 colonies occur in Dorset, especially near the south coast, and in the Isle of Wight. Elsewhere, it is much rarer. Sadly, no more than 15 small colonies survive along the North Downs, and populations are few and far between on the South Downs, in the Chilterns and along the coast of south Wales. North of these areas, the butterfly is now extremely scarce and has disappeared from the large majority of former sites. It is probably extinct in the Peak District and may be reduced to one colony in the border country of south-east Scotland, although small populations persist on warm undercliffs along the north-east Scottish coastline. In Ireland, recent surveys have located a series of previously unknown colonies, especially along the coast of Donegal. Indeed, it occurs quite frequently along the western coastline, for example in County Clare, Limerick and Kerry. However, there are very few records of inland colonies from Ireland in the present century.

Silver-studded Blue

Plebejus argus

The Silver-studded Blue is one of two characteristic butterflies of English lowland heaths, and has become a symbol of this diminishing habitat. Unfortunately, the butterfly declined enormously during the 20th century and is now virtually absent from four-fifths of its former range. The stronghold, as ever, remains the Hampshire Basin, a sandy exposure of Tertiary deposits that encompasses all the heathland of the New Forest and its westward extension into south-east Dorset. Here, the Silver-studded Blue can still be seen by the thousand, fluttering and shimmering above furze and heathers through the hottest days of high summer.

Colonies are not confined to heathland, however. In south-west England, Silver-studded Blues breed on several coastal dunes and cliffs, and the butterfly was once locally common on northern mosses, principally in the old county of Westmorland, where it was considered to be a separate subspecies, *masseyi*, owing to the remarkable blue wings of the females. More surprisingly, scattered colonies once bred on the chalk downs of Kent, Surrey, Hampshire and Dorset. This was the *cretaceus* race, characterised by its slightly larger size and paler blue males – like *masseyi*, this lovely form is extinct. There are also two areas of limestone where the Silver-studded Blue has been known since Victorian days. One is the Isle of Portland, which supports some exceedingly large colonies on its rocky terrain of abandoned quarries and undercliffs. The other is the beautiful limestone stacks of Great Ormes Head, on the north coast of Wales.

North Wales colonies

The Silver-studded Blues of Portland have no sub-specific name, but those on Great Ormes Head are sufficiently distinctive to have been classed as a separate form, called *caernensis*. There, the adult butterfly is unusually small and the females have a noticeably bluish tinge. As on Portland, they emerge two to three weeks earlier than other populations, typically appearing in mid-June. These diminutive Silver-studded Blues have survived well. For whereas the butterfly has disappeared from almost every heath north of Berkshire, *caernensis* populations have increased, in part thanks to A J Marchant's release in 1942 of 90 adults into the Dulas Valley, a region of rough limestone grassland some

Distribution
A declining species now rare outside southern English heathlands. Still abundant on many Dorset and New Forest heaths.

13km east of Great Ormes Head. From this small introduction they spread slowly along the valley over the following 60 years, at roughly 1km a decade. They eventually colonised 15 additional sites in the Dulas Valley, which together support nearly 100,000 adults.

Regional races

Much recent knowledge about this beautiful butterfly derives from studies by Chris Thomas, Diego Jordano and colleagues. They found that our three surviving subspecies or races – *argus* on dunes and heathland, the blues of Portland, and *caernensis* from Wales – are genetically distinct, and differ slightly in appearance and considerably in their ecology. We illustrate *argus* and *caernensis* adults, along with the extinct *cretaceus* and *masseyi* forms. In truth, most physical differences are minor, the only consistent geographical variation being that females have bluer wings the further one travels north.

Sedentary colonies

Like most blues, this species lives in tight, close-knit colonies, which the adults are extraordinarily reluctant to leave. Although both sexes fly readily on sunny days, this is a slow, fluttering affair, seldom more than a few centimetres above the ground, with frequent turns when an obstacle is encountered. Chris Thomas, Mike Read, and Neil Ravenscroft independently marked adults in several populations in Wales, Devon and Suffolk during the 1980s. Each found that the average lifespan of both sexes was four or five days, and that few individuals moved more than 20m during this time, with a lifetime dispersal of further than 50m distance being exceptional. Thus, neighbouring breeding areas separated by a 100m or more of furze or farmland are, to all intents and purposes, isolated, and their populations wax and wane independently of each other.

Longer flights do, of course, occur on rare occasions, sufficient for any new site within 1km of a population to be colonised eventually. Nevertheless, during ten summers working on the chalk downs of Swanage, I have only twice seen a Silver-studded Blue that had strayed from the vast populations on the Dorset heaths just 1km to the north. In north Wales, the butterfly was quite unable to colonise the Dulas Valley naturally from the large populations just 13km to the west; and once introduced, it spread at a snail's pace across a landscape tailor-made for its specialised requirements. These feeble powers of dispersal make the Silver-studded Blue singularly ill-suited for survival in the modern world. Breeding sites are becoming increasingly isolated and, on heathland at least, they seldom remain suitable for more than a few years after a disturbance.

Where it does occur, the Silver-studded Blue can be extremely numerous. The best populations contain tens of thousands of adults, and make a wonderful sight in the early morning, for the butterflies roost together in large groups, often on tussocks of Purple Moor-grass, and bask communally with wings wide open to catch the first sunshine of the day. At the slightest disturbance they flutter away in a cloud of inky blue, brown and silver wings. Most colonies, however, contain between 100 and 1,000 adults, of which no more than a third are flying at its peak. Small ones contain just a few dozen individuals, supported by as little as a 0.1ha of breeding habitat.

The Silver-studded Blue has one adult generation a year. Except on Portland and Great Ormes Head, the adults typically emerge from late June onwards, reach a peak in mid-July, and last well into August. Males make short patrolling flights, fluttering between the sparse patchwork of dwarf bushes that characterises most breeding areas, in a continuous search for newly hatched females. Courtship is then brief, consisting of little more than a short, buzzing flight as the pair weaves between the low vegetation, before the female drops to the ground for mating.

The female is more deliberate when egg-laying. She makes short, fluttery flights over the vegetation before dropping to lay a single white egg, usually on a stalk or piece of tough vegetation. At Great Ormes Head and Portland, the eggs are inserted just under the edges of mats of Common Rock-rose or Common Bird's-foot-trefoil, where these spread onto bare limestone boulders. They remain all winter. On heaths, the egg-laying females concentrate on sparse patches where young Ling, heathers or gorse sprout through the sand in the first years after a clearance. This preference for fresh clearings is accentuated in the north, reflecting the butterfly's need to lay eggs in spots where the ground is several degrees warmer than most of the places where its foodplants grow.

Relationships with ants

It has recently become clear that the Silver-studded Blue has a more intimate relationship with ants than any of our butterflies, other than the Large Blue. The young stages are inseparable from two species of black ant: *Lasius niger* in more moist habitats and *L. alienus* in dry ones. But, unlike the Large Blue, the adult butterfly also interacts with ants. For example, the female can detect *Lasius* black ants as she flutters and tests the rocks and vegetation, and will selectively lay eggs near their nests. This was first suspected in Suffolk, when 'literally hundreds' of eggs were found on the undersides of Bracken fronds, whose nectaries secrete sweet liquids to attract ants. Since then, many observations have been made of the butterfly laying eggs beside a black ants' nest: we illustrate a famous example photographed by

▲ **Male**
Typical coloration of the most common form of the species, found on heaths and dunes.

▲ **Female**
Common form of the female usually has no blue on the wings, unlike the rarer forms.

▲ **Male**
Form *masseyi*, from the old vice-county of Westmorland, now extinct.

▲ **Female**
Form *masseyi*, showing exceptionally blue wings.

▲ **Male**
The smallest form is *caernensis*, found in north Wales.

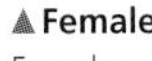

▲ **Female**
Females of the *caernensis* form are noticeably bluish.

▲ **Male**
The extinct *cretaceus* form; slightly larger than the common form.

◄ **Resting adult**
Adult resting on gorse, one of the caterpillar's foodplants on heaths.

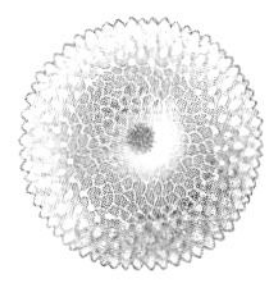

▲ **Egg** [× 22]
Laid on a variety of plants, depending on habitat.

◄ **Caterpillar** [× 2¼]
The caterpillar is tended by ants as it feeds on young shoots.

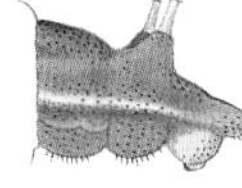

◄ **Tentacles**
Tentacles at the rear of the caterpillar are erected to stimulate the ants into 'milking' the honey-gland.

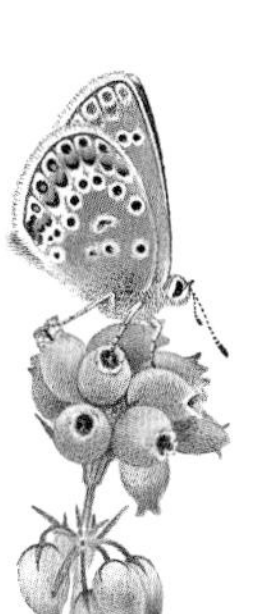

▲ **Feeding adult**
On heaths, Bell Heather is a frequent source of nectar.

◄ **Chrysalis** [× 2¼]
Formed underground, within the chambers of an ant nest.

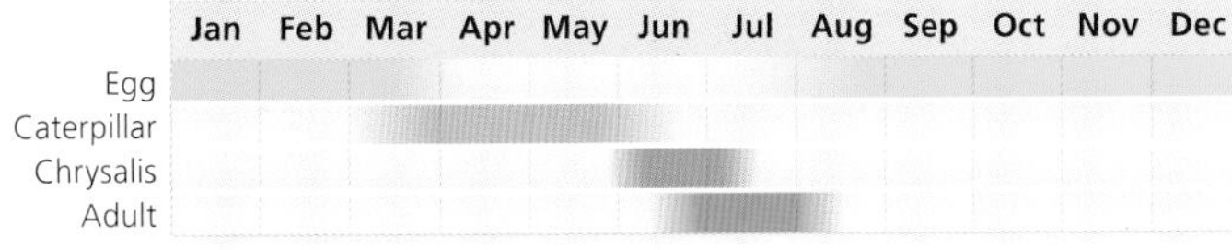

Ken Willmott on Portland (see opposite).

In addition to restricting herself to an ant territory, the female selects spots where the caterpillar's foodplants have room to sprout tender shoots, which are rich in nitrogen, during the following spring. Despite these constraints, she can utilise a wide range of plants growing in this situation, especially Common Bird's-foot-trefoil and Common Rock-rose on limestone sites, and gorses, Ling and other heathers on heaths, all of which are palatable to the caterpillar. The egg hibernates with a perfectly formed caterpillar inside, and is the only stage in the life cycle that is not known to interact with ants.

The Silver-studded Blue egg hatches towards the end of March. The caterpillar is immediately attractive to *Lasius alienus* and *L. niger* black ants as it painstakingly nibbles its way out of the tough overwintering shell. In laboratory experiments, the tiny caterpillar is quickly picked up and carried gently into a *Lasius* nest; the same thing happens when hatching eggs are placed in more-or-less natural situations in the wild. So far as is known, no other European species of blue has a relationship with ants at this early stage, although its close relative, the Zephyr Blue *Plebejus pylaon*, is probably similar. Moreover, unlike both Brown Arguses, Adonis and Chalkhill Blues, whose caterpillars appear equally attractive to a range of red *Myrmica* and black and yellow *Lasius* ant species, the Silver-studded Blue attracts only *L. alienus* and *L. niger*. A foraging worker of the large wood ant *Formica fusca* will sometimes pick up a young Silver-studded Blue caterpillar, but only as food for its colony.

Black *Lasius* ants certainly do not treat Silver-studded Blue caterpillars as fodder. They tend and tap them incessantly with their antennae, and generally surge around in a protective way, as described for the Adonis Blue (see pages 142-3). We do not yet know how the hatchling Silver-studded Blue caterpillar interacts with ants in the wild. In captivity they eat tender plant tissue, and it is possible that this is their food inside ant colonies, for I have often found the roots and blanched shoots of gorses, trefoils and other foodplants twisting through the internal passageways of their *Lasius* nests. But if they do feed above ground at this stage, I suspect they are carried there by the ants, in the same way that workers transport their herds of domesticated aphids from one suitable foodplant to another.

Attracting ants

The caterpillar is more mobile after its first skin moult. By then it has developed additional secretory organs to attract ants, and by the final instar it possesses active tentacle organs, a large dorsal nectary organ and numerous pore cupola (see page 143). While these presumably exude a similar range of sugars and amino acids to caterpillars of Adonis and Chalkhill Blues, I suspect that additional compounds are secreted – probably cocktails of hydrocarbons – that mimic the recognition or appeasement pheromones used by black ants to distinguish their own colonies or species from rival or enemy ants. For if they merely secreted food, Silver-studded Blue caterpillars would surely be as attractive to *Myrmica* ants as they are to black *Lasius* species. Moreover, it is the complex pore cupola of Silver-studded Blue caterpillars that receive special attention from *Lasius*, whereas the sugar-producing dorsal nectory organ is the most attractive area on other blues. Intriguingly, Diego Jordano and Chris Thomas found that the *caernensis* caterpillars from the Great Orme were tended more assiduously by *Lasius alienus* than by *L. niger*, and that the opposite was true with *argus* caterpillars from heathland. In other words, each race or subspecies of Silver-studded Blue is adapted to attract the particular species of black ant with which it cohabits in the wild.

Interactions with ants

There is still much to learn about the relationship between Silver-studded Blue caterpillars and black *Lasius* ants, but it is undoubtedly an intimate one. By day, the caterpillars rest inside their host-ant colonies, generally choosing the outer nest chambers rather than the inner ones that teem with brood, workers and queen ants. Nevertheless, they penetrate deeper towards the heart of the colony than do Chalkhill Blue caterpillars. I have found chrysalises and caterpillars of both blues inside the same *Lasius alienus* nest on several occasions: the Chalkhill Blues rest with a few attendants in the loose earth on the outer edges, while the Silver-studded Blues inhabit the permanent earthen cells and passages patrolled by ants. At dusk, the caterpillars of both blues emerge to feed on Horseshoe Vetch and Common Bird's-foot-trefoil, respectively, each attended by a posse of workers.

The Silver-studded Blue's pupa is formed in the same chambers inhabited by the caterpillars. The ant expert, John Pontin, was the first to find them, lined up along the passages of *Lasius niger* nests in the New Forest. More recently, I have found hundreds of chrysalises in this situation with *L. alienus*, generally 1-3 per ant nest but with up to 20 close together in the best colonies, often accompanied by final and penultimate stages of the caterpillar. The pupae are tapped and tended incessantly by the ants, which seldom desert them, even when exposed to full sunshine after their stone is lifted. The ants show similar loyalty to final-instar caterpillars that are too large to move; the small earlier ones are mostly carried underground with the ant brood into the safety of deeper chambers.

◀ **Laying eggs**
Female laying on foodplant beside a *Lasius* ants' nest.

▶ **Emerging adult**
The adult emerging from an ants' nest is attended by several ants, attracted by the droplets of liquid on its body.

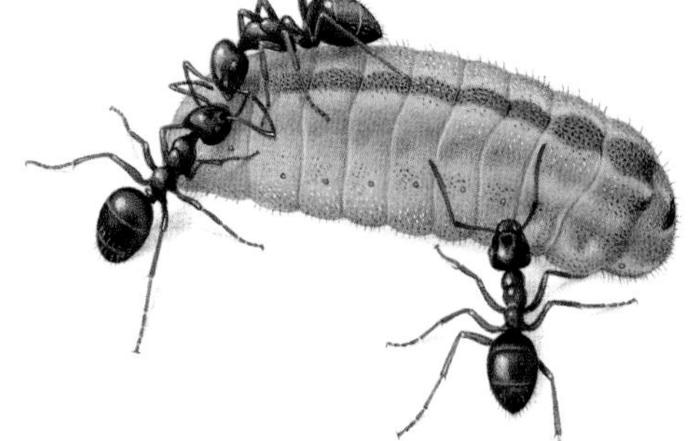

◀ **Developing caterpillar**
Ants attend the developing caterpillar.

▲ **Young caterpillar**
A *Lasius* worker ant carries a young caterpillar.

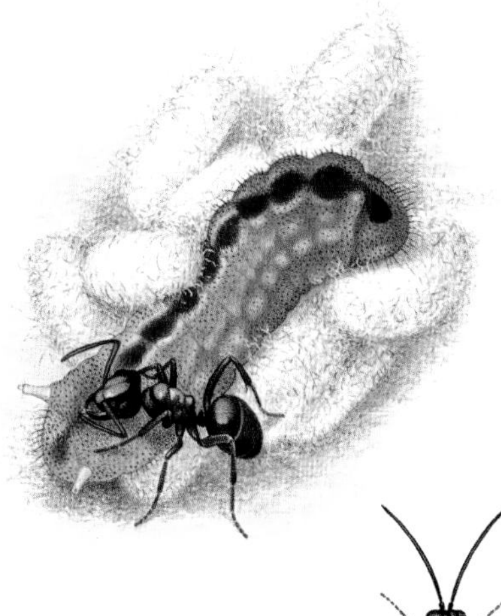

▲ **Ichneumon wasp** [× 2½]
The large parasitoid wasp *Virgichneumon callicerus* (left) emerges from a pupa (shown above as the darker colour). It possesses chemicals that repel ants.

▲ **Braconid wasp** [× 6]
The tiny braconid wasps *Cotesia tenebrosa* (right) emerge from silky cocoons beside the empty husk of the caterpillar, which remains alive, and can attract ants, for a further week.

Emerging adults

Chris Thomas was the first to notice another extraordinary aspect of the Silver-studded Blue's relationship with *Lasius* black ants which, so far as I know, has been reported from no other butterfly. On emergence from its chrysalis, the adult's furry body is wet with droplets of liquid. These, too, are highly attractive to ants, which form an accompanying posse as the butterfly crawls to the surface to blow up its wings. As ever after an intriguing report, this has been confirmed and photographed numerous times across Europe; I have seen scores of ant-attended adults myself. I generally find them between 8am and 10am on a warm, still morning when, in strong colonies, up to three or four adults may be seen being tended simultaneously above the same ant nest, each hanging from a grass stem and accompanied for at least 30 minutes before the workers lose interest and the butterfly flies away. Four to eight ants attend each adult until the wings are inflated and dry. They are constantly active, running back and forth over the head, body and adjoining vegetation, mainly 'licking' the butterfly's thorax and head. Sometimes one slips and may swing from the tip of the antenna, while the butterfly sits motionless, oblivious to this attention. They remind me of cleaning fish nosing around a grouper or a shark, although in this case the ants are presumably protecting the butterfly from enemies during a particularly vulnerable period of its life.

Specialised parasitoids and ants

While it is possible to rear Silver-studded Blues in the absence of *Lasius* black ants, the caterpillars' sticky secretions are so copious that, without constant grooming, they often develop fungal infections and die. It is generally assumed that there is a mutually beneficial relationship with the ants, the latter gaining substantial food in exchange for protecting the caterpillars from enemies, especially parasitic wasps and flies. In fact, Karsten Schönrogge, Kelly Murray and I have found higher infestations of parasitoids in the Silver-studded Blue than in any species of blue butterfly we have studied, other than certain Large Blue populations on the Continent. About half the caterpillars in a British colony of Silver-studded Blue may be parasitised, although a third is more usual.

On the Isle of Portland, we found up to four different species of parasitoid infesting a single population of the butterfly. All were myrmecophilous: that is, they have their own adaptations for living unharmed among *Lasius* ant colonies, especially after emerging from the caterpillar's or chrysalis's body. The two commonest parasitoids are the large ichneumonid wasps, *Virgichneumon callicerus* and *Anisobas cingulatellus*, of which the first is illustrated on page 121. Both are very beautiful, and emerge at the same time as the adult butterflies from pupae inside *Lasius* nests, one wasp per pupa. Although little is known about the natural history of either wasp, we found that they produce obnoxious chemicals that induce aggression and fighting among the ants. It is likely that this distracts or repels the workers, allowing the emerging wasp to escape its *Lasius* nest unharmed. The same strategy is displayed by a rare myrmecophilous tachinid fly, *Aplomya confinis*, which Murray and Schönrogge found in two colonies of Silver-studded Blue on Portland. We do not know whether the ichneumonids re-enter *Lasius* nests the following spring to sting the next generation of caterpillars or soft pupae, or whether they seek their hosts when feeding above ground at dusk. In either case, the possession of chemicals that repel ants would be invaluable.

A less lovely but equally interesting parasite of the Silver-studded Blue is a tiny braconid wasp called *Cotesia tenebrosa* (illustrated on page 121). Like the more familiar braconids of cabbage whites, and Marsh and Glanville Fritillaries, 10-20 minute wasp grubs live inside one caterpillar's body, emerging only when their host is fully grown. At this stage, they bore through the skin to spin cocoons along the sides of the caterpillar, in which they form pupae and later emerge as microscopic wasps, each little larger than a breadcrumb. Remarkably, although it is now an empty husk – all skin with no apparent content – the Silver-studded Blue caterpillar does not die for another week, but remains active enough to produce secretions that attract ants; it is even able to exert its tentacle organs if the workers wander away. Thus, the little cocoons of the parasite are inadvertently attended by protective ants for the whole pupal period, until the adult braconids emerge. There is then a dramatic switch in behaviour. The adult wasps release chemicals that provoke the ants into making rapid biting darts at them. They clearly have a chemical deterrent too, for the ants appear to bounce off the wasps as soon as they reach them, retreating in a state of high agitation. Having watched this strange sequence several times in captivity, our provisional interpretation is that, in wild nests, the braconids drive the ants underground, leaving the wasps free to escape to the outside world from the deserted upper chambers.

Loss of heathland

During the 20th century, far too many of these fascinating communities of Silver-studded Blue, succulent foodplants, black ants and specialised parasitoids have disappeared from the British countryside as a result of the destruction of most areas of heathland outside the Hampshire Basin. Moreover, few of the lowland heaths that survived were farmed in traditional ways, by regular rotational burning or grazing of small patches, generating an annual supply of short-lived breeding sites, each adjoining the previous year's clearings. By the 1990s, disturbances were few and far between. Entire heaths were often abandoned for many years, only to be accidentally burned in fierce summer fires. With its feeble powers of dispersal, this butterfly was ill-equipped to spread to the isolated patches of new habitat that were being regenerated. This was a particular problem in the northern half of the Silver-studded Blue's range, for under cooler climates its breeding sites are smaller, more localised and more ephemeral.

This century-long decline halted a decade ago, thanks to the preservation and renewed management of much surviving heathland. To date, the butterfly has shown no signs of recolonising its former range, and in the Midlands the typical *argus* race is confined to a single fragment, now a nature reserve, sandwiched between two trunk roads. Just a handful of colonies survive in Norfolk, as do five or six on the Suffolk Sandlings. There are stronger populations on some Cornish dunes, but the major colonies of *argus* occur on the southern English heaths. Even here there have been major losses, with no more than a few colonies remaining in Berkshire, Ashdown Forest, or on the lovely pebble heaths of east Devon.

Its status is much stronger on the west Surrey heaths, while in Dorset and the New Forest the Silver-studded Blue can be expected on almost any piece of heathland that looks suitable. Perhaps because of the warmer springs in this region, colonies are less restricted to south-facing slopes or to areas where there has been a clearing within the previous five years. Whilst more or less absent from mature dry heathland, they are frequently encountered on the so-called humid heath – as opposed to the really boggy areas – where the soil is peaty and where the butterflies hover around the clustered bells of Cross-leaved Heath.

The coastal and limestone populations of Silver-studded Blue have also remained stable in recent years, for example in south-west Cornwall, the Bolt Head-Bolt Tail cliffs of Devon, and on Holy Isle, off Anglesey. In addition to the large populations of the Dulas Valley, about ten colonies of the *caernensis* race survive on Great Ormes Head, of which a quarter contain 30,000 or more adults each. Finally, about 30 colonies persist on the Isle of Portland, including several that are carefully maintained on Butterfly Conservation nature reserves.

Brown Argus

Aricia agestis

At first sight, the Brown Argus seems a rather dull butterfly, with both sexes superficially resembling the brown form of the female Common Blue. Yet it has a charm of it own, resulting from a lively behaviour, an unexpected shimmer of silver when it flies, and the rich chocolate and orange colours on the upperwings of fresh specimens. Until recently this was not a particularly common butterfly, having declined and lost about 40% of its populations during the 20th century. By the late 1980s, the Brown Argus was restricted to small, localised colonies on unfertilised southern downland, to several coastal dune systems, and to a scattering of other localities, including southern heaths and woods. Happily, it has increased greatly during the past 20 years, spreading to new types of habitat and recolonising regions further north from which it had been absent for more than 50 years.

Distribution
Until recently, scarce on chalk and limestone downs, cliffs and dunes. Has colonised many grassland sites since the 1990s, and spread north and west.

A neglected species

The Brown Argus was largely neglected by early entomologists, and features in few books of any antiquity. When it does appear it is under a variety of names, including the 'Edg'd Brown Argus' and the 'Brown Blue'. Identifying the butterfly can be tricky. In flight it can be mistaken only for the Small Blue, owing to the silvery reflection as the sun catches the underwings, but the perched butterfly is more difficult. One distinguishing feature is the absence of any blue on the upperwings and, in contrast to Common, Adonis and Chalkhill Blues, there is no spot nearer to the body than halfway-in on the undersurface of the forewing. In addition, the pair of black spots near the outer edge of the underside of the hindwing forms a figure of eight or colon (:) instead of being sideways on (··), as in other blues.

There is some variation in the markings of individual butterflies. In males, the orange spots on the uppersides can be reduced to mere pin-pricks, whereas these merge to form an orange band on some females. In a particularly attractive variety called *subtus-radiata*, the spots on the underwings are distorted to form long streaks.

Colonies and climate change

Little was known until recent decades about the natural history of the Brown Argus, apart from the identity of its main foodplants, the fact that its caterpillars are often tended by ants, and that its populations fluctuate greatly in size. These fluctuations are unusual in that they seldom occur in synchrony on nearby sites. Thus it is common to find that the butterfly is having one of its better years on one downland, while numbers have inexplicably dropped on others.

We now know much about this butterfly, thanks to studies by Nigel Bourn, E J Bodsworth and Rosa Menéndez. Except at its northern edge-of-range, the Brown Argus has two adult broods a year, the first from mid-May to late June,

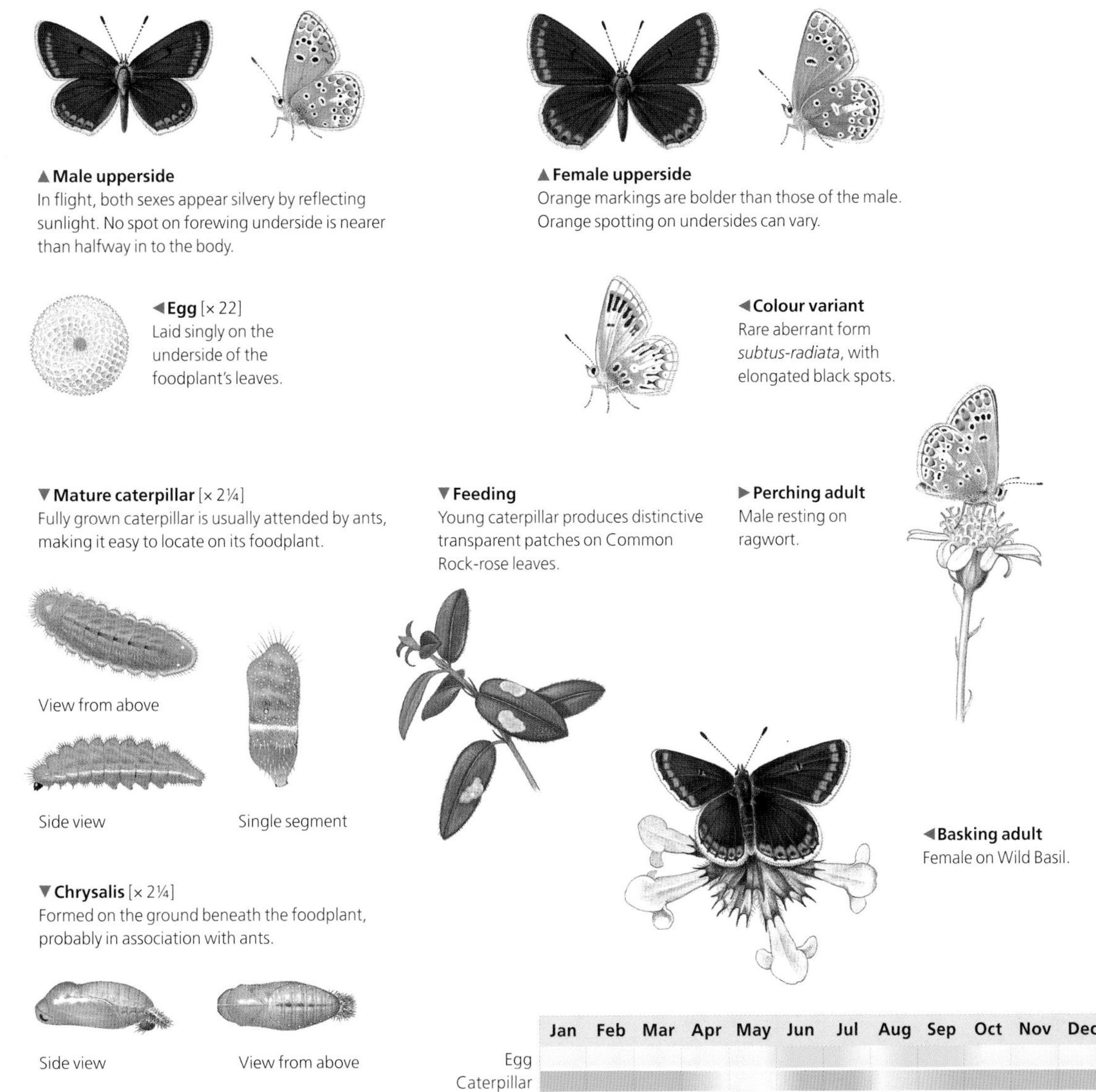

▲ **Male upperside**
In flight, both sexes appear silvery by reflecting sunlight. No spot on forewing underside is nearer than halfway in to the body.

▲ **Female upperside**
Orange markings are bolder than those of the male. Orange spotting on undersides can vary.

◀ **Egg** [× 22]
Laid singly on the underside of the foodplant's leaves.

◀ **Colour variant**
Rare aberrant form *subtus-radiata*, with elongated black spots.

▼ **Mature caterpillar** [× 2¼]
Fully grown caterpillar is usually attended by ants, making it easy to locate on its foodplant.

▼ **Feeding**
Young caterpillar produces distinctive transparent patches on Common Rock-rose leaves.

▶ **Perching adult**
Male resting on ragwort.

◀ **Basking adult**
Female on Wild Basil.

▼ **Chrysalis** [× 2¼]
Formed on the ground beneath the foodplant, probably in association with ants.

the second from July to September. Both are protracted, so much so that although the average lifespan of an individual adult is only four days, it is usual for the earliest males of the second brood to be flying before the last tattered females of the first have died. Most colonies are small, containing just a few dozen adults, and even in good years few sites support more than 500-700 individuals.

Bourn also discovered that although the adults live in self-contained areas, there is considerably more interchange between adjoining sites than is the case, for example, with Adonis or Silver-studded Blues. These latter species travel, on average, only about 10-20m from their birthplaces, while Brown Argus adults move up to ten times further. This explains the notably higher rate of spread of the Brown Argus in recent years, when the warmer climate made many sites suitable for breeding beyond its northern range. By comparison, the Adonis Blue has expanded at a snail's pace, and the Silver-studded Blue not at all.

Courtship and egg-laying

At night, the adult Brown Argus roosts in the lower parts of its site, clustered in groups, head-down on grass stems, often in the company of Common Blues and other blues. At daybreak they bask communally for a few minutes, with wings stretched wide to absorb the warmth of the sun. The rest of the day is spent in solitary pursuits. Males either perch on the ground with open wings, or patrol the lower parts of their site, while any virgin female advertises herself by sitting astride a grass head. A male soon alights alongside, and both then embark on a short and sinuous nuptial flight, meandering just above ground level before settling on a grass clump to mate.

It takes another day or two before the female's egg-load ripens, whereupon she embarks on low, fluttery flights, frequently alighting to search for caterpillar foodplants, twitching and tapping the vegetation with her feet. Until recently, Common Rock-rose was the sole food used in most localities, although Dove's-foot Crane's-bill, Common Stork's-bill and occasionally other geraniums were used on a few sites away from downland.

An egg-laying female becomes agitated when she encounters a foodplant, but tests each carefully before deigning to lay. On downland she will reject most rock-roses outright, instead laying her eggs on a minority of plants that have exceptionally thick, lush leaves and also contain unusually high concentrations of nitrogen. Sheltered, sunny depressions and plants with some bare chalk rubble around them are particular favourites. Few rock-roses meet these specifications, but those that do range from large, mat-like growths to virtual seedlings.

A survey of downs that support Brown Argus colonies revealed that the condition of these plants, rather than their abundance, was of paramount importance to the butterfly. By and large, the species does well during a warm summer, but those colonies that breed on well-drained, thin-soiled downs find that their foodplants become unsuitable for caterpillars during periods of drought, and consequently numbers decline. On sites with deeper soils, the rock-roses remain lush and healthy, with the result that these Brown Argus populations flourish in hot years.

The egg is a pale blue-green disc, and is nearly always laid on the undersurface of a leaf. It hatches after a week or two, and the small caterpillar begins to feed on the soft, rich interior of the leaf, first eating a neat circle of tissue, then larger patches from the underside of the leaf. The upper surface is left intact, creating shiny windows that are easy to spot on leaves in the wild.

Ants and parasitoids

The caterpillar has the same triangular shape in cross section as those of its close relatives, and can be distinguished by the deeper green ground-colour and pink stripes, which blend beautifully with rock-rose shoots and with the pink-veined leaves of stork's-bills. Although it feeds openly by day, the full-grown caterpillar would be extremely hard to spot were it not for its attendant ants. If you scan likely rock-roses and notice an excited group of ants, there will often be a caterpillar beneath them. Be warned, however, that at the slightest disturbance the ants will desert the caterpillar, which will roll off its leaf and is then nearly impossible to find.

In its final stage, the caterpillar possesses all the organs described on pages 142-3. It whips its entourage of ants into a frenzy of excitement by frequently extending the paired tentacles, especially when it walks, and by pumping visible droplets of honeydew out of a large and active honey-gland. In Germany, Konrad Fiedler found that two to three droplets were secreted every hour by the growing caterpillar, rising to five times that rate during the vulnerable period when they stop feeding and hunch up for pupation. In addition, microscopic secretions are exuded like sweat over the rest of the body; there can be little doubt that this caterpillar also chirrups to its ants.

Although constantly providing food for its attendants, the Brown Argus does so at little cost: captive caterpillars reared in the absence of ants form pupae a day earlier than those reared in their company, but on average the resultant chrysalis is slightly smaller. We have yet to discover what advantage the caterpillar gains from this interaction. It may simply be a device to prevent attacks from the ants themselves, for the preferred form of rock-rose invariably grows in spots where red or black ants are abundant. The ants may also deter predators or reduce pathogens.

Ants do not, however, seem effective in reducing attacks by parasitoids. In a fascinating comparison of long-established and newly formed colonies, Rosa Menéndez found no fewer than five species of parasitic wasps and one species of tachinid fly that kill the Brown Argus caterpillars or pupae. Two belong to the same species that we found parasitising Silver-studded Blue caterpillars on the Isle of Portland: the sleek and elegant ichneumonid wasp *Anisobas cingulatellus*, and an inelegant tachinid, *Aplomya confinis,* that looks like a very scruffy blue-bottle. Both parasitoids are specialists at entering and leaving ant nests unharmed, thanks to their ability to secrete obnoxious chemicals that probably drive the ants down into their deepest underground cells while the parasite is in their nest. At least three of the other parasitoids of the Brown Argus, two tiny species of braconid and one slightly larger wasp, have also been bred from

Common Blue or Northern Brown Argus caterpillars, and it seems likely that they will have similar adaptations for manipulating ants.

Menéndez found high rates of parasitism in long-established Brown Argus colonies, such as that breeding on Aston Rowant, the Oxfordshire nature reserve that straddles the M40 where it cleaves the chalk escarpment of the Chilterns. Just over half the caterpillars at Aston and on similar established sites host these fascinating parasitoids, whereas no more than a quarter of them are infested in newly formed colonies.

Changing temperatures and foodplants

Until the 1990s, nearly all Brown Argus sites were on warm, south-facing chalk and limestone downs. Other 20th-century populations bred on cliffs or calcareous sand, for example along the coasts of north Cornwall, south Wales and Norfolk, and in some surviving fragments of Breckland. Not every colony was supported by Common Rock-rose. The butterfly's occasional use of Dove's-foot Crane's-bill and Common Stork's-bill explained a scattering of colonies elsewhere in southern England, for example in woodland glades and rides on clays, and even on acid heaths. Despite both these plants being common and widespread in rough or disturbed ground across Britain, the geranium-using populations were small and, until recently, extremely rare.

No records exist of the earliest distribution of the Brown Argus, but there is little doubt that it was once widespread in the south. Nevertheless, its selective behaviour when egg-laying meant that the species had a much more restricted distribution than its foodplant. The butterfly suffered enormous losses during the 20th century as a result of the elimination of Common Rock-rose from most calcareous grasslands after ploughing or the application of fertilisers. Once lost, rock-rose seldom returns. Between 1811 and 1970, agricultural improvement or conversion to arable fields destroyed 80-90% of semi-natural downland in its stronghold, the county of Dorset. The Brown Argus may, indeed, have suffered greater losses than rarities such as the Adonis Blue, because the gentler slopes and flatter chalklands fared worst. It is exactly this ground, with its more nutritious soils, that supports the lush growths of Common Rock-rose needed by this species. About one third of Dorset's Brown Argus colonies disappeared after the Second World War. Whilst many were the victims of intensive agriculture, others were casualties of reduced grazing from the mid-1950s to the early 1980s, which led to many low-growing rock-roses being swamped by taller vegetation.

A spectacular turn of events occurred in the 1990s, when the Brown Argus started to spread into new grasslands that contained Dove's-foot Crane's-bill and Common Stork's-bill. This occurred throughout its range, but especially in the Midlands and East Anglia, where it had previously been confined to the rock-roses of the Cotswolds, the Chilterns, Breckland and coastal dunes. At the same time, colonies spread about 150km northwards. Today, this charming butterfly can be encountered in many rough but sheltered fields, woodland rides and grasslands, as well as on some road verges, south-east of a line extending from east Devon via the Cotswolds up to the North Yorkshire Moors. A similar, less spectacular expansion occurred along the coastlines of Devon, Cornwall and Wales, but in none of these regions has it penetrated more than a few kilometres inland.

A combination of factors appears to have been responsible for this welcome recovery. In the first place, many potential new breeding sites were generated from 1992 to 2006, when the European Union's Agricultural Policy required arable farmers to set-aside 15% of their land in order to qualify for subsidies. Most set-aside fields were ploughed up at the end of each year, but a proportion was left fallow for up to five years, allowing Dove's-foot Crane's-bill and Common Stork's-bill to develop vast populations. This occurred especially across the arable farms of east and central England.

A more important factor, however, was the warmer climate of the past two decades. E J Bodsworth and Chris Thomas studied the physiology of British Brown Argus caterpillars, and found that whilst they grow better on Dove's-foot Crane's-bill than on Common Rock-rose under laboratory conditions, they thrive only at quite warm temperatures. Since rock-roses nearly always grow in distinctly warmer situations than geraniums in our countryside, and are most abundant on south-facing downland, the large majority of our Brown Argus populations were historically restricted to these warm slopes under the cooler climates that prevailed in the past. Only in a few of the warmest spots where geraniums grew were conditions suitable for breeding. But as the British climate has become warmer, numerous sites with geraniums, that until recently were too cool for Brown Argus, became available for breeding, both within and to the north of its traditional range. Another factor that may have hastened the spread was a build-up in numbers on many traditional rock-rose sites in the 1990s.

Today, the Brown Argus is a locally common butterfly in the south-eastern half of England, and is continuing to spread north at a steady rate. The cessation of set-aside subsidies is likely to result in the loss of many of the new colonies within its current range, but warmer climates and its spread to other patches of grassland that contain wild geraniums make it likely that this charming small lycaenid will continue to flourish for the foreseeable future.

Northern Brown Argus

Aricia artaxerxes

This dusky little insect was probably one of the first cold-hardy butterflies to recolonise Britain when the last Great Ice Age receded, roughly 12,000 years ago. It was discovered around 1795 on Arthur's Seat, overlooking Edinburgh, and in the Pentland Hills to the south, and was named the 'Brown Whitespot' in recognition of the gleaming white mark in the centre of each forewing. In contrast, the spots on the underwings were very faint compared with related lycaenids, with most having no black pupil. In most respects, however, the butterfly resembled the Brown Argus, and for 170 years there was much argument as to whether they were distinct species or mere subspecies, a debate that was heightened by the discovery that some populations in northern England lacked the upperwing's white spot. These latter butterflies, known as *salmacis*, have more clearly spotted underwings and look much more like the Northern Brown Argus on the Continent.

Separating the two arguses

Modern genetic and rearing studies by Kaare Aargaard, F V R Jarvis, Andrew Pullin and others have established unequivocally that there are two separate species. They also showed that our Scottish populations were not the endemic subspecies that was generally supposed, but are closely related to the Northern Brown Argus in Scandinavia, despite being separated from Continental populations for around 10,000 generations. The same studies shed doubt on the status of several *Aricia* populations south of Scotland. Many in the Peak District, Yorkshire Wolds and northern Wales are now known to be Brown Argus, whereas the famous *salmacis* populations in Durham are probably, but not certainly, Northern Brown Argus; further research is required.

The fact that several northern English populations were misclassified stemmed from the belief that the Brown Argus invariably has two generations a year, whereas the Northern Brown Argus has one. For more than a century this was thought to be the one certain way of distinguishing the two species, but is now known to be untrue. It is still a useful rule of thumb that applies to the vast majority of

Distribution
A locally common, but declining, species on northern limestone and alkaline soils.

▲Male upperside
Scottish specimen, with distinct white forewing marks; can also have orange spots on forewing.

▲Male underside
Spots much fainter than those of the similar Brown Argus.

▲Female upperside
Forewings have white marks and more prominent orange margins that extend to tip.

▲Female underside
Underside spots poorly developed, as in male.

◀Male
Form *salmacis*, from northern England, lacks white forewing marks.

◀Female
Form *salmacis*.

◀Egg [× 22]
Hatches within about a week of being laid.

▼Egg on Common Rock-rose
Egg is conspicuously positioned on the upper surface of a leaf.

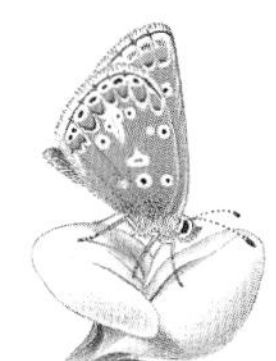

◀Feeding adult
Male of the *salmacis* form feeding on flower of Common Bird's-foot-trefoil, showing more clearly-spotted underwing of this form.

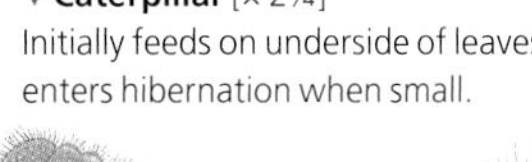

▼Caterpillar [× 2¼]
Initially feeds on underside of leaves; enters hibernation when small.

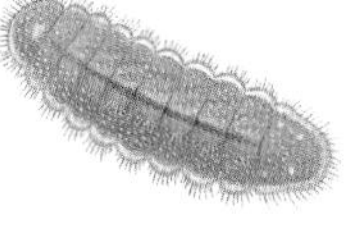

View from above

Single segment

Side view

◀Resting adult
Scottish male resting on stem of Common Rock-rose.

▼Chrysalis [× 2¼]
Tended by ants; less pink than chrysalis of the Brown Argus.

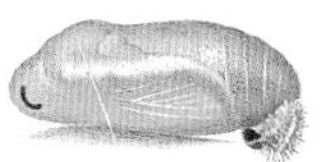

Side view

View from above

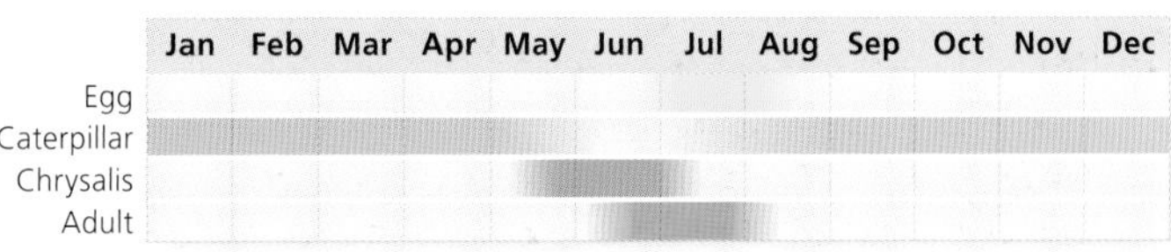

argus populations in the British Isles, but it is now clear that our most northerly populations of Brown Argus have only one annual emergence, except in the warmest of years, for example as occurred in the Peak District in 1999 and 2003.

Sedentary lifestyle

The ecology of this charismatic little butterfly has been well studied by Sam Ellis. The adults generally fly from June until mid-August, peaking from late June to mid-July. Most British colonies are small, containing fewer than 200 adults, of which less than a third will be flying on the peak day. They are also exceedingly sedentary, with individuals generally moving no more than 20-30m during their lifetimes, providing little detectable interchange between neighbouring colonies. The two at Castle Eden Dene, County Durham, appear to be typical: they contained 75-100 adults between them when measured on the peak days of emergence between 1970 and 1972, and no marked individual was found to have flown between the two sites on opposite sides of the River Dene, where the butterflies still breed in exactly the same spots that their predecessors used 170 years ago.

Conspicuous eggs

The egg is laid on Common Rock-rose, but unlike that of the Brown Argus, is placed on the upper surface of a leaf, where it gleams in the sunshine and is extremely easy to find. Like most butterflies, the females choose a small subset of available foodplants on which to lay. The Northern Brown Argus prefers large, bushy rock-roses, with fleshy leaves that are rich in nitrogen and growing in medium-height swards of around 6-10cm tall.

Parasitoids

On hatching, the little caterpillar crawls to the underside of its leaf, where it feeds on the soft tissue, leaving the upper epidermis intact. The little windows of shining epidermis are easy to spot on good sites, making the caterpillar simple to find. Caterpillars hibernate at the base of their foodplant when fairly small, and resume feeding the following spring. They are tended by ants, on the Continent at least, and possess the full array of ant-attracting and appeasing organs (see pages 142-3). The chrysalis also has an association with ants, and chirrups to them in the same way as the Brown Argus. As with most blues, many caterpillars are stung by parasitic wasps, despite being tended by ants. In the case of the Northern Brown Argus, the main parasitoid is a medium-sized wasp called *Hyposoter notatus*, which also accounts for most parasitism found in the Brown Argus. It may, indeed, specialise on *Aricia*, for it is a fairly rare species, yet common where these butterflies occur. At least two-thirds of Northern Brown Argus caterpillars reared from south-west Scotland, and as many in Cumbria, produce this attractive little wasp rather than a butterfly.

Butterfly of the northern hills

The Northern Brown Argus is charming to watch, and its sites are equally beautiful. It is essentially a creature of warm, sheltered, northern hills and mountainsides, and is chiefly found on thin, base-rich or alkaline soils, especially limestones, where large, bushy rock-roses grow in abundance. Typical sites are well-drained, unfertilised grasslands, with patches of bare ground and an uneven sward that is lightly grazed but not cropped very short, intermingled with scattered scrub. A few sites are maintained by erosion or grazing by rabbits, but on most both the butterfly and its foodplant soon disappear in the absence of light winter grazing by sheep or cattle. A mid-successional stage of grassland is optimal for, although Common Rock-rose grows abundantly in short turf, heavy cropping eliminates the butterfly.

In recent years, numerous populations of Northern Brown Argus have been lost as a result of the fertilisation or intensive stocking of northern grasslands, and through the abandonment of many others. In addition, as one of our three genuine northern butterflies, it has suffered much from climate warming, and has retreated on the southern edge of its range. Against this, its ecological requirements are now being regenerated deliberately on many sites, thanks to agri-environment schemes based on Sam Ellis's ecological recommendations.

Overall, this is still a locally common species that occurs in scattered colonies, especially across southern and eastern Scotland, where it remains hugely under-recorded; for example, many 'new' sites were located during recent surveys of the Grampians and south-east Scotland. The same is true of north-west England, where numerous populations breed on the limestone outcrops of north Lancashire and southern Cumbria. Elsewhere, scattered populations survive in south-west Scotland and County Durham.

Common Blue

Polyommatus icarus

As the name implies, this is by far the commonest and most widely distributed lycaenid butterfly found in the British Isles, and the only blue present on most small islands. It was one of the original 18 British butterflies listed in 1634 by Thomas Moffet in *Insectorum Theatrum*, and was familiar to the earliest entomologists, who called it 'The Little Blew Argus'. Happily, it remains common today: any warm patch of waste ground or rough, unfertilised pasture is likely to hold a few individuals, and it is more or less ubiquitous on ancient downland, dunes, sunny flowery banks and heaths.

The brilliant lilac male will be familiar to all naturalists, and is a regular visitor to country gardens, where he sups from flat-headed flowers and establishes small territories, battling with hoverflies and other innocent passers-by. The female is less conspicuous: except when egg-laying or feeding, she remains out of sight, perched among the vegetation until the eggs in her abdomen mature. She also has duller markings, at least in the south.

The colours of wing-scales

As with all butterflies, the eye-catching wing markings and patterns of both sexes derive from minute scales arranged as overlapping tiles across both surfaces of each wing. In the case of the Common Blue, the colours are produced in a combination of ways. Shining physical tints arise from the elaborate internal structure of each scale, consisting of multiple layers of three-dimensional diffraction gratings in the chitin from which they are made. These scatter light to create Tyndall blues, the same effect that produces blue in human eyes and also in the sky. Pigments in each scale make an even greater contribution to wing colour. Many of these are synthesised by the chrysalis during scale development, which occurs a few days before adult emergence during a crucial period when different genes are switched on and off to regulate the intricate patterns and colours of the upper- and underwings.

A second source of pigments in the Common Blue's scales comes from flavonoids, which are obtained from the vetches eaten by its caterpillars. Flavonoids are secondary metabolites that are commonly found in plants: although not essential for vegetative growth or reproduction, they

Distribution
A common species of unfertilised grasslands. Absent only from intensive farmland, mountain-tops and northern Shetland.

confer various health-giving and ecological benefits. Butterflies are unable to synthesise flavonoids, but from an early age the caterpillar lays them down in its fat, a process known as sequestering. The stored flavonoids are passed on to the chrysalis's body-fat, and later released and transferred as pigments to the developing wing pads, where they mix with the colours that are being synthesised.

Although the colours of many flowers derive from a group of flavonoids called anthocyanins, the Common Blue sequesters a different type, the flavonols, which are colourless or pale yellow to our eyes. These are great absorbers of ultra-violet light, which is beyond our range of vision but well within that of a butterfly. Because of the flavonols, the Common Blue perceives very different wings markings on fellow butterflies to those seen by human eyes. Males are especially attracted to females that have high concentrations in their wings, and make persistent mating flights around them, perhaps because these same adults have imperceptibly heavier bodies and longer wings. The concentration of flavonols, in turn, varies with the species and the part of the foodplant that is eaten by caterpillars; it comes as no surprise that legume flowers are the greatest source of flavonols.

Regional variation

Quite apart from their sequestered chemicals, female Common Blues differ greatly in the amount of blue and brown colour on their upper wings, both within colonies and between regions. Some are uniformly brown, although most have a dusting of blue scales towards the base. 'Brown' females predominate in many southern colonies, but are often mixed with others that have variable amounts of blue. The preponderance of blue markings increases the further one travels west and north, culminating in a magnificent form called *mariscolore*, which is regarded as a distinct subspecies by some entomologists. It is found in Ireland and north-west Scotland, including several Scottish isles, and is one of the most beautiful of all British butterflies. Not only are the wings a clear deep blue, but the orange spots are also enlarged almost into a band, and it is distinctly larger than southern specimens. So striking is its appearance that English visitors sometimes mistake it for more exotic species such as the Large Blue or Adonis Blue; the literature on butterflies is bedevilled with such mistakes.

The Common Blue also produces atypical forms or aberrations, although less frequently than Adonis or Chalkhill Blues. For example, the spots on the undersides may vary in size and shape, and the colour of the male's upperwings is occasionally paler, as in the *pallida* form of the butterfly illustrated opposite.

For all this variation, there are a few constant features that distinguish both male and female Common Blues from related species. The outer fringes of the wings are clear white and not crossed with the dark lines or chequering found in the Adonis and Chalkhill Blues. On the underside of the forewing, there is a spot about a quarter of the way out from the body that is absent from the Silver-studded Blue and our two Brown Arguses. Finally, there are orange marks around the edge of the hindwings, which is not the case with Holly, Small and Large Blues.

Variable generations

The number of broods and emergence dates of this butterfly vary in different parts of the British Isles. In the south, there are always two or more emergences, the first from mid-May to mid-June, the second lasting from late July into September. F W Frohawk, who reared many Common Blues in Kent, maintained that the second brood was no more than a partial one, and that many caterpillars enter hibernation in June rather than develop to form adults. On the other hand, the species sometimes manages a third generation in October after an especially warm summer.

Further north, there is time for just one generation a year, which emerges on different dates according to the local climate: the warmer the climate, the earlier the emergence. George Thomson gives an authoritative account in *The Butterflies of Scotland*. Briefly, Common Blues may be flying in late May in south-west Scotland, although a prolonged emergence from June to late August is more usual. But in the north-east, for example on Speyside, they are seldom seen before the second week of July and are really August butterflies. Further south, in Yorkshire, the butterfly may switch between one and two broods a year according to the warmth of the season.

Territories and roosts

The Common Blue lives in reasonably discrete colonies, though adults wander more than most close relatives. Thus the occasional stray has been picked up on lightships, and new patches of habitat tend to be colonised quite quickly. But it is not a true nomad like the Holly Blue, and most individuals fly, pair and lay within the breeding grounds from which they emerged.

Although this is easily the commonest British blue, little was known of its behaviour before studies by Deryk Frazer and Roger Dennis in England, and by Konrad Fiedler and colleagues in Germany. Male Common Blues are distinctly territorial, and either perch or patrol in search of females, frequently skirmishing with rivals and other butterflies such as the equally combative Small Copper. Typical flights are short and rapid in both sexes, as Common Blues flit from one flowerhead to the next. When settled, the wings are

▲**Male upperside**
Brilliant lilac-blue; outer fringes of wings clear white.

▲**Male underside**
Both sexes have a spot on the forewing near the body.

▲**'Brown' female upperside**
Predominantly brown, with varying blue near wing-bases.

▲**Female underside**
Both sexes have orange marks on edges of the hindwings.

◀**Male**
Rare aberrant form *pallida*, with pale upperwings.

▲**Female**
Form *mariscolore*, found in Ireland and north-west Scotland.

▲**'Blue' female upperside**
An extreme example, with brown restricted to the wing edges.

▲**Egg** [× 22]
Laid singly, often on Common Bird's-foot-trefoil.

▶**Roosting adults**
In the evening, adults congregate in small groups on grass heads, where they roost head-downwards.

▼**Feeding adult**
Common Blues feed at flat-headed flowers, such as hawk's-beards.

▲**Caterpillar** [× 2¼]
Feeds during the day on tender leaves.

▲**Chrysalis** [× 2¼]
Formed on the ground; probably tended by ants.

In Scotland, single brooded (flying in June, July or August).

opened fully only in weak light, so as to absorb maximum warmth from the sun. This mainly occurs first thing in the morning and in late afternoon, which are the best times to examine and photograph the Common Blue. Adults are easily approached then, and there is the added bonus that many may congregate on clumps of tall grass, settling on the tops to catch the last rays of the afternoon sun, or waking in the morning from their communal roosts. Roosts tend to occur on banks or in sheltered hollows, where both sexes rest head-down, two or three to a grass head, often accompanied by other blues.

Egg-laying

The lovely females come into their own when laying eggs. They make short, fluttery flights just above ground level, frequently alighting to crawl over low-growing herbs, twitching and drumming their feet to test the quality of different growths. Each dips her antennae when she finds a potential egg-laying spot, and tests the leaves by rubbing them with the tip of a half-curved abdomen. Eventually a suitable plant is found, and the abdomen is bent double before it recoils like a buffer, leaving a tiny pale green egg on the leaflet.

I have watched females lay many times, and have found thousands of their eggs in the wild. They use a wide range of leguminous plants, but the commonest by far is Common Bird's-foot-trefoil. Other favourites, that can support an entire colony of this butterfly, are Greater Bird's-foot-trefoil, Lesser Trefoil, Black Medick and various restharrows.

Konrad Fiedler has shown that the types of plant chosen by the females have important consequences for her offspring, not just because of the different flavonoids that can be sequestered from different species or parts, but also because of their effect on growth and survival, and on the caterpillar's ability to produce secretions that attract ants. More important than the species of plant eaten is the luxuriance of its growth, and whether the plant will develop flowers. Eggs are invariably laid on the soft growing tissue of the youngest pods or leaflets, whilst tough old growth is ignored. Females select tissue that has a high content of water and of nitrogen, both of which are greatest in flowers. This can lead to different plants being used in the two generations. In Cheshire, Roger Dennis found that the first brood was laid principally on Lesser Trefoil, while the second brood switched to Common Bird's-foot-trefoil in August, when the Lesser Trefoil was more withered.

Searching for eggs

Common Blue eggs turn white as they dry, and are easy to find. Search the midribs near the base of tender leaflets, especially on the upper surfaces of small, sheltered plants that are sprouting back into leaf after a disturbance. The Common Blue is simple to rear. It emerges from the egg after a week or two, and starts feeding on the undersurface of its leaflet, excavating mouthfuls of lush mesophyll (the interior of the leaf). As with most blues, the upper epidermis is left intact, creating silvery blotches on the leaves that are easy to find. The older caterpillar eats whole leaves, flowers and pods, and, unless it enters hibernation, completes its growth in six weeks.

Caterpillars and ants

The caterpillar is green and furry in its final stage, and beautifully camouflaged on the foodplant. It feeds by day, easing its slug-like body over leaves and chirruping a tuneless song, too soft for human ears, that is probably caused by the rhythmic compression of air in its abdomen. The sounds synchronise with the protrusion of its shiny black head, which pops in and out as it walks, and almost certainly attract ants. Nevertheless, although this caterpillar possesses the organs described on pages 142-3, its powers of attraction are weak. Only 19 of the 25 undisturbed caterpillars that I once watched for long periods in the wild were being tended, and then usually by just a single worker ant rather than the clusters that smother Adonis and Chalkhill Blue caterpillars when they emerge to feed. At Oxford, James Rawles found that Common Blue caterpillars attracted just half as many black *Lasius* ants as those of the Adonis Blue. This is perhaps because the Common Blue produces few amino acids in its secretions, relying mainly on a sugary cocktail of sucrose, melezitose, glucose and fructose to reward ants.

I have only once seen ants showing real enthusiasm for a Common Blue caterpillar. This was in Devon, where both the red ant *Myrmica sabuleti* and the wood ant *Formica rufa* were common. I watched the caterpillar several times during the day, and it was always being milked by one or other of these species. Twice they fought for it: the first time, a red ant was in possession and easily saw off a feeble challenge from the more excitable wood ant. Later, the opposite occurred: by now a wood ant was in control, and there were frequent skirmishes as a red ant, in its typical sneak-thief way, scurried back and forth under the foodplant, slinking up from beneath the caterpillar to take quick licks of its body. The wood ant was perched squarely on the back of the caterpillar (which continued to browse Common Bird's-foot-trefoil, oblivious to the battle), and at each attack reared back and up on its hindlegs, spraying formic acid at the red ant. It is tempting to think that the ant in possession held some territorial advantage over the intruder, but more observations are needed to confirm this.

The chrysalis also attracts ants, both through secretions

that ooze from its microscopic pore cupola, and its ability to crackle into song. Like hairstreaks and other blues, it generates long bursts of noise, less staccato than those of the caterpillar but equally tuneless, like the drumming of a male snipe in flight. It seems likely that chrysalises are tended in the wild, and I once found one beneath a slate inside a red ant nest. This stage lasts roughly a fortnight before the adult butterfly emerges.

Numbers and drought

Colonies of Common Blue exist in a wide variety of places where its foodplants are common. Each typically consists of a few tens or hundreds of adults, in contrast to the thousands of Adonis, Chalkhill and Silver-studded Blues that are found on their best sites. Common Blue numbers also fluctuate greatly from one year to the next. By and large, after a warm, moist spring and summer in the south, there is high breeding success in the second brood, and this in turn leads to large numbers the following year. On the other hand, populations plummet during a hot, dry summer. A spectacular crash occurred after the great drought of 1976. I spent most of that summer in Devon, and remember one south-facing slope where almost all the foodplants had shrivelled by the time second-brood females were flying. These unfortunate mothers had to search for the few green wisps of trefoil that grew in deeper soils and were shaded by gorse. Even these were parched and miserable, but became peppered with eggs. I suspect that most later died, for Common Blues were virtually absent from the area the following year, and took two to three seasons to recover to normal numbers.

Our most widespread blue

In most years, the Common Blue is a widespread and common butterfly in the British Isles, so much so that it is simpler to list the areas where it does not occur. It is absent, for example, from land higher than about 550m in Scotland and Wales, and although colonies are known from the warm southern sand dunes of Shetland, it appears to be missing from the rest of those islands. Elsewhere, it may be found wherever its foodplants are abundant, even on islets less than 0.5ha in area. The largest colonies inhabit warm chalk and limestone downs, with those breeding on cliffs, undercliffs and dunes a close second. Numbers tend to be lower elsewhere, but the butterfly is nonetheless common on banks, cuttings and roadsides, on lowland heaths with moderately rich soil, and on patches of wasteland. Colonies are also found on much heavier soils, in unfertilised pasture, marshy areas and along ditches, where whole populations appear to be supported by Greater Bird's-foot-trefoil.

Although primarily a butterfly of open grassland, the Common Blue is frequently encountered in the glades and sunny rides of woods, or breeding on leguminous plants along the edges. This, indeed, is almost the only habitat in which the butterfly survives in parts of Cambridgeshire and East Anglia, where intensive agriculture has eliminated so much grassland that its foodplants have become rare in the open countryside. Despite being one of the commonest butterflies in Britain, there is little doubt that countless populations have been lost from lowlands as a result of the efficiency of modern agriculture and the general tidying-up of the countryside. There was, moreover, a time when trefoils were deliberately sown in pasture to sweeten the hay and enrich the soil with nitrogen – a far more sensible way of fertilising leys and hay crops than adding synthetic nitrates.

Chalkhill Blue

Polyommatus coridon

As its name implies, this dazzling butterfly inhabits unfertilised chalk and limestone downs, where the caterpillar's foodplant, Horseshoe Vetch, blooms in abundance. No-one can fail to identify the male's milky blue wings, although when flying in strong sunshine their hue is so pale that they can look surprisingly like a Marbled White. The chocolate-brown females are a different matter and give rise to much confusion. Note that there is a distinct chequering to the white fringes along the wing edges which distinguishes this species from all except the Adonis Blue. Both females fly together on several sites, and are hard to separate: a good way is to examine the spots around the margins of the upperside, which are edged with white in this species but with blue on the female Adonis Blue.

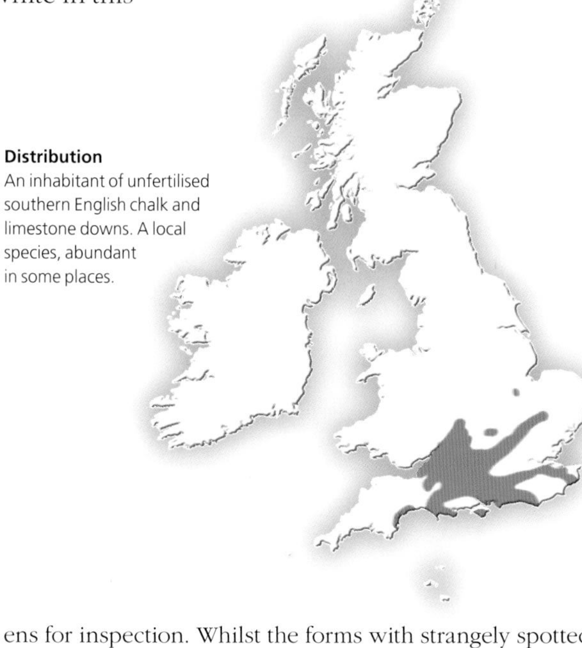

Distribution
An inhabitant of unfertilised southern English chalk and limestone downs. A local species, abundant in some places.

Prized aberrations

Although most Chalkhill Blue colonies consist only of typical adults, several contain aberrations, of which we illustrate a few. These were much prized in the heyday of British butterfly collecting, so much so that in 1938 an entire book, *Monograph of the British Aberrations of the Chalkhill Blue*, was devoted to the countless varieties assembled by Percy Bright and H A Leeds during a lifetime's work. In *The Aurelian Legacy*, Michael Salmon quotes their contemporary, Sidney Castle Russell, after a successful foray on the South Downs with the Revd J A Marcon. It epitomises a more innocent era when butterflies were plentiful, when the impact of gentlemen collectors was negligible, and when the thrill of the chase was paramount:

> 'We made quite a good bag at Shoreham of *coridon*, a green male, *obsoletas* and *caecas* ... The amusing thing about this bug is it found its way into a tea-cup and laid down and died ... Marcon had remarkable luck here too – a dozen fine vars. and *gynandros* [gynandromorphs] and six green males in *coridon* ... he is as Wells says, "A ruddy Marvel!" I should think his Shoreham vars. are well worth £50.'

The most famous population, however, was at Royston golf course, near Cambridge. There, *semi-syngrapha* and other forms were common for several years before the last war, and the scramble to net them led to accusation and counter-accusation between rival collectors. One, to general disgust, simply sat in the car while his chauffeur caught dozens for inspection. Whilst the forms with strangely spotted undersides are intriguing, by far the commonest varieties of Chalkhill Blue are the blue *tithonus* and *semi-syngrapha* females. Curiously enough, it is unusual to find both within the same population.

Chalkhill Blues live in discrete isolated colonies, from which males occasionally wander and can be found far from the nearest breeding site. There is one generation a year, with the first adults usually seen in mid-July, reaching a peak a month later and often lasting well into September. Numbers vary considerably between sites and in different

▲ **Male**
Milky blue upperwings are unlike those of any other blue, and look very pale in full sunlight.

▲ **Female**
Very similar to female Adonis Blue, although slightly larger; spots around the margins of the upperwings are edged with white.

▲ **Colour variant**
Female of the *tithonus* form, with blue wings.

▲ **Colour variant**
Form *caeca*, from Watson Collection, British Museum.

▲ **Colour variant**
Form *antiextrema*, from Watson Collection, British Museum.

▲ **Colour variant**
Form *flavescens*, from Watson Collection, British Museum.

◀ **Egg** [× 22]
Laid on or near the foodplant; hibernates and hatches in spring.

▶ **Feeding adult**
Chalkhill Blues visit Kidney Vetch and many other flowers.

◀ **Resting adult**
Female resting in dull weather on flower-bud of Common Knapweed.

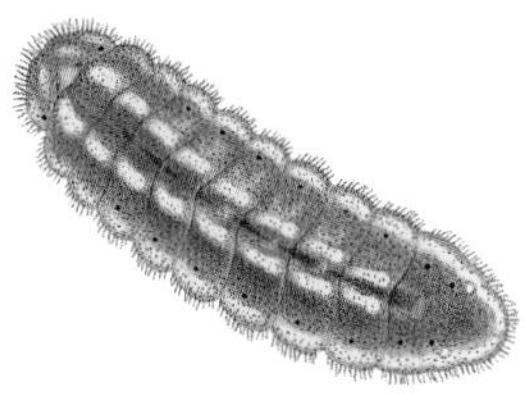

▲ **Caterpillar** [× 2¼]
Similar to the caterpillar of the Adonis Blue, but rather lighter.

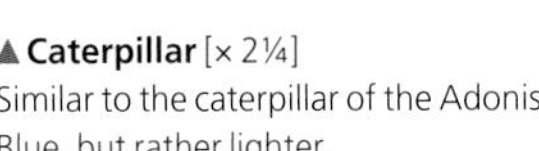
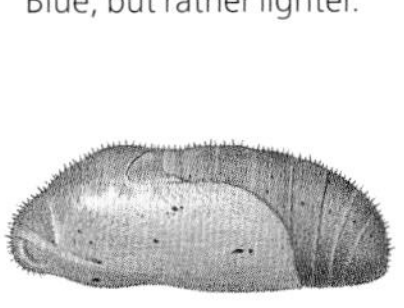

▲ **Chrysalis** [× 2¼]
Formed underground inside an ant nest, or in a cell moulded by ants beneath the foodplant.

	Jan	Feb	Mar	Apr	May	Jun	Jul	Aug	Sep	Oct	Nov	Dec
Egg												
Caterpillar												
Chrysalis												
Adult												

years. Two colonies of no more than medium size contained around 10,000 and 18,000 adults when measured, both breeding in less than 2ha of land. Many populations are much smaller, and here only a few individuals will be seen. The largest examples, on the other hand, probably contain hundreds of thousands of adults in the best years, and the turf shimmers with their milky-blue wings. But such outbreaks are unusual: I have seen them twice in Britain, at Fontmell Down, Dorset, in the 1970s, and on the slopes of Old Winchester Hill, in Hampshire.

Adult behaviour

Adults behave similarly to the Adonis Blue, gathering to roost by night on tussocks of tall grass near the base of a hillside. Often two or three sit together on a grass stem, with perhaps a score in the same small clump. They make a lovely sight when the sun wakes them in the morning, for all bask together for a few minutes, with wings wide open to absorb the maximum warmth.

The main daytime occupation of males is to flutter inches above the sward in an incessant search for mates. Females fly less often, generally only to feed or lay eggs. Chalkhill Blues visit a wide variety of flowers, but on most sites scabiouses and knapweeds are common and frequently used. Males also gather to drink water; on the Continent I have often seen hundreds grouped on pebbles near alpine streams.

When ready to lay eggs, the female Chalkhill Blue flutters and crawls over the turf, searching for Horseshoe Vetch. Each egg is laid singly, usually on stalks and tough woody parts, as well as on neighbouring vegetation. She is much less fussy than the female Adonis Blue, which chooses only the shortest available vetch plants. Chalkhill Blues use these too – I have often found eggs of both species on the same plant – but they select more overgrown vetches as well. Yet they are not entirely undiscerning. Lush, nitrogen-rich plants appear to be favoured, as are large vigorous mats of the vetch, sprawling the size of a dinner plate over crumbling chalk scree. The butterfly is often common on sites where grazing has been temporarily relaxed, releasing the close-cropped stumps to produce a luxuriance of delicate leaflets.

Eggs and early development

As befits an egg that will endure the winter, the shell is robust and heavily sculptured; even without a hand-lens it is easily distinguished from the delicate tracery of the Adonis Blue egg. Chalkhill Blue eggs are quite simple to find on the best sites; I often encounter them while working on Adonis Blues, finding several hundred in a good season. The search is best made in late summer, for an increasing number fall off during the autumn to complete their hibernation on the ground, where they are almost indistinguishable from the tiny chippings of lime or chalk.

The Chalkhill Blue caterpillar is fully formed by late summer, but remains within its eggshell until late April. At first, it behaves very like the similar-looking Adonis Blue. Apart from the difference in timing and size – Adonis Blue caterpillars are almost fully grown when Chalkhill Blue eggs hatch, and there is little overlap in the second generation – Chalkhill Blue caterpillars can be distinguished by their slightly paler bodies and their decidedly whiter bristles. They are also nocturnal during the final stages of growth, emerging at dusk to browse on vetch leaves.

Ants and proteins

Despite their nocturnal habits, the caterpillars are as attractive to ants as the Adonis Blue's, and possess similar secretory organs. I have never seen a Chalkhill Blue caterpillar browsing at dusk that was not being attended by ants. On some sites, these include the yellow ant *Lasius flavus*, a subterranean species that will nevertheless follow caterpillars above ground after dark. In Germany, Konrad Fiedler found that the large dorsal honey-gland across the top of the ninth segment exudes about 40 droplets of honeydew an hour during the caterpillar's final stages of growth. Each drop is a 15% solution of sugars, primarily sucrose and often some glucose. More importantly, it contains no fewer than 14 different amino acids (the chemical building blocks of proteins), especially leucine. This differs from the honeydew secreted by the Common Blue, which contains more sugar but a very much weaker solution of just six amino acids, and perhaps explains why Chalkhill Blue caterpillars are more attractive to ants than those of the Common Blue.

In a fascinating experiment involving the ant *Tetramorium caespitum*, Fiedler estimated that Chalkhill Blue caterpillars alone provided between a quarter and a half of all the energy needs of this ant's colonies on downlands that support large populations of the butterfly. He demonstrated, for the first time, that the secretions of a lycaenid butterfly provide a tangible reward to ants, thus establishing that the relationship between the two partners was one of genuine mutualism. Similar benefits are likely to accrue to the red *Myrmica* and black *Lasius* ants that typically milk caterpillars in British colonies of Chalkhill Blue. The feast lasts for only a few weeks, but is preceded and followed by comparable densities of Adonis Blue caterpillars on our finest downs, resulting in a near continuous supply of liquid ant food from April to August.

The chrysalis of the Chalkhill Blue is hard to find, but well worth seeking in June, when it lives concealed within an earthen cell moulded by ants beneath vetches, or in the outer chamber of an ant nest. It is also adaptated to attract ants. It possesses secretory pores which exude a film of amino acids over the cuticle, and a formidable sound organ, formed like that of the Purple Hairstreak from two toughened segments of the abdomen that rub together as a tooth-and-comb (see page 92). The vibrations and chirruping sounds it produces clearly stimulate and agitate the ants, particularly when an emerging adult struggles to free itself from the pupal case; a veritable posse of protective ants quickly assembles to surround it.

Decline and recovery

Sadly, this interesting and beautiful butterfly became much scarcer during the 20th century, when many gentler slopes were ploughed and fertilised, and most steep downland was abandoned as grazing land. The loss of rabbits as effective grazers from the mid 1950s to early 1980s was a further blow. Nevertheless, the Chalkhill Blue experienced nothing like the severity of decline seen in its warmth-loving relative, the Adonis Blue, and generally survived on traditional sites so long as Horseshoe Vetch persisted. The story of the past 25 years has been a happier one. By the 1990s, about four-fifths of Chalkhill Blue sites had been designated as statutory protected areas, and many entered agri-environment schemes whereby the landowner was subsidised to farm sites to enhance wildlife, including specifically this butterfly. As a consequence, none of the 161 Chalkhill Blue colonies that were regularly monitored since 1981 has been lost; on the contrary, adult numbers on them increased by roughly three-fold on average, in response to burgeoning Horseshoe Vetch populations. For reasons as yet unexplained, Chalkhill Blue numbers have recovered more strongly on sites in east and south-east England than in the south-west. They also increased disproportionately on sites being managed under agri-environment schemes, although a Butterfly Conservation survey found that this was more closely correlated with the return of rabbits than with the controlled regimes of cattle or sheep grazing. A series of warm, moist summers has also helped to raise numbers.

Twyford Down

Despite these welcome increases, there has been virtually no spread by the Chalkhill Blue to new sites. One exception is at Twyford Down, Winchester, where the decision to construct a motorway through what was perceived to be its last core breeding area in central Hampshire turned the Chalkhill Blue into a *cause célèbre* in the 1980s. Whatever one's views of the scenic impact of the M3 cutting east of Winchester, its route was very properly altered to avoid the prehistoric artifacts of the Dongas, where this butterfly bred. In fact, Twyford Down had been ploughed and lost its colony of Chalkhill Blues in the 1930s; the national protest to save the butterfly was in reality 50 years too late.

In the 1980s, Twyford Down consisted of two large arable fields, 9ha of which were off-route and available as compensation land, together with 2km of the old Winchester bypass, that was to be buried beneath several million tonnes of chalk rubble from the new cutting. With Lena Ward and Rowley Snazell, I had the pleasure of designing new restoration sites, with the intent of creating optimum habitat for this and other downland butterflies, including the Small Blue. No reconstruction of a chalk grassland from arable had previously been attempted on this scale, but by carefully varying the micro-topography, soil depths and the nutrient status, and by sowing millions of locally sourced downland flower seeds, we obtained within a few years a herb-rich sward that was of similar quality to nearby Sites of Special Scientific Interest. After an absence of 60 years, the Chalkhill Blue soon colonised the restored grassland, and increased to high numbers; it also spread along the former route of the Winchester bypass to form a third new colony nearby. Today, five populations of Chalkhill Blue flourish east of Winchester where once there were two, each linked by the exchange of a few adults that fly annually across the motorway.

Southern strongholds

In the distant past, Chalkhill Blue colonies bred as far north as Lincolnshire, but there has been no northwards spread under recent warmer climates. Today, its north-western limit is marked by some fine colonies in the Cotswolds, while in the east it survives on fragments of unspoiled chalk grassland near Cambridge. It becomes more common, although still distinctly local, on herb-rich downs in the Chilterns, and is present on most sites south of this where Horseshoe Vetch survives in reasonable abundance. Thus there is a string of large colonies both along the South and North Downs of Sussex, Surrey and Kent, and many isolated populations breed on the chalk hills of Hampshire. It is frequent, too, on many unfertilised parts of Salisbury Plain, especially in the dry valleys around the edges. Other strongholds are the Isle of Wight and the escarpments of Dorset, where it is still to be expected in any unfertilised valley that supports Horseshoe Vetch. These include the limestone coast and the Isle of Portland, where some magnificent populations survive.

Adonis Blue

Polyommatus bellargus

The Adonis, along with the Chalkhill Blue, is the quintessential butterfly of southern English chalklands. It is still very localised as colonies recover from a calamitous 30-year decline, yet today it can be seen by the hundred on certain warm downs from Dorset to Kent, where the brilliant blue butterflies hover over a turf yellow with Horseshoe Vetch. This spring flight is followed by a second emergence in late summer, providing a magnificent finale to the season when the browns, skippers and Chalkhill Blues are either tattered or have disappeared.

Brilliant wings

The sheer brilliance of the male's wings has long attracted attention, and collections exist containing drawer upon drawer of this one species. It was all the more collectable because the wing markings vary considerably in some colonies. The male's blue, for example, is almost violet in some cases and turquoise in others. This is most apparent in the wild, where the strength of sunlight and the extent to which scales have been lost also affect appearances: there are always a few individuals which, in flight, are almost indistinguishable from the Common Blue. At rest, the latter is more violet, while the Adonis Blue has fine black lines that cross the outer white fringes and just enter the body of the wing.

These chequered fringes are also a good way to distinguish the female, although the Chalkhill Blue is similar. She is a lovely chocolate-brown on most sites, but much given to variation on some. The commonest aberration, which I see by the hundred in the Isle of Purbeck, is called *semiceronus*. Here, the ground-colour is wholly or partly blue, resembling a male, but with orange spots around the border. Other varieties have underwing spots that are distorted or reduced.

Distribution
A recovering species, but still a rare inhabitant of unfertilised, well-grazed, south-facing downs in southern England, principally in Dorset, the Isle of Wight and Wiltshire.

Sedentary populations

Adonis Blues live in close-knit colonies, with little or no movement between those on adjoining downs. With David Simcox, I have marked thousands of adults over the years, and found that they fly freely over open, herb-rich downland, but almost always turn back when they encounter a barrier or a ploughed field. I have yet to detect any mixing between four study sites separated from each other by just 50-100m of chest-high scrub. Of course, the odd individual must sometimes migrate, but this is a rare event and has hampered the recovery of this species after its habitat was restored to English downs from the 1980s onwards, following the introduction of conservation grazing and the return of rabbits after myxomatosis. For a decade, new colonies were restricted to sites a mere stone's throw away from existing populations. The butterfly gradually spread, however, by stepping-stone colonisation from one down to the next, eventually reaching more distant sites, so much so that most former localities with suitable habitat had been reoccupied by the time of the last national survey of Adonis Blues,

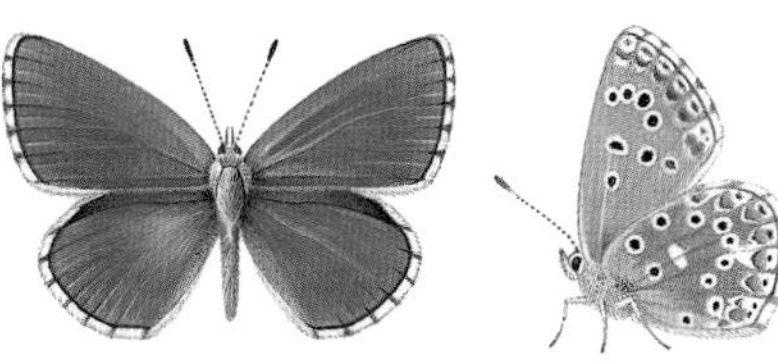

▲ **Male**
The brilliant colour of the male is brighter than that of any other blue.

▲ **Female**
Similar to the female Chalkhill Blue, but pale hindwing scales are blue, not white.

▶ **Male colour variant**
A rare and extreme form caught at Ranmore, Surrey, in 1972.

◀ **Female colour variant**
Aberrant form *semiceronus*, common on some sites.

◀ **Feeding adult**
Second-brood adults often feed at the flowers of Wild Marjoram.

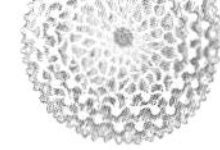

▲ **Egg** [× 22]
Laid singly on Horseshoe Vetch; many may be found on the same plant.

◀ **Adults on Horseshoe Vetch**
Horseshoe Vetch provides nectar for adults as well as being the caterpillar foodplant.

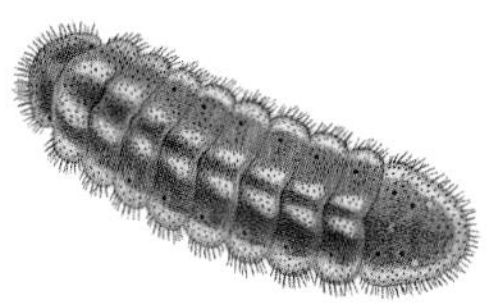

▲ **Caterpillar** [× 2¼]
Made conspicuous by the constant presence of ants.

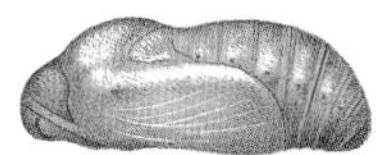

▲ **Chrysalis** [× 2¼]
Formed on or under the ground, in association with ant nests.

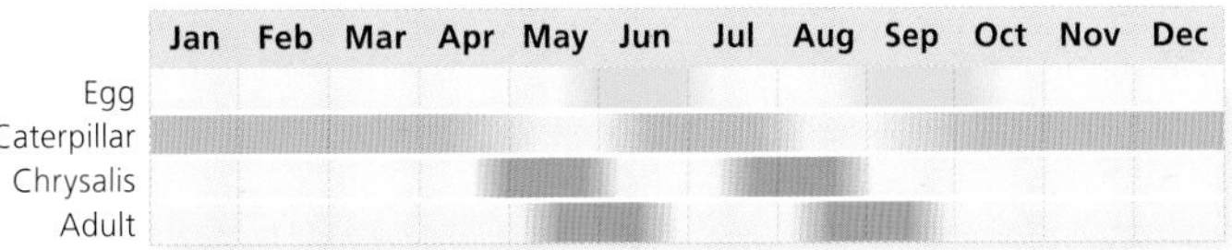

Ground temperatures under Horseshoe Vetch

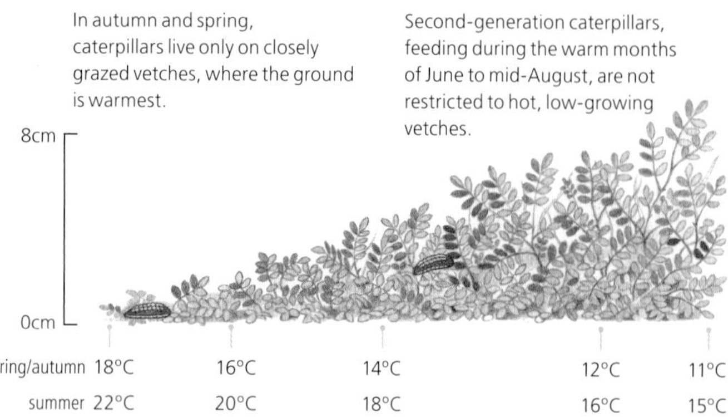

organised by Nigel Bourn around the Millennium. There were still some isolated sites that had not been reached, but the main reason why certain downs supported the butterfly while others did not lay in the quality of their grassland sward for breeding.

Seasonal shifts in egg-laying

On every site, the male Adonis Blue is the more conspicuous sex because, in addition to his vivid markings, he spends long periods hovering just above the turf, slowly flying up and down the hillside in a perpetual search for virgin mates. Females are pounced upon almost immediately after they emerge, and are often swamped by males before their wings are dry. This accounts for the many individuals seen with slightly crumpled wings.

After pairing, the female spends long periods resting on the ground. Typical Adonis Blue sites contain a short, broken turf, with numerous little pits and depressions that catch the sun, and it is in these that she lingers, waiting for her egg-load to ripen. When it has, she flutters slowly over the turf, dropping frequently to crawl over the caterpillar's foodplant, Horseshoe Vetch. The female is highly selective in her choice of plants, especially during the second flight in late summer, when she chooses spots that will remain sufficiently warm for her caterpillars to develop during their autumn and early-spring phases of feeding. At this time of year, she usually rejects large clumps of vetch in favour of small, short sprigs growing in turf that is just one or two centimetres tall (see diagram), especially when situated in a warm depression or hollow, such as an old hoof-print on a steep, south-facing slope. She has more latitude in the May to June emergence, for during the warmth of summer the caterpillars can develop successfully in taller, relatively cooler turf. Thus, from May to August the breeding area temporarily expands to double or treble the size of the autumn and spring sites.

David Roy has studied this seasonal shift in the breeding patches available to Adonis Blues, and rightly considered that it provides an explanation as to why the springtime emergence of adults is usually much smaller than the summer one. Despite the additional areas available to summer-feeding caterpillars – which have become even greater under the warmer climates of recent years – every colony goes through an annual bottleneck during autumn and spring, when eggs and caterpillars are restricted to a minority of short-cropped plants, growing in sheltered hotspots, that provide sufficient warmth for survival.

In addition to favouring south-facing downs, the females often lay eggs on warm boundary banks. Indeed, a remarkable number of Adonis Blue colonies became restricted to ancient fortifications during the nadir of this butterfly in the late 1970s. Spectacular examples included the battlements of Maiden Castle and Hod Hill, in Dorset, and the fascinating Bockerly Dyke, which snakes its way between the boundaries of Hampshire, Dorset and Wiltshire, supporting high densities of Adonis Blues on a ribbon of turf no wider than a metre or two in some places.

The white pin-head-sized eggs are found mainly on the underside of terminal leaflets, and are easy to spot if you gently lift the branches with a pencil. Although laid singly, there may be 30-40 on the best plants in a good year. They hatch after a week or two, and the tiny caterpillars start nibbling at the undersurfaces of the leaves. They are tiny, translucent and hard to see at first, but their presence is obvious from characteristic feeding damage, visible as numerous little pale circles on the leaflets, each with the opaque upper cuticle left intact. Later, entire leaves, shoots and fruits are devoured. Caterpillars feed throughout the day, and are camouflaged from their first moult onwards. They would be very difficult to find were it not for the large clusters of ants that often surround them. This is a common feature of blue butterflies, although the interaction with this species is unusually strong.

Tended by ants

I have had the pleasure of watching several hundred Adonis Blue caterpillars being tended by ants in the wild. The relationship begins after the first moult, and continues almost incessantly until the butterfly has emerged from the chrysalis. The caterpillar possesses three distinct organs that attract ants. The most obvious is the honey-gland. This consists of a slit across the tenth segment back from the head, which forms the opening to a complicated gland housed deep in the body. Fleshy, slightly raised lips surround this slit, which on magnification are seen to contain bizarre clusters of knobs and plates, looking rather like medieval clubs. These are mechano-receptors, and the ants drum them with their antennae, prompting the honey-gland to release sweet droplets of honeydew.

The caterpillar also possesses microscopic pores over the whole body, and an extraordinary pair of tentacle organs situated either side of the honey-gland, one segment back. These are normally kept hidden within its body, but are periodically unfurled, like the expanding horns of a snail. At the tip of each is a circle of hairs that finally spring open, looking very like a chimney sweep's brush. They are extended by the caterpillar particularly when it crawls from one leaf to another, or when the ants have wandered off. It is thought they release a volatile chemical that agitates the ants: their extrusion certainly causes ants to run around in great excitement, and usually results in them finding the caterpillar from which they had become detached.

Ants attending an Adonis Blue caterpillar

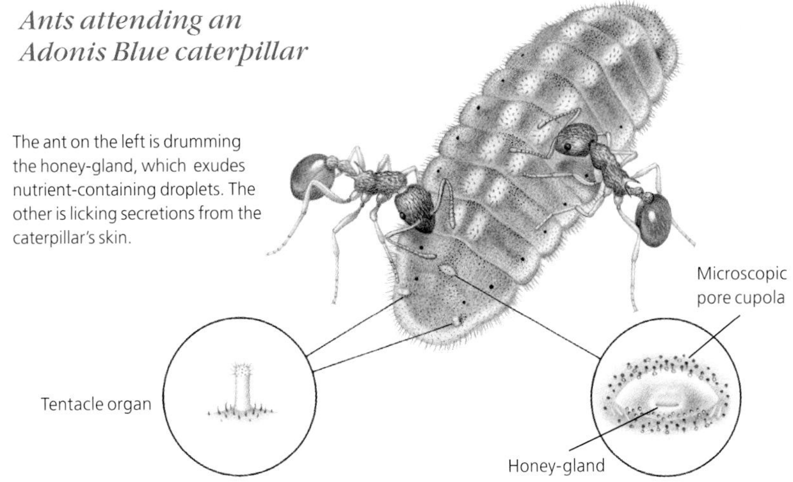

The ant on the left is drumming the honey-gland, which exudes nutrient-containing droplets. The other is licking secretions from the caterpillar's skin.

Another extraordinary feature of the caterpillar was discovered in 1989 by Phil DeVries, the great American butterfly expert. He had already found that the caterpillars of certain riodinids (relatives of the Duke of Burgundy) produce strange songs by rubbing two ribbed tentacles behind the head against striations on the head capsule. These arouse and possibly pacify ants, which themselves make various noises, for example to spread alarm when faced with danger. No such organ has been found on a blue's caterpillar, but DeVries knew that a few American lycaenids sang. It was therefore with some excitement that we tested the Adonis Blue. We found that not only did its caterpillar sing, but that the tune was loud compared with other species, although below our feeble threshold for hearing. It has an eerie loveliness: a sort of rasping, barking noise that is especially apparent when the caterpillar crawls, reinforced by regular high-pitched squeaks that over-score the tune. We later recorded other British blues and found similar sounds. More recently, Karsten Schönrogge found the same phenomenon in Brown and Purple Hairstreaks.

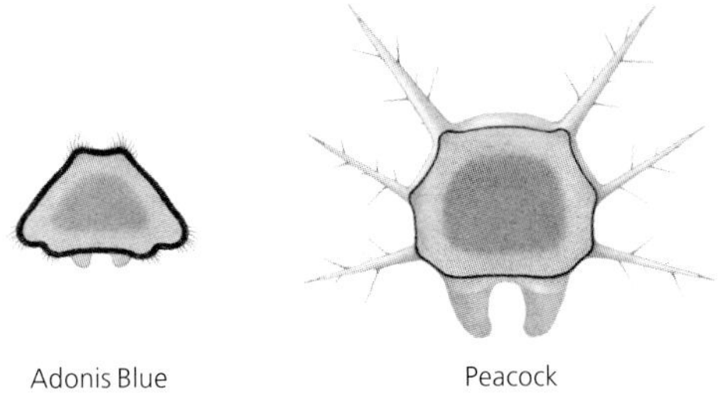

Caterpillar cross-sections
The protection afforded by ants and the thick rubbery skin (indicated in black) of the Adonis Blue caterpillar means it has no need to arm itself with devices such as spines and long hairs, as shown by the Peacock caterpillar.

Obsessive attention

It appears that any species of ant may attend Adonis Blue caterpillars, but on English chalk it is usually the red ant *Myrmica sabuleti* or the black ant *Lasius alienus*, which are the commonest species in the spots where this butterfly breeds. The ants are obsessive in their attention, and the caterpillar presumably obtains protection from pathogens, invertebrate predators and other carnivorous ants. The armed guard extends around the clock, for ants follow the caterpillar down to the soil surface when the day's feeding is complete, frequently burying it for the night in a pile of loose earth. They often work this into a firm, crusty cell, which may contain up to eight Adonis Blue caterpillars together, with perhaps a dozen ants doing sentry duty around them. Cells are also constructed around caterpillars while they moult their skins, a process that can take two or three days.

The chrysalis has an equally intimate relationship with the ants, relying on a combination of sounds and amino acid secretions to attract them, in the same way as the Purple Hairstreak. Adonis Blue chrysalises are not easy to locate. Some enter crevices and cracks in the ground, where ants quickly bury and tend them. I have found an equal number inside ant nests, resting among the brood in the warmest upper chambers. Whether they are attracted to or carried to the nests, or merely enter them because of the soft soil, I have yet to determine. In either case, the chrysalis receives constant attention until it emerges three weeks later. With a crackling burst of song, it breaks open the pupal case amid a sea of frenzied ants, before crawling to the surface to inflate its wings.

It is unlikely that an Adonis Blue colony could survive were it not for the protection of ants. This is true of some other blues, but is perhaps one reason why this species is so fussy in its breeding requirements. Whereas the Chalkhill Blue needs ants only from mid-May to August, Adonis Blue caterpillars must be tended from the earliest days of March until the end of October. For ant colonies to be active so early and so late in the year demands an abnormally hot environment, and the caterpillar itself may need warmth. Whatever the precise reason, Adonis Blue colonies are restricted to the hottest localities in England – south-facing southern hillsides – and then only to sites where the soil is thin and the turf so sparse or closely grazed that the sun can bake the ground.

Collapse and recovery

Adonis Blue populations have endured a roller-coaster ride in abundance over the past century. They were always scarce. This was, indeed, one of the last British butterflies to be discovered, and for many years was known as the 'Clifden Blue' or 'Deptford Blue' after two early localities. The main populations were probably always in Dorset and the Isle of Wight, where the chalk is warmest. These areas, with Salisbury Plain, became its stronghold, for they suffered less than other regions when most unfertilised downlands were abandoned from the beginning of the 20th century, or when most sites became overgrown after the disappearance of rabbits in the 1950s. The impact of myxomatosis triggered a spate of extinctions as swards became too tall for breeding. Colonies being monitored in Kent declined from a few thousand adults to extinction in just two years after rabbits disappeared. This was repeated throughout southern Britain, and by the 1970s only 2% of former sites outside Dorset still supported the butterfly, even though Horseshoe Vetch remained common on most.

By the late 1970s, the Adonis Blue was at an all-time low. After one survey, I estimated that about 75 colonies survived, of which half were in Dorset and most were very small. It seemed likely that the species would be lost to Britain around the turn of the century if the decline continued unabated. This, happily, has not been the case. Most nature reserve managers started managing their sites more appropriately for this butterfly. This has included grazing many of the downlands owned by the National Trust, for example along the beautiful Hog's Back of the Purbeck Hills and its extension between the Needles to Ventnor on the Isle of Wight, where there is a string of protected populations. Private landowners have also played a part, such as Maddy Pfaff's textbook restorations on the chalklands of the Weld Estate, in west Purbeck. In addition, the widespread return of rabbits, and the introduction of agri-environmental schemes that include Adonis Blue habitat as an objective, have meant that many other downs that had been ungrazed for 25 years returned to some sort of short turf in the 1980s. The warmer springs and summers of recent years may also have hastened the revival.

The mid 1980s saw the first evidence of a recovery. It was painfully slow at first, and involved little more than the expansion of the strongest colonies onto neighbouring downs. By 1990, the English colonies had doubled in number to about 150, reaching 250 populations by the Millennium, still centred around its core areas of Dorset and the Isle of Wight (100 colonies), Wiltshire (90), and the South Downs of East Sussex, from the Seven Sisters to Lewes and Brighton (40). Butterfly Conservation's most recent survey of 2000-2004 showed a further increase in numbers on existing sites and an expansion to new ones, again mainly to sites close to a current colony. These include increases from the much smaller refuges of the butterfly in the Chilterns, and along the escarpment of the North Downs near Dorking and in east Kent. Adonis Blues have also reappeared and spread in the Cotswolds after an absence of about 40 years, but these are believed to have resulted from introductions.

Although the Adonis Blue has yet to recover its status of the 19th century, it has been a heartening experience to witness its expansion over the past 25 years. Today, one can chance upon a colony on most southern downs, and expect to find it in any abandoned quarry or stretch of escarpment in Wiltshire, Dorset or the Isle of Wight that is steep, south-facing, unfertilised and grazed. On the best sites – especially those scarred with cattle hoofs during winter – the populations are huge, containing tens of thousands of adults in the second generation of a good year. The only shadow of concern results from a comparison of the DNA found in 19th-century museum specimens and in current populations of Adonis Blue, which shows that much genetic variation was lost during the bottleneck years of 1955-1980. There has been no obvious loss of vigour in the species – far from it, considering its current expansion – but the butterfly might be less able than its Victorian ancestors to adapt to future changes to its environment.

Mazarine Blue

Cyaniris semiargus

Of the five British butterflies to have become extinct since the mid-19th century, the Mazarine Blue is the least spectacular and has attracted the least attention. To my knowledge, there has been no serious attempt to reintroduce it, in contrast to the considerable efforts made over the Large Copper, Black-veined White and Large Blue. Occasional records of single specimens still crop up, suggesting that a few surreptitious attempts may have been made. But it is more likely that 20th-century records stem from accidental releases, for the eggs of this butterfly are easy to find on the Continent, and I know several naturalists who have brought them back to rear in captivity. The resultant adults are then photographed on wild flowers, and a few inevitably escape.

Habitat and foodplants

This is an attractive, medium-sized blue, looking rather like a large Silver-studded Blue from above, with lovely cinnamon-coloured underwings, spotted more like those of a Small Blue or Holly Blue. It lives in small, discrete colonies on the Continent, as it clearly once did in Britain. Our populations, like those in north Europe, had one generation a year, with the main adult period lasting from mid-June to mid-July. My experience of the species in Scandinavia, central France, the Alps and the Pyrenees suggests that it lives mainly in flower-rich meadows on damp soils. This appears also to have been the case in Britain, or at least in its stronghold of Dorset, where most recorded colonies were on the heavy clays of the Blackmoor Vale, apart from a few on sandstone and chalk. The British foodplant of the Mazarine Blue is unknown, but was presumably Red Clover, for some of its sites were clover meadows and this is its principal food on the Continent. Kidney Vetch is also eaten on drier limestone sites.

The females lay their eggs on flowers of both foodplants, fluttering around clovers, and often laying three or four eggs on the pale bases of the firm young tubes of the petals. In central France, the butterfly is particularly attracted to the clovers that regenerate in steamy, warm hay meadows, about three or four weeks after the first crop has been cut.

Former distribution
A grassland species, extinct for over 100 years. Its main strongholds were Dorset, south-east Wales and Gloucestershire.

The eggs are quite easy to find if you part the petals of the youngest clover buds. They hatch after one to two weeks.

The caterpillar lives first in the flowerheads, boring into the flowers with an ugly little head on an extensible neck, rather like a young Holly, Small or Large Blue. Pupation, in north Europe, takes a further nine months, for the caterpillar hibernates when half-grown, and resumes feeding on sprouting clover shoots the following spring. In its later stages, it possesses all the ant-organs described on pages 142-3, and is attractive to ants in captivity. It is almost certainly tended by them as it feeds, and although no-one to my knowledge has studied the association, it is likely to be a strong one in view of the powerful interactions found in its close relatives. The chrysalis, too, attracts ants, and is probably buried by them in the same way as other blues.

Colonies in Britain

In Britain, the Mazarine Blue has always been one of the rarest butterflies. It was first recorded in 1710, and was subsequently found in 22 counties during the next 150 years, mainly in the south but with a few records extending to Yorkshire. It is impossible to say how widespread it really was, for butterfly records were then extremely patchy, and travel was often difficult before the second half of the 19th century. On the other hand, some records may be false, for frauds were quite common among certain Victorian dealers, and Continental Mazarine Blues were imported and sold as British from as early as 1860.

There are three main areas where colonies definitely occurred. The most famous is in Dorset, where the butterfly was 'formerly widely distributed and locally common' according to E R Bankes, a leading local 19th-century entomologist. Almost everything known about these populations comes from the collection and diaries of J C Dale, squire of Glanvilles Wootton, in north Dorset. Dale's most famous colony was in a meadow near his home, where he collected or recorded the butterfly in 27 of the summers between 1808 and 1841. It was clearly common in at least ten of those years, including 1808 when the 17-year-old Dale began his famous diaries. Nearly 300 adults were seen or collected from this colony; some survive in mint condition in the Hope Museum at Oxford, where I have had the pleasure of examining them. Three are included in the famous first plate of E B Ford's *Butterflies*, and others were loaned to Richard Lewington for the illustrations in this book.

There were at least eight other colonies of Mazarine Blue in Dorset at that time, but these were further afield and seldom visited. When and why they disappeared is unknown. The Glanvilles Wootton population 'suddenly disappeared' in about 1841, and no colony has been reported from Dorset since.

The other localities where the butterfly was seen regularly and often were near Hereford, in Gloucestershire, and in Glamorgan. In Gloucestershire, it was found in various places up to 1865, and survived 12 years longer in Glamorgan, where it had previously been known 'in plenty' near Merthyr between 1835 and 1837. The last authentic British colonies were at Penarth and Llantrisant, near Cardiff, where the butterfly was seen from 1871 to 1877. These became quite famous as a source of specimens. A local entomologist, Evan John, 'took it every year and once saw about twenty in a field'. He invited collector friends from Bristol to come and watch the butterflies at the sites, and had the edges of the hay meadow scythed so as not to impede their progress. This colony also disappeared, and there have been only a few single sightings since, although it is rumoured that a colony survived on the north Lincolnshire border until as late as 1903.

An unexplained disappearance

No-one can say why this attractive blue disappeared from its scattered sites in the 19th century. One suggestion is that it was the victim both of farmers sowing Red Clover in hay meadows, which lured egg-laying females in from safer pastures, and of changes in the timing of haymaking, which resulted in clover fields being cut exactly when the eggs and caterpillars were still present. This is speculative but plausible: the single-brooded north European colonies of Mazarine Blue are much more vulnerable in this respect than the multi-brooded ones in the south. In central France, colonies appear to move between the patchworks of little hay meadows that are still mown over a range of dates.

Today, most former British sites are greatly altered – certainly in the Blackmoor Vale where I live – although there are still a few unfertilised humid meadows where clover grows luxuriantly and is seldom cut. It would be interesting to try a reintroduction, perhaps using Scandinavian livestock. The relationship with different ants, and their availability on sites, should be studied first, for I suspect ants play a greater role than has been recognised in the survival of this vivid, subtly coloured little blue.

▲Male
Deep blue coloration of upperwings is distinctive; underwing markings resemble those of the Small Blue.

▲Female
Upperwings have a bronze sheen shortly after emergence from the chrysalis.

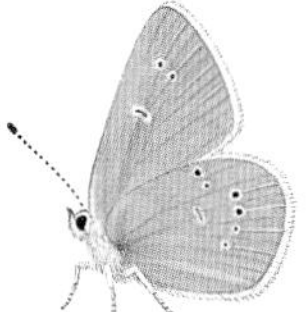
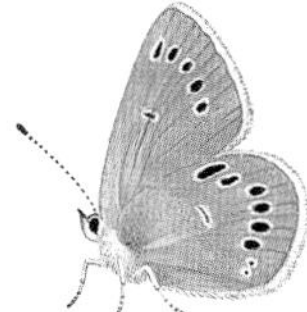

▲Colour variants
Two aberrant forms, with extremes of spotting, caught in the 19th century at Glanvilles Wootton, Dorset.

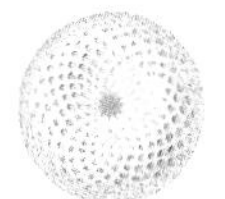

▲Egg [× 22]
Pale blue-green when laid, hatching after about ten days.

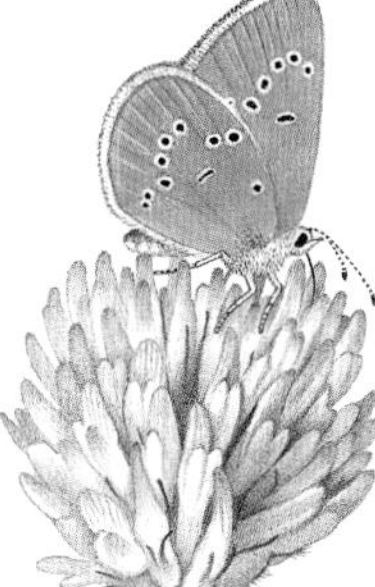
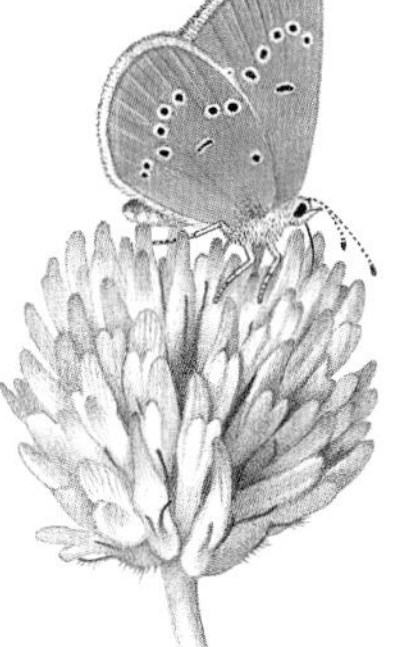

◄Feeding female
Both adults and caterpillars feed on Red Clover.

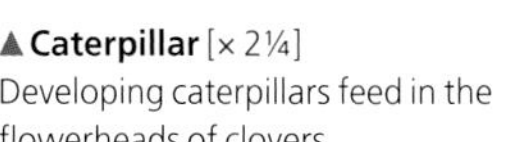

▲Caterpillar [× 2¼]
Developing caterpillars feed in the flowerheads of clovers.

►Feeding male
Kidney Vetch is used as a source of nectar by the adults and as a foodplant by the caterpillar.

	Jan	Feb	Mar	Apr	May	Jun	Jul	Aug	Sep	Oct	Nov	Dec
Egg												
Caterpillar												
Chrysalis												
Adult												

Holly Blue

Celastrina argiolus

The Holly Blue differs from its close relatives in being a butterfly of shrubs rather than grassland, and in roaming the countryside instead of living in tight-knit colonies. It is a regular visitor to gardens, and can be found in the heart of most cities in southern England. The largest population I have seen was in the grounds of the John Innes Park at Merton, a small oasis of walled gardens planted with Ivy and cultivated hollies, among the concrete and asphalt of south London.

Flashing blue and silver

This is a species that can confuse beginners, who often glimpse a flash of blue as it flits over a treetop or wall. While it is reasonable to assume that any blue butterfly seen flying high among bushes could be a Holly Blue, it is not an infallible guide. The Common Blue ascends to considerable heights, and even settles on treetops to drink aphid honeydew.

Identification is easier when the butterfly is at rest. The silver, black-spotted undersides, without the slightest trace of orange, can be confused only with those of the Small Blue, which is scarce and very much smaller. Furthermore, while the upperwings of the male Holly Blue are similar to those of a Common Blue, its fringes are distinctly chequered around the forewings, whereas those of the Common Blue are clear white. The female Holly Blue has especially lovely uppersides that are heavily tipped with black, particularly in the second emergence in midsummer. These are distinctive in flight, giving an inky blue impression very similar to a smallish Large Blue. Indeed, the Holly Blue is responsible for more false reports of Large Blues than any other butterfly.

Distribution
Numbers fluctuate in 4- to 6-year cycles. Seen almost everywhere within its range in peak years, but scarce at other times.

Two generations

There are two emergences of Holly Blues a year, the first from mid-April to June, the second in late July and August, with an occasional third brood in hot years in the south. In both generations adult Holly Blues behave rather like hairstreaks, resting among bushes at night and by day skipping in the sunshine around the canopies of bushes. They settle quite often on leaves, where they fidget in the sunshine with wings firmly closed, preferring to drink honeydew rather than nectar. Only in weak sunshine do the wings open, and then seldom more than 90°, giving a glimpse of the

▲Male, first/second brood
Males from the first (spring) and second (summer) broods are similar.

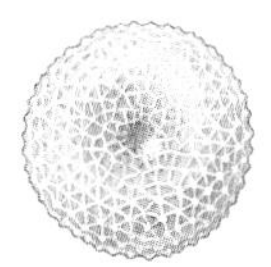

◄Egg [× 22]
Laid singly at the base of flower-buds, usually of Holly and Ivy.

▼Caterpillar [× 2¼]
Camouflaged markings vary greatly, but plain green is the most usual form.

▼Caterpillar on Ivy [× 1½]
Damage to the developing flower-buds makes the caterpillar easy to find.

◄Chrysalis [× 2¼]
Second-brood chrysalises hibernate, producing adults the following spring.

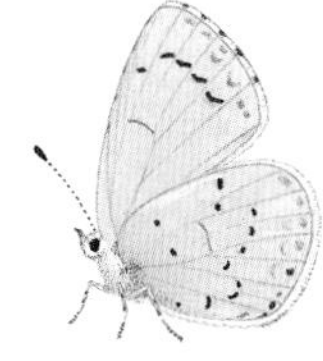

▲Female, second brood
Second-brood females have broad black margins to the upperwings.

◄Female, first brood
First-brood females have less black on upperwings than in second brood; underside identical.

►Basking male
Male basking on Ivy, the foodplant of the second-brood caterpillars.

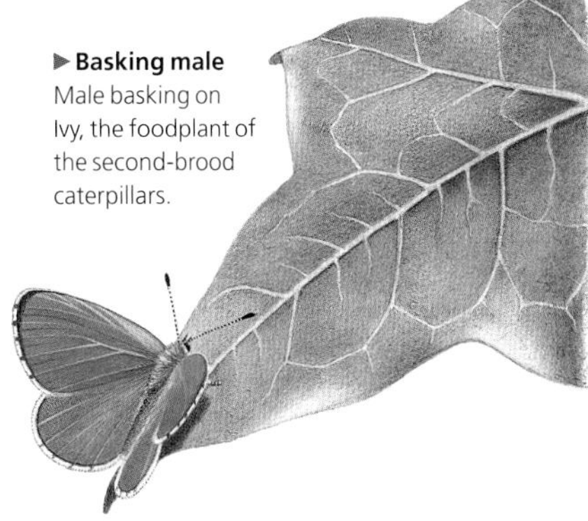

►Parasitic wasp
The grubs of *Listrodomus nycthemerus* kill large numbers of developing caterpillars.

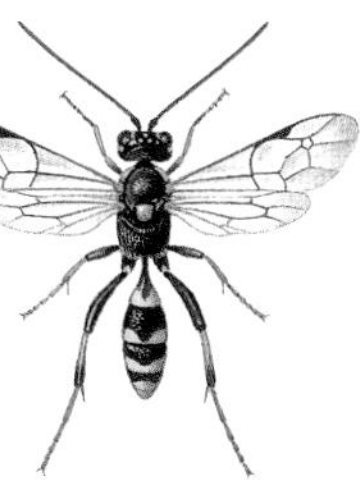

	Jan	Feb	Mar	Apr	May	Jun	Jul	Aug	Sep	Oct	Nov	Dec
Egg												
Caterpillar												
Chrysalis												
Adult												

beautifully marked uppersides. They resemble hairstreaks, too, in being tame, and can be closely approached during their frequent descents to ground level. Males, in particular, congregate to feed avidly on the salts in patches of mud; this occurs most frequently during midsummer droughts, when I have seen half-a-dozen drinking together on a muddy stream-bed. Like many butterflies, they are not averse to dung, benefiting from the minerals that it contains.

Egg-laying

Egg-laying females are often seen in gardens, fluttering slowly around the contours of shrubs and pausing to hover over any potential foodplant. Eggs are laid singly at the base of flowerbuds or the very young fruit of various shrubs, and less often on fresh leafbuds. In spring, by far the commonest host-plant is Holly, whereas most midsummer eggs are laid on the flowers of Ivy. However, Spindle, dogwoods, snowberries and heathers are also used, and I know of whole colonies in the West Country where the sole food in both generations is gorse flowers.

Eggs are easy to find in good Holly Blue years, when favoured bushes are peppered with the tiny white discs. You will notice that most are laid on somewhat prominent bushes growing in warm sunny positions, especially beside a wall or in a hedge or wood edge. Richard Revels, who has an unrivalled knowledge of this butterfly, found that particular Ivy plants are favoured year after year, and that the eggs themselves tend to be laid on half-hidden Ivy flowerbuds hanging in the shade. If you have a garden in southern Britain, it is well worth growing one or more foodplants, for this lovely blue is enchanting to watch as its hops around the bushes. Cultivated and variegated varieties of female Holly trees and Ivy are perfectly acceptable, so long as they are sheltered and produce fruit.

Feeding on Holly and Ivy fruit

The egg hatches after a fortnight and the little caterpillar attaches itself to the side of the young fruit. It then bores a small hole in the skin and eats the contents, scouring out the inner surfaces with a head that is mounted on a long extendable neck. It switches to fresh berries as it grows, leaving distinctive round holes on the withered fruit. The damage is easy to find on both Holly and Ivy, as shown in the illustration. Larger caterpillars may devour a complete Ivy berry, but the cup remains as evidence, and there is often much bleeding of sap. In either situation, this conspicuous damage should soon lead to the discovery of the caterpillar, marvellously camouflaged against the side of a fruit. Most are a clear fleshy green, but a few come with a variety of markings ranging from pink to maroon stripes down the back and sides; we illustrate an extreme example.

The fully grown caterpillar has the complete set of ant-attracting organs, although they are used intermittently in Britain, probably because few tree-climbing ants inhabit places where Holly Blues breed. In captivity, the caterpillars are lavishly attended, and in Somerset I have seen one being milked for its sugary secretions by the red ant *Myrmica ruginodis* on a low wall draped with Ivy. In Bedfordshire, Richard Revels finds them attended by black *Lasius* ants.

Parasitic wasps and cyclical populations

The caterpillars could do with some protection, for they are frequently attacked by two species of parasitic wasp. The braconid *Cotesia inducta* may kill appreciable numbers of

Cyclical populations driven by wasps

Holly Blue numbers undergo dramatic cycles every 4-6 years that are shadowed by those of its parasitoid, the wasp *Listrodomus nycthemerus*. When numbers of both insects are low, the Holly Blue recovers quickly to reach peak densities in 2-3 years (4-6 generations). The wasp lags a year or two behind, but soon exploits its host to such an extent that almost every caterpillar may be parasitised. This causes the butterfly population to crash, and in turn the number of wasps plummets in its wake.

Data from Richard Revels in *British Wildlife*, August 2006. To smooth seasonal fluctuations between the two annual generations, counts of caterpillars (+ 0.1) are presented on a log scale as 3-generation running averages.

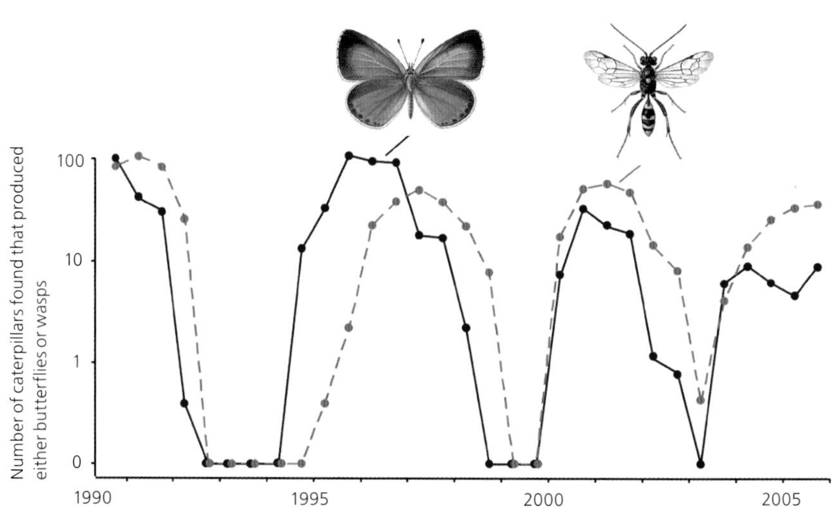

young caterpillars, whereas the older ones are attacked by a large ichneumon, *Listrodomus nycthemerus*, which cruises patiently around Holly and Ivy blooms, seeking its hosts. It punctures the young caterpillar with a long stiletto of a sting, and injects an egg into its body. This develops inside the caterpillar as it grows, later emerging as an adult wasp, one per Holly Blue chrysalis. This beautiful parasite has no other host than the Holly Blue, and can kill the vast majority of caterpillars in some years.

The mature caterpillar leaves its shrub to pupate, and acquires a purple mottling rather like the pre-pupal Brown Hairstreak. No-one, to my knowledge, has found a Holly Blue chrysalis in the wild, but its speckled brown colour and strong ant-attracting capabilities make it likely that pupation occurs on or near the ground, where it is probably tended by ants. Captive ants certainly find the chrysalis attractive, due to its secretions and, perhaps, a mimetic song (see page 92). Springtime chrysalises hatch after three to four weeks, whereas those from the midsummer brood hibernate.

Adult Holly Blues are considerable wanderers, first appearing in a district, increasing for a few years and then quickly declining. Permanent colonies exist throughout its range, especially in more wooded habitats, but even so there may be much movement between neighbourhoods. Populations fluctuate greatly over time, in a distinct cyclical pattern, with the butterfly building up in numbers and then crashing every four to six years. Richard Revels has painstakingly measured these cycles in Bedfordshire, by counting the caterpillars and rearing them for *Listrodomus nycthemerus*. He found that numbers of both the host butterfly and its parasite cycled in synchrony, but with the wasp generally lagging one or two generations behind. This provides the most convincing evidence to date that the dramatic oscillations in Holly Blue populations are driven by its parasite. When the wasp is rare, many caterpillars survive and the butterfly temporarily achieves high densities, but in a year or two the wasp catches up and becomes so abundant, through breeding on its plentiful host, that the number of butterflies plummets. This in turn precipitates a crash in the wasp population, and so the cycle continues.

The garden blue

At the peak of its cycles, the Holly Blue is widespread and plentiful, although always in lower numbers at any one spot than is the case with colonial species of blue butterflies. The best places to see it are in sheltered, sunny gardens and woods. In good years, the adults will be seen fluttering in ones and twos almost anywhere along the hedgerows of southern England and lowland Wales, with more localised concentrations further north, in the Lake District and the Isle of Man. It occurs even in our most intensive southern agricultural regions, such as the East Anglian fenlands, although here it is restricted to overgrown churchyards and gardens.

Modest expansion

The Holly Blue is one of the few British butterflies that held its own during the 20th century, although in the 19th century it was more widespread in north-east England. It has fared even better in recent decades. Superimposed on the cyclical oscillations in numbers has been a gradual trend upwards over the past 30 years, no doubt due to warmer weather conditions. During this period it expanded to many new localities in the northern midlands and Wales, where hitherto it had been exceedingly localised, and it has gradually extended its range north, reaching Scotland in appreciable numbers in its peak years of 2002 and 2004. A similar, more modest expansion has occurred in Ireland, where it was historically more localised and confined mainly to coastal and wooded areas.

It is reasonable to conclude that the Holly Blue is as numerous as it has ever been since its first record as a British insect, in 1710, by James Ray. It appears to have declined at certain periods in the 19th century, and the early entomologists gave no indication that it was common. Indeed, few mentioned it at all, and when they did they often created a new name. It was, for example, first known as 'The Blue Speckt Butterfly', with the female called 'The Blue Speckt Butterfly with Black Tips'. 'Azure Blue' and 'Wood Blue' were alternative names in the 19th century, and it was only comparatively recently that 'Holly Blue' gained universal acceptance.

Large Blue

Maculinea arion

About 18,000 species of butterfly are known from around the world, but none has a more curious lifestyle than the European large blues. These five lycaenids have evolved far beyond the mutually beneficial relationship that normally exists between ants and blue butterflies: they have instead overturned it. For whereas ants usually protect caterpillars while they browse upon their foodplants, and in return drink nutritious secretions from the larval honey-gland and pores, the caterpillars of large blues live like cuckoos inside red ant *Myrmica* spp. nests, eating the ant grubs or, in two Continental species, being fed directly on regurgitations by worker ants who are tricked into mistaking the intruder for their own brood. With both types of feeding, a large blue generally destroys its host ant colony.

Sadly, this small group, the *Maculinea*, are among the rarest and most endangered insects in the world. We in Britain have only one species, and that became extinct here in 1979. Since then, an indistinguishable race has been reintroduced from Sweden to a few carefully restored nature reserves, and has spread to a further 30 sites. Today, the Large Blue is more numerous in the UK than at any time in the past half a century.

Our largest, darkest blue

Despite its name, the Large Blue is not particularly big: the wingspan of a typical adult only slightly exceeds that of a large male Chalkhill Blue, and miniature specimens the size of a Small Blue are apt to emerge on every site. Their markings are always lovely, although seldom constant. The male Large Blue generally has finer black wing margins and fewer and smaller upperside spots than the female, but the heaviness of these marks varies considerably between individuals and, to a lesser extent, between sites. The colours merge into a deep ink-blue blur when the butterfly flutters across its grassland sites, in a jinking flight that is slower and jerkier than that of any other British blue.

I have had the luck to work with these butterflies for nearly 40 years, and to live among them for six summers on one site. I can therefore confirm and extend the remarkable observations made by Dr T A Chapman, Capt E B Purefoy and F W Frohawk, who discovered the species' association with ants nearly a century ago.

Distribution

Reintroduced colonies breed on about 25 sites in Somerset and on isolated restoration areas in the West Country and in the Cotswolds.

▲ **Male**
Smaller and generally less heavily spotted than the female, with finer black wing margins.

▲ **Female**
Forewings heavily spotted with black – a characteristic unique among British blues.

▲ **Colour variant**
Aberrant *alconides* form, found most often in males.

▲ **Colour variant**
Aberrant *imperialis* form, found most often in females.

◀ **Egg-laying**
The egg is laid deep in the flowerhead of Wild Thyme.

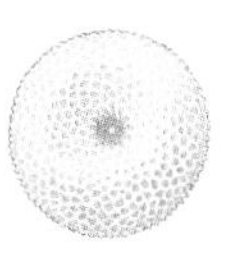

◀ **Egg** [× 22]
Laid singly on the flower-buds of Wild Thyme.

▲ **Young caterpillar** [× 2¼]
At first, the young caterpillar feeds on Wild Thyme flowers.

▶ **'Milking' a caterpillar**
An ant taps a caterpillar with its antennae, to stimulate production of secretions.

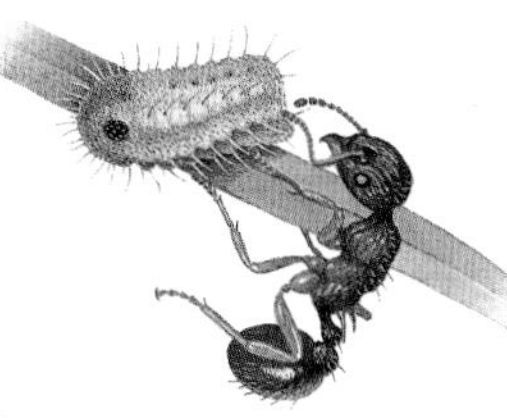

◀ **Adoption by an ant**
Deceived by its resemblance to an ant grub, an ant siezes the caterpillar and returns to its nest.

▼ **Feeding on grubs**
After being carried inside the ant nest, the caterpillar attacks and eats the grubs.

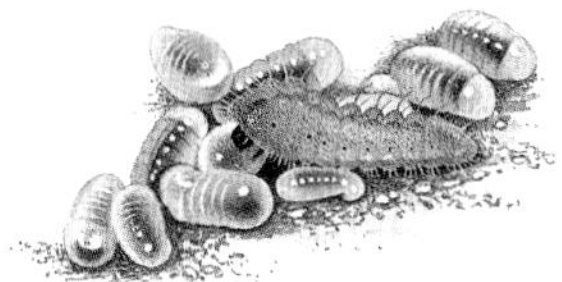

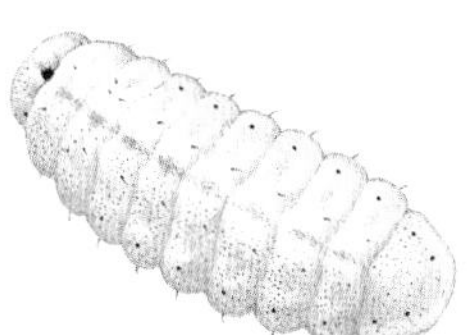

▲ **Mature caterpillar** [× 2¼]
The caterpillar can live for up to two years in the ant nest, feeding on their grubs.

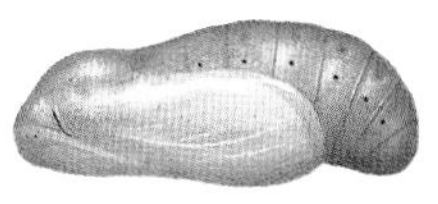

◀ **Chrysalis** [× 2¼]
Formed within an ant nest, where it produces sugary secretions.

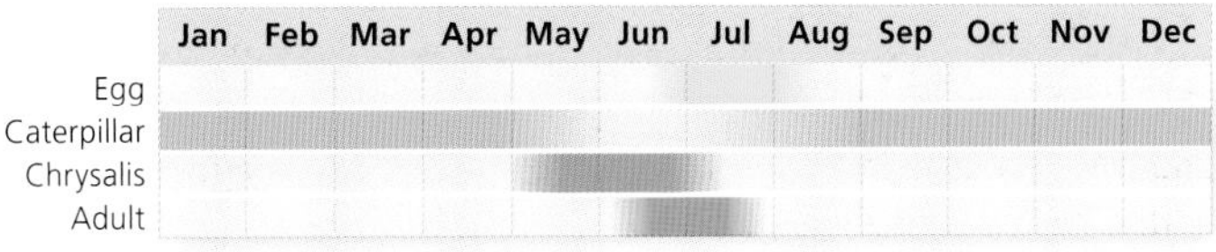

Singing in the ants' nest

Adult Large Blues emerge from early June to mid-July and live, on average, for five days. Before taking to the air, the adult must first escape its ant nest. The chrysalis, like the caterpillar, lives underground, usually just below the soil surface in a warm earthen cell. It is attended by a dozen or more ants, which incessantly lick secretions off its body. The chrysalis also possesses a sound organ similar to that of the Purple Hairstreak, but the calls that it makes are different, for they mimic the acoustics of adult red ants. All *Myrmica* ant species possess a stridulatory organ consisting of a plectrum and a file located on opposite sides of the first segments of the abdomen (the petiole and the gaster), which they rub together to produce vibrations and squeaks that are just audible to the human ear. When amplified, it is clear that queen ants produce calls that differ from those of their workers. These induce high levels of attention from the worker ants, which cluster around a calling queen, providing the enhanced protection worthy of the most valued member of their society. Although rasped from a very different organ, the 'song' of the Large Blue chrysalis is remarkably similar to that of a red ant queen, so much so that the workers treat it, too, like royalty.

Emergence and mating

Although constantly attended by ants for the three weeks that it exists as a chrysalis, the Large Blue produces a greater burst of sound when the adult splits open the pupal case and struggles to shake itself free. Emergence occurs between 8am and 9.30am, when the ants are rather sluggish, but the noise soon whips them into a frenzy, and they accompany the butterfly as it crawls up the narrow passages to the outside world. Once above ground they mill excitedly around, while the butterfly makes its way up a tussock and inflates its wings. The adult rests for about 45 minutes while the expanded wings set hard. It then flies to the lowest part of its site. Typical British breeding grounds are south-facing hillsides, and the males spend the rest of their lives near the bottom, patrolling in the morning sunshine, dipping regularly to investigate tussocks that might hide an emergent female. This activity diminishes around noon on hot days, when the males roost in the shade. They reappear in late afternoon to drink nectar from Wild Thyme.

Virgin Large Blue females are sometimes mated while they rest above their ant nests with drying wings, accounting for the slightly crumpled individuals seen in high-density colonies. Most, however, flutter down the slope and are soon intercepted by the males. There is a brief aerial courtship before the two butterflies land on a clump of grass and pair for about an hour. They then part, he to feed before resuming his patrol, she to hide until her eggs are ripe.

Egg-laying

Some females lay eggs on the afternoon of their emergence, but the main egg-laying session occurs on the second day of life. Each flutters slowly up and down the breeding slope, hovering just above the short turf before alighting on Wild Thyme. After twitching and rotating on a flowerhead, the butterfly curves her plump abdomen almost double before plunging the tip between tight young buds. A single egg is injected into the inflorescence before the abdomen slowly recoils. Many females are killed by spiders, dragonflies or birds before laying a single egg, but those that survive can produce two or three hundred each. By the end of the season, an average of about 50 eggs will have been laid per female in each colony.

The eggs hatch after five to ten days. The minute caterpillar burrows into a thyme flower to feed on the pollen and developing seed. Although the eggs are laid singly, a large flowerhead growing on the edge of a thyme plant often receives four or five during the few days that it is in bud, and I have found over 100 eggs on a single plant. Most are destined to die, for the caterpillar is cannibalistic before its first moult, and no more than one survives on each flowerhead. This is beneficial rather than harmful, for it thins the caterpillars out in spots where numbers are high, and reduces the chance that ant nests will adopt more caterpillars than they can support.

Adoption by ants

The caterpillars develop quickly on thyme, but put on little growth. After two to three weeks, each weighs roughly a milligram, yet has already completed all its skin changes and developed the organs needed for the next phase. These include a minute honey-gland that exudes sweet droplets, and high densities of microscopic pore cupola (see pages 142-3) which, in other species of blue, produce the amino-acids that are particularly attractive to ants with brood to rear. It is probably the pore cupola of Large Blues that secrete the cocktail of hydrocarbons that mimic the recognition chemicals of the ant colonies that they will shortly infiltrate.

I have watched several hundred Large Blue caterpillars being adopted by red ants, but the sight never loses its thrill. The tiny caterpillar completes its final skin moult at any time of day, but remains hidden in the thyme flower until 5-7pm. It then flicks off the flower and drops to the ground, where it hides beneath a leaf or in a crevice. By doing this, it enhances the chances of being found by a red *Myrmica* ant by at least a hundred-fold, for their workers forage mainly in early evening and very close to the soil. Other ants climb above the turf and forage during the heat of the day.

The caterpillar does not search for *Myrmica*, but on a good site it is soon discovered and the ant taps its body,

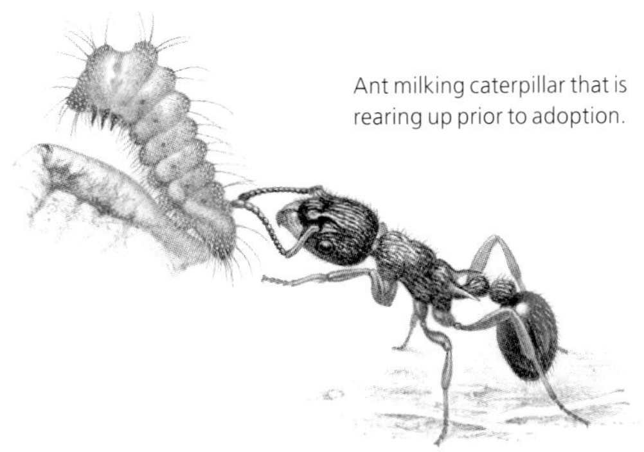
Ant milking caterpillar that is rearing up prior to adoption.

stimulating the honey-gland to produce a minute drop of liquid. This causes great excitement. After drumming the gland with its antennae for a few minutes, the ant rushes away to recruit up to a dozen nest-mates. These crawl all over the little caterpillar, milking its gland and licking and prodding its body with their mandibles. After around ten minutes they wander away, leaving the caterpillar with the original ant that discovered it. She is far more possessive; I have seen fights to the death when an ant from a different colony tried to muscle in on the feast.

Milking may last for up to four hours, but eventually the caterpillar rears up to signal its readiness for adoption. This causes a frenzied response from the ant, which is now tricked into believing that the caterpillar is one of its own grubs that has somehow escaped from the nest. In the dark of the brood chamber, red ants recognise their young by a combination of stimuli, including size, hairiness, scent and how firm the skin is to touch. The tiny Large Blue caterpillar already possesses the size and hairiness of an ant grub. It inevitably acquires its distinctive smell through being milked by the ant, for the odour of the nest is dissolved in waxes that coat the body of every ant that belongs to it, and this rubs off onto the caterpillar. The final stimulus is provided by the rearing up and distortion of the caterpillar's body, which causes its normally flabby skin to balloon out and become taut, very like the firm body of an ant grub. The caterpillar almost certainly releases chemicals of its own at this moment which, in close relatives of the Large Blue, mimic the scent of red ant grubs. The end result is always the same. The ant immediately snatches the rearing caterpillar in its jaws, and runs with it back to the nest. It then carries the caterpillar underground and places it among the ant brood.

Cuckoo in the nest

Once inside the nest, the caterpillar scrabbles among the brood, trying to puncture the skin of a grub. When this eventually bursts, the caterpillar slowly eats the soft tissues. Between feeds, it rests on a small pad of silk spun on the wall of an empty cell, near to the main brood chambers. The caterpillar grows quickly in the nest, and soon turns into a bloated white maggot that dwarfs both the ants and their grubs. It crawls into the deep recess to hibernate, and resumes feeding near the surface in spring. By now it is large enough to glide over the ant grubs, rather like a starfish eating its prey, keeping the vulnerable underside hidden from the worker ants, while exposing only its thick blubbery upper surface to nips or attack. Even so, preying on ant brood is a dangerous activity which the caterpillar reduces to a minimum. It rests on its silk pad for days at a time before gliding into the brood chambers to binge-feed on grubs, then returns to digest them in its haven for perhaps another week.

Being found by the right ant

There is growing evidence that some Large Blue caterpillars spend two years in an ant nest, in a manner similar to its relatives, the Alcon and Rebel's Blues of continental Europe. Whatever its age, by the time it pupates in late May the caterpillar is nearly 100 times heavier than its weight at adoption, having consumed up to 1,200 ant grubs. In fact, most caterpillars die long before this stage. The few that survive are those fortunate enough to have been adopted by one particular species of red ant, *Myrmica sabuleti* and, moreover, into a particularly large nest or a cluster of smaller nests that contain few, if any, queens during the first weeks of its life.

Up to five different species of *Myrmica* ant forage beneath Wild Thyme on British Large Blue sites, and all adopt its caterpillars with equal readiness. Survival, however, is over five times higher in the nests of *Myrmica sabuleti* than with any other species. The other four large blues on the Continent are equally specialised, but in each species or region, a single – and usually different – species of *Myrmica* is exploited. As yet, the reason for this discrimination is understood only in

A mature caterpillar in an ant nest; before pupation the caterpillar is nearly 100 times heavier than when it was adopted.

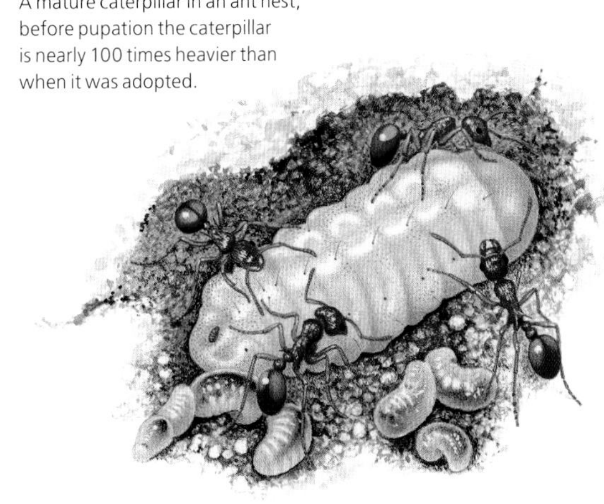

two related species of large blue, but undoubtedly applies also to *Maculinea arion* in Britain.

Above ground, foraging worker ants cannot distinguish between the grubs of different *Myrmica* species if these are artificially placed in their territories, and the same applies to the mimetic caterpillars of Large Blue butterflies. All are recognised just well enough to be carried to the 'safety' of the brood chambers, and into the domain of nurse ants. These are much more discriminating. If food is plentiful, and the ant colony is in a state of benign tolerance, all may be well. But if they become stressed or underfed, the nurses turn xenophobic. The ants accept grubs or caterpillars only if they smell like their own species (or even colony), and butcher imperfectly matching intruders to feed to their grubs. Inside the nest, the caterpillars of each species (or race) of large blue secrete a cocktail of chemicals that sufficiently resembles one species of *Myrmica* ant for it to be accepted as a society member, even in times of stress. The downside is that the closer a caterpillar mimics one type of *Myrmica*, the less it resembles any other. Thus, although often carried into the nests of *Myrmica rubra*, *M. ruginodis*, *M. scabrinodis* or *M. schencki* red ants, the British Large Blue seldom survives with them because its glands release the odour of *M. sabuleti*.

A hazardous lifestyle

Life in an ant nest is full of danger, even for Large Blue caterpillars fortunate enough to be adopted by *Myrmica sabuleti*. Many are killed by worker ants which, acting under the influence of their queen, kill large ant grubs that might develop to become her rivals. Caterpillars are vulnerable only during the second to fifth week after adoption into nests, after which they are either too large to be mistaken for potential rival queens or too rubber-skinned to be harmed.

Starvation is another threat, particularly in nests that adopt two or more Large Blue caterpillars. In fact, only a minority of nests contain sufficient brood to rear a single butterfly, even though the caterpillars exploit their prey in the most efficient way possible by eating just the largest available grubs; they also break hibernation a month later than the ants, allowing the workers to fatten up more brood before it is killed. If the caterpillar still eats all the grubs in a nest, the ants typically desert, leaving it behind. This allows a neighbouring *Myrmica* colony to bud off and invade the apparently vacant nest site, carrying in a fresh supply of ant brood. With its remarkable ability to fast, the caterpillar merely waits for up to three weeks for fresh ants to arrive. Despite these adaptations, starvation in ant nests is a major cause of death that determines the ultimate size to which a population can grow on most of the Large Blue's British sites.

Parasitoids that manipulate ants

As if these hazards were not enough, many caterpillars on the Continent are killed by a beautiful ichneumonid parasitoid, *Neotypus melanocephalus*, which can parasitise no other species than the Large Blue. I have found it a few times in Dordogne and the Rhône valley of France, but nothing is known of its behaviour except that the wasp emerges from the butterfly's chrysalis inside an ant nest. In appearance it resembles *Neotypus pusillus*, the specialist parasitoid of the Dusky Large Blue. But whilst I have frequently watched the latter stinging the tiny caterpillars of its host shortly before they drop off their early foodplant, Great Burnet, in the fens and marshes of eastern France, I have never seen the parasitoid of our Large Blue sting its host. It is possible that it attacks caterpillars at a later stage, after they have been adopted into ant nests, as occurs with the remarkable parasitoids of Rebel's and Alcon Blues. If so, it probably follows the same procedure. The wasp first identifies a nest that contains caterpillars, then sprays the ants with noxious chemicals that induce such fierce fighting between nest-mates that most ants become too tied up attacking one another to protect their nest. This allows the heavily armoured parasitoid to force its way in and sting the Large Blue caterpillars, before escaping relatively unscathed. Although it is known from only a handful of sites in Europe today, Victorian records exist of *Neotypus melanocephalus* on British sites. My ambition is that one day this fascinating parasite may also be reintroduced to British Large Blue colonies.

Decline to extinction

With such a specialised life cycle, it is not surprising that the Large Blue has always been rare in Britain. It was discovered near Bath late in the 18th century, and about 90 colonies were found during the next 150 years. Most were in remote parts of southern England, at least a day's horse ride from the nearest railway station. This gave the butterfly a tremendous cachet among Victorian collectors, second only to that of the Large Copper.

The first British site where collectors could almost guarantee to secure specimens was at Barnwell Wold, in Northamptonshire. This sustained heavy collecting for about 20 years, but in 1860 about 200 adults were taken at rest by one dealer in the course of a wet summer: the colony never recovered. With the demise of the Barnwell colony, entomologists turned their attention to two other areas where it had recently been discovered: three small colonies in the limestone Polden Hills of Somerset, and some large ones along the Devon coastline south of Salcombe, the Bolt Head to Bolt Tail cliffs. Here, enormous numbers were caught. The Revd F R Elliot wrote of a friend who 'in the summer of 1859 took as many as a hundred specimens in one day... which were later sold for half-a-crown apiece'.

There were bitter debates about the wisdom of such collecting, and many appeals were made for restraint. But this had little effect, and when the south Devon populations crashed in 1875, collectors turned to the Cotswold colonies that had been discovered 17 years earlier. As had occurred elsewhere, the Cotswold populations all but disappeared in the 1880s, and remained in very low numbers. They were, in fact, to recover and become locally common again in the 1920s and 1930s, but entered a terminal decline in the mid-1950s, becoming extinct ten years later.

It was not until 1891 that the finest British colonies were discovered, along the Atlantic coast of Cornwall and Devon. This came as an enormous relief to Victorian entomologists, who believed the butterfly to be on the verge of extinction elsewhere. They responded by arriving in droves, and contemporary reports claim that over 1,000 butterflies, 'and probably double', were taken each year from 1895 to 1914 between Tintagel and Millook, with no apparent harm to the populations. Unfortunately, a considerable decline set in during the 1920s, at least in the traditional collecting grounds south of Bude. Thereafter the butterfly was seen in insignificant numbers, disappearing finally from Crackington Haven in 1963.

There was still one stronghold, however, along the coast north of Bude, where the Large Blue bred in almost every valley up to Clovelly in the 1950s. It was exceptionally abundant on some sites, with individual valleys probably supporting up to 5,000 adult butterflies. But, as in the Cotswolds, a sharp decline started in the mid-1950s, and within 18 years the butterfly was extinct here, too. This left one small colony on Dartmoor; that disappeared in 1979.

Despite the large numbers taken by collectors, almost all these extinctions were caused by agricultural change. Nearly half the old breeding sites were simply destroyed, mainly by ploughing and seeding. The remainder appeared superficially unchanged, with most possessing a sward of fine grasses dominated by mats of Wild Thyme. But these had altered in a more subtle way. All had been heavily grazed when they supported Large Blues, but farmers gradually abandoned these unproductive pastures for flatter, richer land. As described on page 144, the effect was not serious until the 1950s, for large populations of rabbits maintained a close-cropped turf. But once myxomatosis was introduced, the turf grew dense and tall.

Tall, shaded turf is as fatal to this butterfly – or rather to its ant host – as it is to the Adonis Blue. The cooling of the soil that occurs when a short sward grows just a few centimetres taller results in the rapid disappearance of *Myrmica sabuleti*, whose nests are often taken over by other, more tolerant species of red ant. Since these will not rear the Large Blue, the butterfly disappears as quickly as its host.

Research and reintroduction

Knowledge of this butterfly's specialised life cycle came just too late to save it from extinction in the UK, but conservationists soon attempted to recreate suitable conditions on former sites. Hillsides were grazed in targeted ways that allowed the essential ant, *Myrmica sabuleti*, to return in extraordinarily high densities to places where it had been undetectable in the 1970s. By 1980, a site on Dartmoor owned by the National Trust seemed ready again for the butterfly. This prompted a programme of re-establishment, carried out in collaboration with David Simcox – who has led the project since 1999 – and many other dedicated ecologists and conservation organisations.

After much searching in northern Europe, we found colonies in Sweden that appeared identical to the former British races. These were introduced as a trial in 1983, with immediate success. The butterfly increased for five years before experiencing a slight fall in 1989, caused by a severe summer drought. Happily, it recovered, and has persisted on this and a neighbouring site, albeit in low numbers, for over 20 years.

A conservation success

Spurred by this initial success, a further ten organisations joined the partnership to produce a powerful team dedicated to the conservation of this curious butterfly. At the time of writing, restorations are at an advanced stage on about 100 sites in four former landscapes inhabited by the Large Blue. Much the most successful has been the reintroduction to the Polden Hills, where more than 40 grasslands were restored. The first introduction there occurred in 1992, to the Somerset Wildlife Trust's reserve at Green Down. Within five years this population had grown to a few thousand adult butterflies, a level that has persisted into 2013, 22 years after introduction. Three other introductions were made to far-flung sites in the Polden Hills, including the National Trust's Collard Hill, where a large colony has also developed and where visitors are welcomed throughout the flight period.

More encouraging still has been the Large Blue's natural spread to about 30 other Polden Hills conservation sites, including some which now support densities that match that achieved on Green Down. Among the most exciting are five medium-sized to large colonies breeding alongside railway lines, two in new cuttings created from scratch, to David Simcox's design. In comparison, the colonies in other regions remain small and isolated, providing a challenge for us to repeat the success of Somerset along the Atlantic coast of Devon and Cornwall, and in the Cotswold Hills, where two flourishing colonies already appear to have been established, thanks to introductions made by Simcox and Sarah Meredith.

Duke of Burgundy

Hamearis lucina

Few British butterflies have experienced a more rapid or more worrying decline than the Duke of Burgundy. Although never particularly numerous, it once inhabited most large woods in the south, where it fed on half-shaded Primroses that grew in regenerating clearings. All but 20 of these colonies have gone – victims of the decline in coppicing – leaving fewer than 200 populations in the country as a whole. Most breed in scrubby patches on north-facing downs. This was historically a poor alternative as a habitat, and all but three of these sites support a mere handful of adults. All must be considered vulnerable.

A link with the New World

It would be sad, indeed, were we to lose this little insect, for it is the only European representative of that wonderful tropical family, the metalmarks or Riodinidae, that finds its greatest diversity in Central America. Our Duke, to be sure, is one of the dullest and least interesting members of the group, with little of the subtle beauty of the adults of most related species, and none of the extraordinary adaptations that many metalmarks have evolved to attract ants.

It is a charming butterfly, nonetheless, with much of the perky behaviour of a hairstreak or blue, combined with the chequered wing pattern of a fritillary. It was, indeed, known as the 'Duke of Burgundy Fritillary' for two centuries, after originally being named 'Mr Vernon's Small Fritillary' in memory of the collector who first found the butterfly in 1696 in White's Wood, Gamlingay, in Cambridgeshire. Its markings still give rise to confusion. The distinctive features that separate it from the fritillaries are two clear-cut bands of white marks running parallel down the underwing, and spots around every wing edge.

Distribution
A rapidly declining species, approaching extinction in most of its range; frequent only in the Cotswolds and on the edges of Salisbury Plain.

Territorial males

The Duke of Burgundy lives in small, close-knit colonies, typically consisting of a few dozen individuals, although there are three sites where one or two thousand adults occur. Not all will be flying on the same day, and you will do well on a typical site to see more than five or six adults on each visit. An average adult lifespan is five to seven days, with the first males emerging in late April or early May. Emergence then continues in ones and twos, building up to a peak towards the end of May and dwindling away by mid-June.

▲**Male upperside**
Both sexes have spots around all wing edges.

▲**Male underside**
The forelegs of the male are reduced and not used for walking.

▲**Female upperside**
Orange markings in the female are generally brighter than in the male.

▲**Female underside**
Both sexes have two bands of white marks on the underwings.

▲**Colour variant**
Markings may be heavier on some individuals.

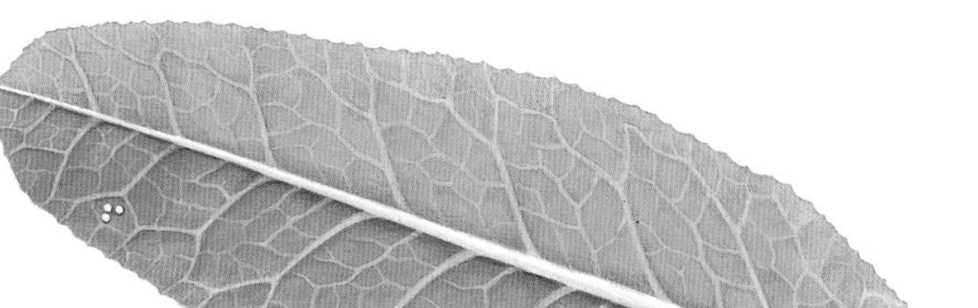

▲**Egg-laying**
Small groups of eggs may be laid on the same leaf by different females.

▼**Egg** [× 22]
As the egg develops, the hairs of the young caterpillar may be seen through the shell.

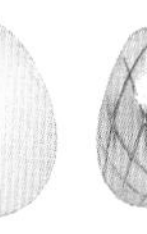

Newly laid

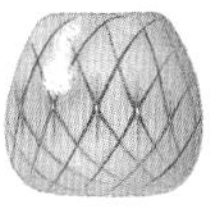

Prior to hatching

▲**Caterpillar** [× 2¼]
Feeds nocturnally on the leaves of *Primula* spp., leaving distinctive damage.

▼**Chrysalis** [× 2¼]
Formed in vegetation or on the ground; lasts for nine months.

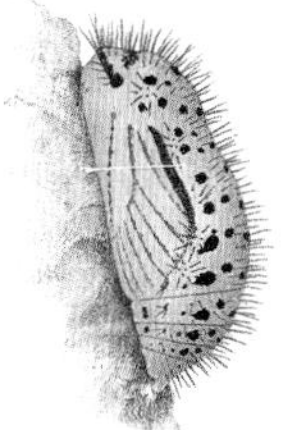

Side view

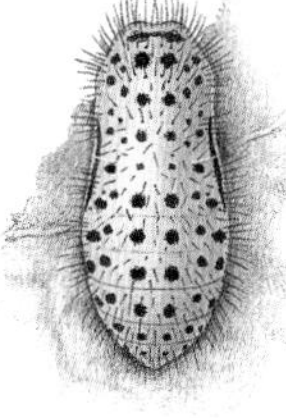

View from above

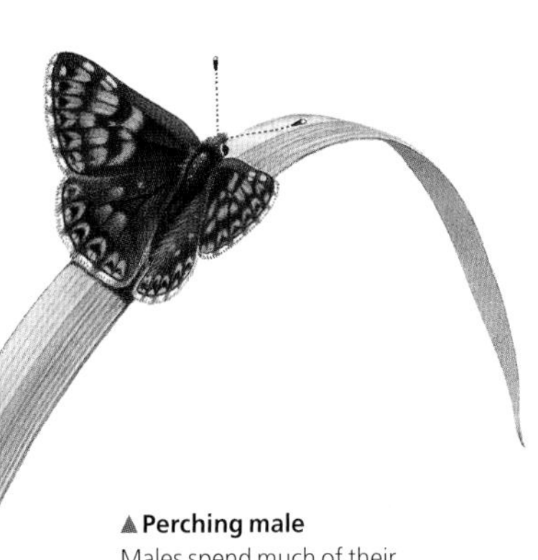

▲**Perching male**
Males spend much of their week-long lives defending territories against rivals.

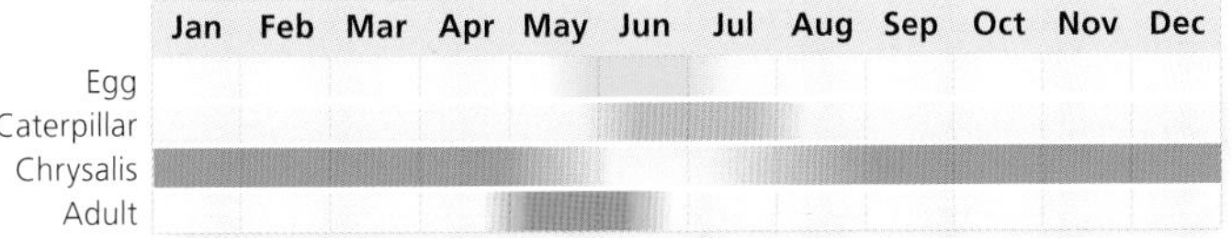

The sexes behave rather differently. Males are much the more conspicuous sex and can be located with ease, even on sites with low numbers, owing to their distinctive territorial behaviour. They congregate in leks in quite small areas, where each male selects a small bush or tussock and fiercely defends it against rivals through a series of spiralling dog-fights. Territories are generally established in clearings towards the base of a hill, especially on the edges of shrubs that jut out into sheltered recesses and catch the morning sunshine. In woods, there is a strong preference to perch at the junction of two rides, or in small sunny hollows near recent clearings. Each perch is used by a sequence of males throughout the season and, very often, from one year to the next. Here, the male sits for long periods on a leaf, with wings half-open, pointing backwards towards the sun, ever on the alert for intruders to his space. A highly pugnacious butterfly, it soars up to engage any passing insect in prolonged, violent flights. But courtship itself is minimal, with none of the elegant behaviour seen in the browns and vanessids. Mating typically occurs on a low shrub within one metre of the ground.

Fortunately for the persistence of this species, the females fly considerably further than the males. In one study, more than half the marked females had shifted more than 250m when they were recaptured, and new sites up to 5km from an existing breeding area have been colonised in recent years. They are also more elusive than males, and are generally seen only when laying eggs. This involves fluttering a few centimetres above the undergrowth in a series of jerky hops, punctuated by short glides. It is often said that neither sex visits flowers, but this is not so: I have watched them several times feed from the blooms of Hawthorn, and Matthew Oates reports that buttercups and other yellow flowers are visited.

Searching for Primula

Duke of Burgundy eggs are laid on the undersurfaces of *Primula* leaves, with Primrose used mainly in woodland, and Cowslip in scrubby grassland. Extraordinary care is taken to choose a plant that is growing in a precise situation which, once learnt, makes the delightful glassy eggs very easy to find. Primroses and Cowslips need warm, bare ground for their seedlings to germinate, and are often abundant in the year or two after woodland has been cleared, or on downland that is trampled and scarred in winter by cattle – hence the name 'cow-slip'. However, *Primula* live for a good many years, during which time it is common for them to become swamped by regenerating shrubs or by tall, abandoned grassland. Eventually, they persist as anaemic non-flowering plants, whose large, flabby leaves lie limply on the shaded ground, waiting for the next clearing to occur.

The female Duke of Burgundy invariably selects a *Primula* about halfway through this succession. In scrubby grassland, she chooses the widest and lushest green leaves of large, flowering plants, preferring vigorous Cowslips that are half-shaded by an encroaching shrub or tussock, yet not so swamped as to prevent the long vertical leaves from protruding above the sward. This is an unusual growth-form on most downs, where Cowslips generally occur in more open turf and have a tight rosette of small, oval leaves that lie flat against the ground. The butterfly always avoids these, as it does heavily shaded plants with lank, limp leaves.

A similar situation pertains in woodland. There, Tim Sparks found the eggs predominantly on mid-sized to large Primroses, growing as part-shaded plants in young conifer plantations or along broad rides in deciduous woods, typically those that ran east-west. As in grassland, there tends to be a short window in the vegetational succession during which a colony can thrive: the butterfly generally colonises clearings that are becoming slightly overgrown for fritillaries, then moves on to another patch a year or two later. Thus in coppice woodland, Martin Warren found that adults were absent one year after a cut, but became abundant two to three years later, shortly before the canopy closed again. By the time a coppice panel had five years' regrowth, the butterfly had disappeared. Breeding patches are therefore confined to very small areas within a wood at any one time, although large numbers can briefly develop within them. I remember in Grovely Woods, near Salisbury, there were scores of this species in the late 1990s, thriving in a five-year-old plantation.

Eggs, caterpillars and chrysalises

When egg-laying, the female Duke of Burgundy perches on the top of a leaf and reaches around and below with her abdomen. The eggs are conspicuous glassy spheres, usually laid in groups of two to four, although I have found up to ten together, and there may well be two egg-batches per plant. They are an attractive opaque creamy-yellow when laid, developing a curious criss-cross pattern of lines shortly before hatching, caused by the long black hairs of the little caterpillar being visible through the transparent shell.

Duke of Burgundy eggs hatch after one to three weeks, and the tiny caterpillar crawls to the base of the stem. It rests in this position by day, emerging at night to rasp deep grooves in the fleshy *Primula* leaf. The damage is quite distinctive, especially when the caterpillar grows older, for it eats large holes, leaving the veins and midrib intact, until the leaf looks like a moth-eaten blanket. The final-stage caterpillar is a particularly voracious feeder that can be found on warm July evenings after dusk if you search likely Primroses and Cowslips by torchlight. This is a critical period of growth, for many *Primula* wither and die back around this time of year, and the caterpillar feeds only on lush green leaves. Matthew Oates reports that they will sometimes wander several metres to find suitable plants at this stage, browsing on any fresh *Primula* seedlings encountered *en route*. Finally, after six weeks of feeding, the fully grown caterpillar deserts its foodplant and forms a hairy, speckled chrysalis in dry nooks, such as 30-60cm up in the heart of a tussock of fine grass, among chalk scree, or even in the empty cases of beech nuts. Although it remains there throughout the winter, the chrysalis is extremely hard to find.

A tenuous survival

Today, the Duke of Burgundy is largely confined to grassland sites in central southern England. The large majority of these are on sheltered, north-facing downland slopes, with the butterfly generally breeding towards the base of the hill, where the soil is deeper and the Cowslips less prone to summer drought. South-facing downs are almost never used. Although this is one of our best-monitored butterflies, colonies of Duke of Burgundy are sometimes overlooked because of their small size and the local shifts in breeding sites. It is still worth searching suitable spots for new colonies within its surviving regions.

These, alas, are few and far between, for most of the well-known colonies have disappeared in recent decades. The butterfly has long been extinct in Scotland, Wales, East Anglia and most of northern England. Surrey and the east Midlands forest belt of Oxfordshire and Northamptonshire stretching up to Castor Hanglands in Rutland must perhaps be added to the list. It survives, but is at the lowest of ebbs, in east Kent, the Chilterns, Dorset and Somerset, and Butterfly Conservation's 2000-2004 survey showed that the Duke had also fared poorly at its two northern outposts, the Morecombe Bay limestones of Cumbria, which support 11 colonies, and the North Yorkshire Wolds, where perhaps 15 populations survive. The large majority of remaining colonies, representing a tenuous stronghold for this butterfly, survive in the limestone grasslands of the Cotswolds and on the chalk downs of Wiltshire, Hampshire and West Sussex. There is, however, some hope for the future, as targeted management programmes led by Butterfly Conservation bear fruit. Already, unusually high counts of Duke of Burgundy were made on several monitored sites in 2010, 2011 and 2013, and a major restoration project has started in the Cotswolds.

White Admiral

Limenitis camilla

The White Admiral is a butterfly much loved by naturalists across its global range, from England to Japan. It has an elegance unmatched by other species, and strange but attractive young stages. It is, moreover, one of our few uncommon butterflies that has been on the increase, in distribution if not in numbers. Although still scarce enough to arouse excitement, it is not so rare that every locality is known or protected.

No account can do justice to the White Admiral's dainty movements, or convey the character of a creature so ideally suited to gliding in and out of dappled shade among the branches of mature woodlands. The undersides are particularly beautiful, and seen close to, there is no other butterfly with which this can be confused. Hopeful beginners sometimes mistake high-flying adults for Purple Emperors, but the White Admiral is considerably smaller, has a silhouette of rounded rather than pointed wings, and a more graceful, flitting flight. It hugs the contours of the trees as it skims around the woodland canopy, gliding through still air instead of rising high and battling with the breeze above the treetops.

Colour forms

Adult White Admirals are usually constant in appearance, but have two named varieties that are encountered in some years with reasonable frequency, and in certain woods more than in others. One of these is the *obliterae* form, in which the white markings are considerably reduced. The other is the scarcer *nigrina* variety, in which the upperwings are universally black. Both types can be obtained by chilling captive chrysalises in a refrigerator, and it is possible that they are produced in the wild from caterpillars that pupate in frost pockets.

Distribution
Has spread to reoccupy many former sites in the past 50 years. Mainly seen in large woods, especially in Hampshire, Sussex and Surrey.

Life in the canopy

There is typically one brood of the White Admiral a year. It begins in mid-June after a warm spring, and usually reaches a peak in the second or third week of July, with tattered adults lasting into August. Very occasionally, a few individuals emerge in September after a warm summer, suggesting a partial second brood. This is essentially a woodland butterfly, and the adults are normally seen in ones and twos. That is not to say it has small populations, rather that the adults

▲ **Male upperside**
Slightly smaller than female, with forewings more pointed.

▲ **Male underside**
Wing patterns are similar in both sexes.

▲ **Female upperside**
Females are most easily seen when they descend to ground level to lay eggs.

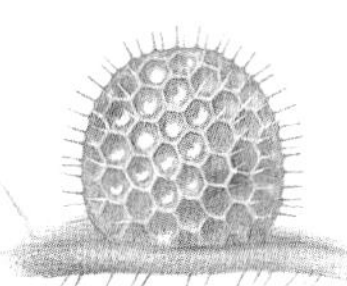

▲ **Egg** [× 15]
Hairy, with a shell divided into cells; laid singly on Honeysuckle leaves.

▲ **Young caterpillar**
Rests on midrib of Honeysuckle leaf, and strips the leaf on either side.

▲ **Colour variant**
Aberrant *obliterae* form, in which the white bands are obscured.

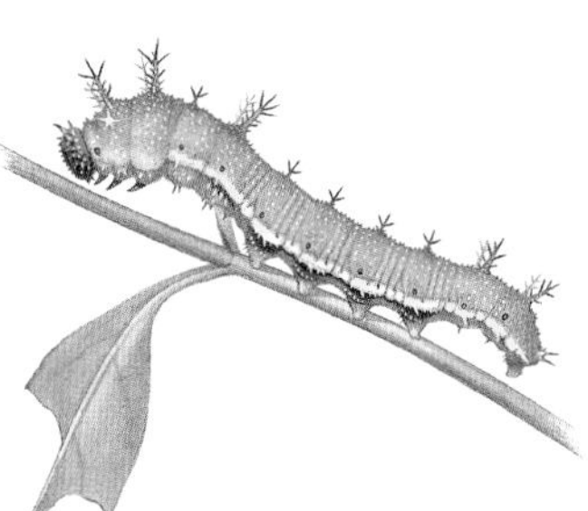

◀ **Mature caterpillar** [× 1½]
After emerging from hibernation, the caterpillar feeds from April until June.

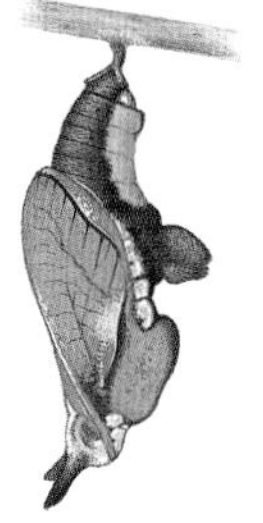

Side view

View from above

◀ **Chrysalis** [× 1½]
Suspended from a pad of orange silk attached to a stem or leaf.

▲ **Resting adult**
Adult at rest on a leaf of Honeysuckle.

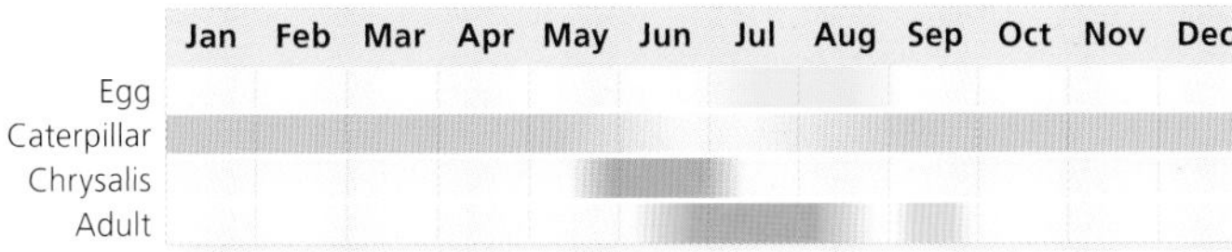

are elusive and spend much of their lives out of view in the canopy, where they bask on oak leaves and drink aphid honeydew. From time to time, some are attracted to ground level, where they cluster on the blossoms of Bramble in sunny rides, or sip at water or the dissolved salts in dung in the early hours of the morning.

Both sexes have a swift and elegant flight, consisting of minor whirrings of the wings punctuated by long, graceful glides. Their manoeuvrability is remarkable. The Victorian entomologist, C G Barrett, described this better than most:

> 'This special grace seems to arise from the habit of the insect of sweeping down over the trees to near the ground, then rising a little, gliding into every opening, taking the curves of the branches and high bushes with the perfection of ease, sweeping rapidly away, or soaring over the trees, to return in a few minutes to the same spot.'

The males, especially, seem to soar in this way, launching themselves off large trees and other vantage points in sunny glades and rides. The females are more retiring and circumspect, especially when immature. Their mating is seldom seen, but I understand it occurs out of sight in late afternoon, high on the woodland canopy.

Egg-laying and young stages

When the time is ripe, fat fecund females descend to ground level where they flutter slowly through shadier woodland, searching for spindly growths of Honeysuckle on which to lay eggs. They choose straggly, trailing wisps that do not flower – the ones that dangle in the half-light around a tree-trunk or beneath a bough, that clamber in the vegetation overhanging ditches, or weave through the shrubs bordering rides. Each female lays very quickly once she has found a suitable spot. She alights with a flutter sideways onto the leaf, arches her abdomen until it touches the opposite edge, and then glues a single glassy green egg to its upper surface.

White Admiral eggs have a curious cellular shell with projecting hairs, and look rather like miniature sea-urchins. They are fairly easy to find, and hatch after one or two weeks. The young caterpillar is even simpler to spot, and is probably the easiest stage in which to find this butterfly. It feeds in late summer and autumn by nibbling the egg-leaf back at right angles from the tip, but leaving the midrib intact and projecting. It rests for most of the day on the far tip of the midrib, a tiny, crusty brown object hidden on a silk 'faecal cushion' that is adorned with its droppings, with further knobs of excrement trapped in a skein of silk over its back. This unsavoury disguise is abandoned after a week or two, when the little caterpillar sits exposed on its midrib.

As autumn progresses, each caterpillar manufactures a shelter, called a hibernaculum, out of the Honeysuckle leaf, and it is in this that it spends the winter. It first fastens the stalk to the stem with silk to prevent the leaf being shed, and then chews off the outer two-thirds of the leaf, before folding the remainder over the midrib.

The small caterpillar emerges to resume feeding in spring, resting at first on the base of a rosette of sprouting leaves, where its spiny brown body is well camouflaged against the leaf-scales. It turns green after the final moult, and is extraordinarily hard to spot as it rests, with head and tail end raised in a curve, on a tender Honeysuckle leaf. The chrysalis is even more beautiful, and hangs suspended like a half-dead leaf, twisted and beginning to brown, and bedewed with beads of water or rain. It is not difficult to find, dangling beneath the Honeysuckle on which it has recently fed. It hatches after two to three weeks.

Numbers and climate

Little was known about the ecology of this lovely butterfly until Ernie Pollard studied the colony at Monks Wood, in the 1970s. He found there was one key period in the life cycle that determined whether numbers would be high or low in any year. This hinged mainly on the weather in June: large numbers of both chrysalises and full-grown caterpillars are devoured by birds, but in warm seasons these stages are short, so fewer are killed. Conversely, in cool summers they remain vulnerable for very much longer, and only a few survive.

Pollard's research, coupled with the known egg-laying preferences of the females, does much to explain the extraordinary fluctuations in this species over the centuries. When first recorded it was quite widely distributed in southern England, extending as far north as Lincolnshire. It was, however, distinctly local in most counties – for example, only four colonies were known in Dorset in all the years up to 1913. It was plentiful enough in some large woods, notably the New Forest, but only in the major timber-growing regions is it likely that conditions were shady enough for breeding. The White Admiral does not thrive in freshly cut coppices, the condition of most woods in Victorian days and before.

An expanding habitat

A sharp contraction of range occurred in the second half of the 19th century, perhaps partly due to the very cool summers in some of those decades, and by the early years of the last century its status was at an all-time low. A happier state of affairs was to follow. During the 20th century, commercial coppicing virtually ceased, and a great many broadleaved woods and old coppices were neglected or left to grow as mature timber crops. Although the sun-loving fritillaries rapidly disappeared under this lack of management, the abandoned woods provided ideal breeding grounds for an insect of partial shade. Moreover, the massive conifer plantations of the past 75 years have provided good breeding conditions for the White Admiral, albeit for a relatively short period in their middle age, when the trees are beginning to close up but before the woodland floor is cast into deep shade.

The White Admiral responded swiftly to capitalise on this increase in habitat, at least in some parts of the country. By 1942, White Admirals had expanded northwards to Lincolnshire, and had more or less reoccupied their former range. The spread in other directions was more gradual. In Dorset, colonies were reported to be steadily increasing from 1913 to 1923, yet by the 1950s only 47 new sites had been located, with just one in the Blackmoor Vale, its current stronghold. By the late 1970s and early 1980s, however, we found the butterfly in about 100 localities in Dorset, and it is still spreading slowly westward. In the past decade, its range also expanded in East Anglia, Leicestershire and Lincolnshire.

This is not to say that the White Admiral has done nothing but expand in recent decades. Numbers have been very low in some years, notably those with cool summers, and the long-term trend has been for population sizes to become lower in established colonies over the past 30 years, despite wonderful numbers in the warm summer of 2013. It is probably being shaded out of many conifer plantations as these mature, and further losses are likely. Against this, the butterfly's fate seems secure in many southern broad-leaved woods and coppices. At the time of writing, it is quite widespread, and to be found in almost every slightly shaded wood of reasonable size, from Dorset to Kent and as far north as Oxford, with scattered, more localised colonies breeding in Devon, Somerset, Gloucestershire, the southern Midlands and East Anglia. This represents a most heartening turn-around from the White Admiral's status a century ago, and some recompense for the loss of most sun-loving butterflies from English woodlands.

Purple Emperor

Apatura iris

There are few greater thrills for any naturalist than to stand beneath an oakwood in high summer and watch male Purple Emperors soaring and wheeling above the canopy. They are large, pugnacious butterflies that glide effortlessly through the air with short flicks of their wings, flashing every shade of purple in the sun. But spectacular though they may be, these are scarce and elusive insects. Only seldom does the casual naturalist see one, even in its strongholds of south Wiltshire and the western Weald.

The 'noble Fly'

The combination of beauty, rarity and elusiveness quickly elevated the Purple Emperor to become a great prize among early lepidopterists. Butterfly literature is littered with hyperbole about 'this noble Fly' and 'Royal Game'. And understandably so, for no-one forgets his first sighting of this insect or can fail to be delighted by its young stages. Moses Harris, in the 18th century, was more restrained than most when he wrote of his 'unspeakable pleasure' when, having received a caterpillar from 'an ingenious Aurelian', the resultant chrysalis hatched out as a male Purple Emperor. Later, in Victorian times, the Revd Morris was less inhibited. This extract is from two pages of suitably purple prose that were devoted to the capture of his first three specimens:

> 'The 19th of July, 1852, must ever be the most memorable one... for on that day did I first see the Emperor on his throne – the monarch of the forest clothed in his imperial purple... One! two!! three!!! "Allied Sovereigns"... I hope that Her Most Gracious Majesty [Queen Victoria] has no more profoundly loyal subject than myself, and I may therefore relate that, while plotting and planning "an infernal machine" against His Imperial Majesty's liberty and life... in the shape of a fifty-foot net, and without any reference therefore to what is now going on in France, or any allusion to the career of Louis Napoleon, my toast that evening after dinner was (with as much sincerity as in the minds of the French), *"Vive l'Empereur!"*.'

There was no incongruity to the mind of a Victorian collector about his love of this butterfly on the one hand and his very strong desire to catch and kill it. 'You can't think how I put my whole soul into egg and butterfly collecting when I'm at it, and how I boil over with impotent rage at not being

Distribution
A scarce and elusive woodland species that is often overlooked. Mainly found in Wiltshire, Hampshire and the western Weald.

▲ **Male upperside**
Purple sheen visible only when wings are at certain angles to the light; colour variation rare.

▲ **Female**
Lacks the purple sheen; white markings are bolder than in the male.

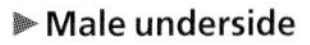

► **Male underside**
The undersides of both sexes are similar.

▲ **Egg** [× 15]
Laid singly on upper surface of a sallow leaf.

▼ **Caterpillar** [× 1½]
After hibernation, green coloration is gradually resumed.

◄ **Hibernating caterpillar**
Young caterpillar hibernates beneath a bud or a forked twig.

▼ **Feeding adult**
Males sometimes feed on the salts in moist ground or in animal droppings.

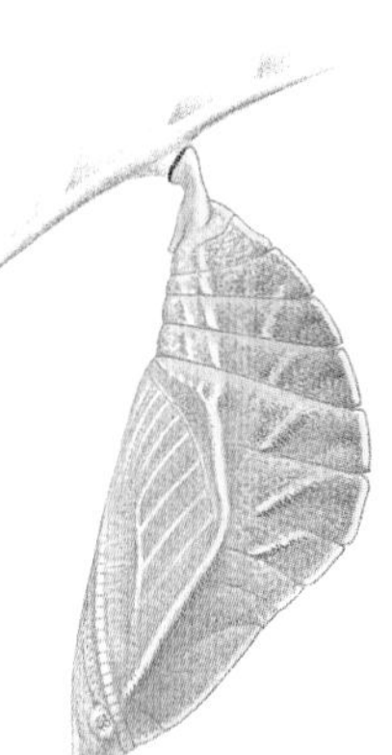

◄ **Chrysalis** [× 1½]
Camouflage strongly resembles the leaf of a sallow.

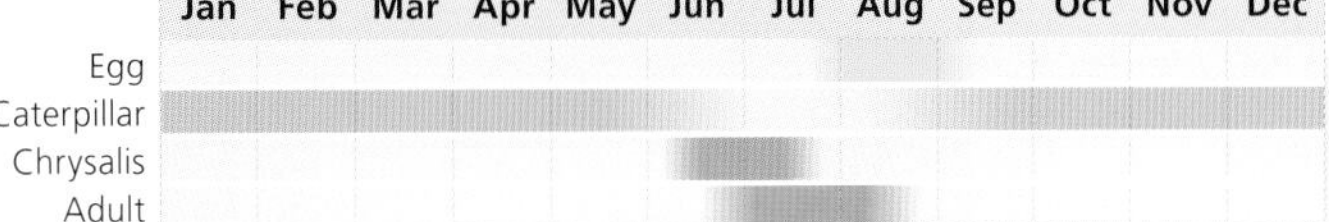

able to attain the object of my desires,' wrote Frederick Courtney Selous when rain had prevented him from collecting Purple Emperors, some years before he found fame as a big-game hunter and as the prototype of Alan Quatermaine. 'I... raced across and took him easily. Strange to say, as soon as I had caught him, I felt a pang of sorrow I hadn't got *two*,' wrote another Victorian on catching his first Emperor. But most that was written concerned methods of catching this butterfly. Of particular interest were the respective merits of using a long-handled net to sweep males one by one off their favoured perching posts, or of capitalising on their love of putrid flesh and excreta by luring them down. For it was soon discovered that, in some years at least, male Purple Emperors descended to gamekeepers' gibbets, and could 'easily be captured while enjoying the luxurious juices of a dead cat, stoat or a rabbit, or of a seething mass of pig's dung'. There was much discussion about the allure of different baits. It was generally held that fox and dog droppings were much superior to those of deer and horse, and a large number of male Emperors was caught in this way.

Honeydew, sap and salts

Butterfly-collecting has largely been replaced by watching and photography, but the agreeable eccentricity lives on. A glance, for example, at the July 2009 blog on the website www.thepurplempire.com shows a breakfast feast set in a ride in Fermyn Wood, Northamptonshire, to lure down male Emperors. Five tables are bedecked with tablecloths and laid with bowls of pungent attractants, including Pimms, horse manure, Stinking Bishop cheese and seven types of shrimp paste. *Plus ça change...*

Although we welcome the shift to butterfly-watching, the collectors, even in their heyday, are unlikely to have inflicted serious harm on this species. For it is impossible to catch more than a small proportion of a population, and the adult Purple Emperor is sufficiently mobile to spread back from one wood to the next. It is not, however, a migrant or a vagrant. Rather, a single colony breeds over a very large area of land, which generally encompasses several woods and copses.

Adult Purple Emperors emerge from late June onwards, and survive well into August. Although they hatch at extremely low densities, the sexes soon meet by flying to high points in the neighbourhood, where the males establish territories or leks. Mating occurs in the late morning or afternoon, leaving the early morning and evening for other activities. Neither sex visits flowers, but instead drinks the aphid honeydew that coats many broadleaved trees in summer. Honeydew is merely the part-digested sap of plants, and both male and female Purple Emperors will also fly long distances to major bleeds on trees, where the sap weeps and congeals from cankers and damaged tree-trunks.

Only the males descend to feed on other matter. This occurs mainly in the morning, especially between 10am and 11am. Their love of dung and rotting flesh has been described, although they are by no means always attracted to these. I have more often found males drinking dew along rides or nearby roads, from which they probably obtain vital salts. Puddles are a particular attraction, especially during dry weather, when they probe deep into the mud with their sturdy yellow proboscises. This is a sight worth seeking in our better Emperor woods, for although you are unlikely to find more than one or two individuals in a morning, they usually drink for several minutes before taking off, and are always extraordinarily tame.

Colours by refraction

Mud-puddling males drink with closed wings if the sun is bright, displaying beautiful undersides with an eye-spot and pattern in the outer half that must look frighteningly like a raptor – say a Hobby – to any smaller bird hopping near it on a treetop. But in weak light the wings are held open, exposing their purple sheen. This is usually shown over the whole wing in illustrations, but in the wild the wings often appear black, or have just a small portion gleaming purple. For it is only when light refracts through the scales at certain angles that this glorious colour appears. The exact tint varies with the intensity of light and with the angle of the wing to the sun. Males battling high above the canopy in strong sunshine emit a cascade of purple and blue flashes, quite unforgettable to anyone who sees it.

Male Emperors take to the air from about 11am onwards. They first indulge in comparatively low-level soaring in and out of the upper branches of trees. Large distances are covered in the process, and they often move from one wood to the next, if possible flying along tall hedgerows and shelter-belts. They gradually ascend as they fly, and by noon begin to gather in the highest-lying woodland in the neighbourhood, often but not always on the summit of a hill. There they establish territories high up on prominent trees. One tree, or a small group, is always occupied throughout the season, year after year. This is the so-called 'master tree' or 'master oak', which in reality is seldom a single tree and can be any species; fewer than half those in current use are oaks.

Master trees

There is a considerable mystique – and much flummery written – about master trees, a good deal of it summarised in a fascinating book, *Notes and Views of the Purple Emperor*, by Heslop, Hyde and Stockley (1964). Many conflicting claims have been resolved in recent years, thanks to the patient observations of Ken Willmott, on which much of this account is based. It was widely believed, for example, that no colony of Purple Emperors could survive the felling of its master tree. This is certainly not the case, but there may be a grain of truth in that the males may move to a different focal point in their neighbourhood or wood if no suitable canopy structure is left to attract them.

Everyone who knows this butterfly has his favourite master tree, which is often a closely guarded secret. The earliest collectors soon learned that not only is the same tree used every year, but that particular branches, or even leaf clumps, are continuously occupied. Matthew Oates recently examined more than 50 examples, and considers the term 'master tree' a misnomer; he instead suggests 'sacred grove', in view of the fact that male leks are generally situated on a group of three or more trees, typically situated near hilltops or on minor summits. These are rarely the tallest trees in a wood, for those are too windswept. Instead, the males prefer elevated nooks in the canopy that provide some shelter from the wind, especially those situated along rides or on the north- or east-facing edges of a wood.

Most master trees are classed as secondary territories, and are used intermittently by males. In contrast, primary territories are occupied daily in suitable weather, usually by more than two males and sometimes by half a dozen. There, each male perches with wings half open in a sheltered but prominent nook, facing outwards and slightly downwards, ready to launch himself at females or intruders. Silver-washed Fritillaries are rapidly seen off, but the battles with rival males are remarkable. The combatants climb high into the sky, clashing their wings and flashing purple all the while.

Males also swoop down to investigate other shiny objects, and are well known for striking car windscreens that gleam upwards in the sun. I was once the object of attack in a glade beneath a master oak in Surrey. I was wearing a white top at the time, and a male immediately descended to fly in tight circles around my chest, encircling me four or five times before soaring back to his perch. It was a memorable experience, for the sun was shining and the colours that accompanied every wing-beat were breathtaking when seen so close.

Each male occupies the same perch all afternoon, although he regularly abandons it to make brief circuits around the treetop or wood. But if he disappears for too long from a key point, another male soon arrives to take his place. This, too, was known to early collectors. Barrett, a century ago, wrote of a 'Mr Tugwell [who] has taken half a dozen in succession from the same branch, and has known dozens to be secured from it in different years'.

The female Purple Emperor is more elusive than the male. She lacks his purple colour and skulks among the treetops, seldom being seen even when egg-laying. On her first morning of adulthood, she flies gradually towards the highest point in the neighbourhood, ending up at the master tree if she has not already been accosted *en route*. In practice, many are discovered at an earlier stage, for in the late morning males indulge in 'sallow searching', which may involve flying hundreds of metres from the master trees to patrol the favourite stands of sallows used for breeding. Wherever she is found, the female receives excited attention and, if receptive, leads the male to a suitable platform to pair, high in the canopy.

Egg-laying

After mating, the females disperse throughout the woods and copses of their breeding sites, in search of sallows on which to lay. They are pestered by males when they fly over high ground, but elude them by drifting straight down to the ground. The male spirals round each female as she falls, but soon loses interest and regains his perch, leaving her to resume the search. The eggs are laid mainly on the broad shiny leaves of Goat Willow (or Sallow), although the narrower leaves of Grey Willow are also often used. Certain types of tree are much preferred. These are always in accessible situations in woods, such as on the edges of rides and glades, or in the undulating canopy of a fairly young plantation. The females prefer the larger sallows in particular and, above all, trees that are partly, but not entirely, shaded. Many that are chosen are on the north side of a clearing. Few sallows fit these criteria, and the females fly swiftly through the shaded branches hunting them out.

Once she has found a suitable sallow, a female 'strikes' the tree (as old collectors used to say) by flying at it and disappearing deep inside. Perhaps a dozen eggs will then be laid, singly and well spaced on the upperside of shaded leaves, usually near the crown of the tree. The great majority are laid above eye-level. The long hours spent searching for eggs are repaid when you eventually find one, for they are highly attractive, each resembling, as one Danish naturalist wrote, a rum pudding. Although the egg is green when laid, it soon develops a purple band around the base, and hatches after about ten days. The tiny caterpillar then crawls to the tip of its sallow leaf and sits facing inwards towards the stalk.

Early development

Caterpillars adopt the same position on a leaf for the rest of their lives. They are extraordinarily well-camouflaged after the first moult, when an attractive pair of horns appears. At first they nibble the edges of the leaf beside them, causing characteristic damage that is quite easy to recognise in late

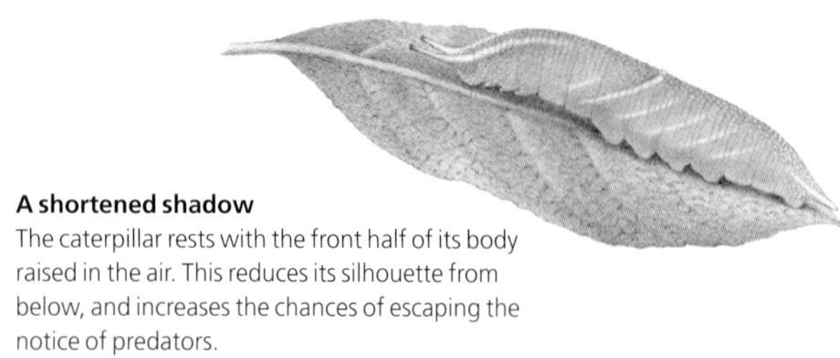

A shortened shadow
The caterpillar rests with the front half of its body raised in the air. This reduces its silhouette from below, and increases the chances of escaping the notice of predators.

summer. By early November they make their second skin change, and soon turn a muddy brown. It is at this time that they desert the leaf for the winter, to hibernate on a pad of silk spun over the crotch of a forked branch, where again they are extremely well camouflaged. Caterpillars resume feeding in spring, first on expanding leaf tips, then adopting the characteristic perch at the tip, as illustrated, once leaves are large enough to support them. Finally, in early June each crawls some distance from its 'seat leaf' and forms a beautifully camouflaged pale chrysalis beneath the undersurface of another leaf.

Emperor woods today

The main localities for the Purple Emperor are in southern England. Scattered colonies once occurred in the more wooded districts of Wales, East Anglia and south Lincolnshire, but these declined in the 20th century and the butterfly is probably extinct in all those regions. It is, however, greatly overlooked, and has recently been rediscovered by patient searching for adults on master trees in many former woods in Middlesex, Hertfordshire and Kent, and in Rockingham Forest, in Northamptonshire. We can be confident in writing, however, that although the Purple Emperor survives, and may recently have expanded, in certain woodlands in east Devon, Somerset, Northamptonshire, Nottinghamshire and Kent, it is a very local and scarce butterfly in these peripheral counties to its core range.

The stronghold of the Purple Emperor is central southern England. Its extensive breeding grounds encompass virtually every wood and copse in the western Weald, including many on the Surrey sandstones and chalk, in addition to the famous populations on Wealden clay. The species is not, of course, uniformly abundant throughout this area, but nevertheless breeds almost everywhere that suitable sallows grow. This includes many conifer plantations for the first 20 years or so after planting.

The populations of the west Weald extend more widely into east Hampshire, where again the butterfly can be found in almost every large wood or heavily wooded district. This area extends to the New Forest, although numbers are extremely low there, as was ever the case. Far superior are the woods a little to the north, around the southern and western borders of Salisbury. These currently support some of the highest densities of Purple Emperor in the country, and there are several spots where one can guarantee to see the males. Fortunately, some of the largest populations in this region breed in nature reserves, where the requirements of Purple Emperors are a key feature of the management policy. Finally, fine populations exist north-east of Oxford, extending well into Buckinghamshire, breeding in the patchwork of conifer plantations and deciduous woods that form the remnants of Bernwood Forest.

Camberwell Beauty

Nymphalis antiopa

The Camberwell Beauty is one of the most spectacular butterflies to be seen in the British countryside, and one that has always elicited great excitement. Formerly known as 'The Grand Surprize', it is a scarce and irregular immigrant from the Nordic countries, although for about 150 years the belief prevailed that it was native. British specimens came to be regarded as the greatest of all prizes among early dealers and collectors, and fierce debates raged about their distinguishing features. For over a century it was held that true British Camberwell Beauties had 'a superior whiteness of their [wing] borders' compared with their counterparts on the Continent. Cream-bordered specimens were considered to be foreign and valueless, with the result that at least one 'British' specimen in the Natural History Museum was found to have had its borders painted white.

Antiopa *years*

Although the Camberwell Beauty has a reputation for excessive rarity in Britain, roughly 2,000 sightings have been reported since the first two adults were seen in 1748, flying around willow trees along Cold Arbour Lane, near Camberwell, then a village rather than the urban sprawl of today. It has been a regular visitor ever since, with at least one sighting in nine years out of ten since 1850. Nevertheless, it is unusual for this butterfly to occur in any numbers. The great '*antiopa* years' of 1789, 1793, 1820, 1846, 1872, 1947, 1976, 1995 and 2006 caused a sensation partly because they were so unexpected. In 1995, more than 500 Camberwell Beauties were seen; in 2006 the tally was 273, although both were years when many naturalists were on the alert for butterflies, surveying for *The Millennium* and later *Atlases*.

Distribution
Most sightings are in late summer, on the east coast of England and Scotland. Seen far inland in exceptional years.

Occasional
Rare

Climate and breeding

Despite being the second-most frequent of our vagrant species, the Camberwell Beauty is the only one for which there is little or no circumstantial evidence of breeding in the British Isles. The nearest we have is the following account by William Lewin, at Faversham:

'The middle of August, 1789, I was surprised to see two of these elegant flies, near Feversham [*sic*], in Kent; one of which I thought it great good fortune to take; but in the course of that week I was more agreeably surprised with seeing and taking numbers of them in the most perfect condition. One of my sons found an old decoy pond, of large extent, surrounded with willow and sallow trees, and a great number of these butterflies flying about, and at rest on the trees, many of which appearing to be just out of the chrysalis, left no room to doubt, that this was a place where they bred.'

Later authors, myself included, consider it more probable that Lewin was fortunate enough to encounter a spot where a flight of pristine immigrants had landed, for this is a butterfly that migrates in flocks. In *The Aurelian Legacy*, Michael Salmon cites evidence of another mass flight, in 1820, when William Backhouse reported 'vast numbers of this species strewing the sea-shore at Seaton Carew [near the mouth of the Tees], both in a dead and living state'.

It is nonetheless puzzling that very few of the adult Camberwell Beauties that reach our shores hibernate successfully, and that probably none goes on to breed. It may be that our winters are too mild and wet for more than a handful of individuals to survive. Thus the vast majority of British sightings are made in August and September, with small numbers often seen in April following an *antiopa* year. An exacerbating factor may well be that, like all species which hibernate as adults, the Camberwell Beauty does not pair until the following spring. There may simply not be enough survivors within a particular region for any to find a mate.

There appears, at first sight, to be an abundance of suitable breeding sites for the Camberwell Beauty in the British Isles. The butterfly lays its eggs in large clusters around twigs, like its close relative, the Large Tortoiseshell. Grey Willow is the main foodplant in Scandinavia, although other willows, poplars, elms and birches are also used. The caterpillars are gregarious and highly conspicuous. They wander to form solitary chrysalises in late July, and the adult emerges about three weeks later. It is both the latest and largest of the northern European tortoiseshells to appear, and loops among the trees in a bold, flitting flight, similar to that of a Peacock. Most of the butterflies enter hibernation quite early, and although they roam over the entire countryside the following spring, numbers fluctuate enormously, even in Scandinavia.

Arrivals in Britain

British immigrations appear to be an extension of these periodic build-ups, wanderings and expansions. Records in the British Isles were certainly few and far between from 1954 to 1972, when Camberwell Beauties were absent from Denmark, whereas in 1995 large numbers were recorded simultaneously from the Netherlands and Denmark.

The great influx of 1995 is believed to have originated in Poland, whereas the 2006 immigration came – as is more typical – from Norway. The flight patterns of this most recent expansion have been carefully monitored and reconstructed by Chris van Swaay, Richard Fox and Jaap Bouwman. It began with a substantial build-up of numbers in Norway during July 2006, a period of exceptional warmth. The weather changed abruptly in August, when cool northerly winds helped to blow thousands of Camberwell Beauties south. The first were spotted in the Netherlands, where 102 sightings were made on 6th August, with five being seen on the same day in Britain. The wind then veered eastwards for a few days, presumably taking the next wave of emigrants to Sweden and Poland. A further shift to north-easterlies brought the main influx to Britain: up to 20 individuals a day were reported between 15th August and the end of the month, predominantly from the east-coast counties, especially Norfolk and Suffolk. During the last third of August until early October, many butterflies spread inland, and single sightings were made in Ayrshire and County Down during September. By the end of the season, a total of 273 Camberwell Beauties had been reported from 33 British counties, with roughly half of them from Norfolk, Suffolk or Yorkshire. This compares with 654 and 131 sightings, respectively, in the Netherlands and Denmark during the same period. True to form, although a few individuals successfully overwintered in Britain and were seen the next spring, there is no evidence of breeding, and numbers in 2007 were comparatively low.

Despite several unsuccessful attempts to establish this species in Britain, the vast majority of sightings are clearly of immigrants. As in 2006, most are seen in eastern counties. These typically encompass the entire coastline, including Scotland, but the majority are from Kent and East Anglia, with Norfolk having the highest tally. The Camberwell Beauty does, however, wander inland more than most other rare immigrants: as in 2006, it was recorded as far west as Ireland in 1995, when there was also a record from Shetland.

▲**Male upperside**
Pale wing margins vary in colour from lemon-yellow to white. Sexes are similar.

▲**Male underside**
Dark underwings camouflage the butterfly when hibernating in hollow trees.

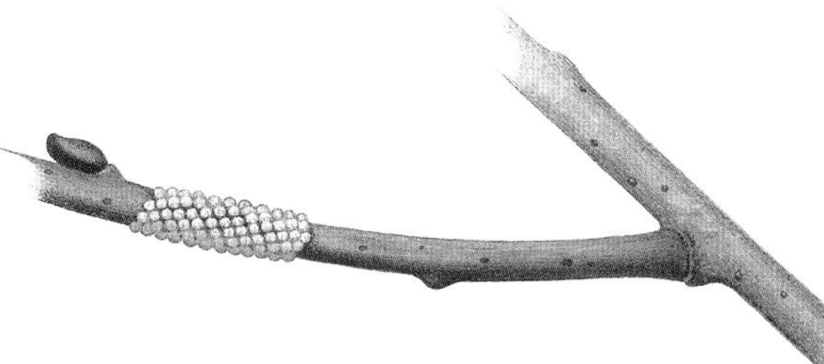

◀**Egg batch** [× 1½]
Batches of eggs encircle the twigs of willows, poplars, elms and other trees.

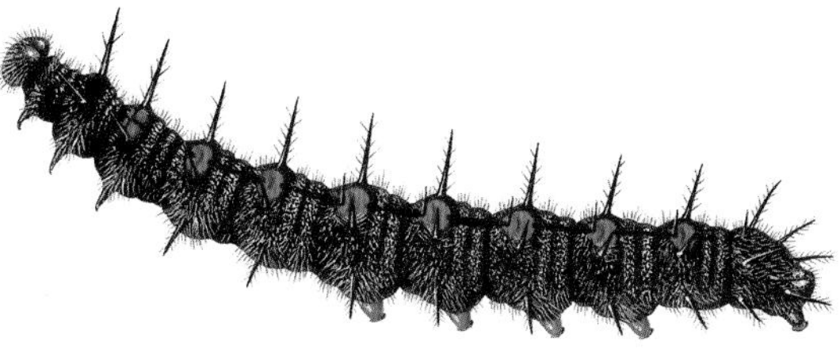

▲**Caterpillar** [× 1½]
Conspicuous, spiny caterpillars live gregariously in groups of up to 150.

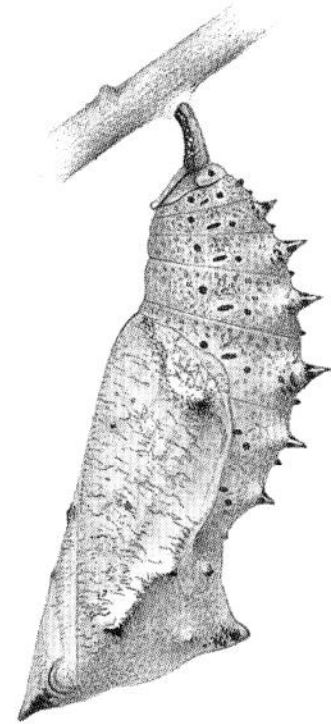

◀**Chrysalis** [× 1½]
Suspended from a stem or twig, some distance from the foodplant.

	Jan	Feb	Mar	Apr	May	Jun	Jul	Aug	Sep	Oct	Nov	Dec
Adult												

Red Admiral

Vanessa atalanta

This familiar insect is one of the largest and strongest-flying butterflies in our countryside. It is not, however, a resident, and the individuals that remain in the British Isles die in all but the mildest winters. The vast majority of Red Admirals seen in spring and early summer are immigrants. These arrive in varying numbers every year from southern Europe to breed and spread throughout the country, reaching the more distant Scottish isles and the remotest mountain chains in good years. Numbers increase throughout the summer until, one or more generations later, there is a return migration to the butterfly's winter breeding grounds in southern Europe. Although this is a powerful insect, with strong wings and a broad, muscular thorax, it is nevertheless extraordinary to consider that the black-and-scarlet butterfly seen gliding around a British garden may have flown there from the Mediterranean coast.

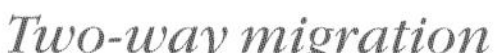

Two-way migration

Much has been learnt about the two-way migration of Red Admirals in recent years, thanks to admirable studies made in different parts of its European range by Constanti Stefanescu, Oskar Bratttström, Rebecca Nesbit, Robin Baker, Ernie Pollard and others. Until recently, it was widely believed to have two broods in central and northern Europe, and then to migrate south where it hibernated as an adult. That is largely incorrect. For example, 'Pollard counts' of adult numbers in Britain and north Europe show that there is little synchronisation of broods, such as occurs with the Small Tortoiseshell. Rather, there is a gradual build-up through the year, starting with rare sightings on warm January and February days, with real numbers being seen from May onwards, culminating in a peak during the autumn. Then, as temperatures fall, the fresh adults switch behaviour and start flying south without mating, an event that varies greatly in timing from one year to the next, but ranges from late August in cool seasons to mid-October after an Indian summer. Vast numbers can be seen near the south coast at this time of year. In October 1904, F W Frohawk counted up to ten Red Admirals to the square yard, stretching across a mile of Sussex downland, each drinking nectar from Devil's-bit Scabious to build up resources before crossing the Channel.

Distribution
A regular immigrant from southern Europe; common throughout the lowlands of England, Wales and Ireland in most years.

▲ **Male upperside**
Sexes are similarly marked on both upper- and underwings.

▲ **Male underside**
Brightly marked forewings contrast with camouflaged hindwings.

▲ **Colour variant**
Variation in the amount of red and white occasionally occurs.

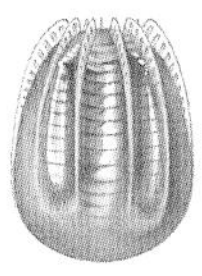

◀ **Egg** [× 22]
Laid singly on the upper surface of a leaf.

▼ **Caterpillar** [× 1½]
Solitary, usually found on nettles; lives within a protective tent of leaves.

Black form

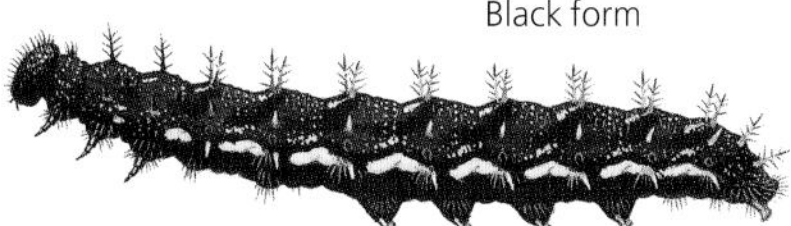

Yellow form

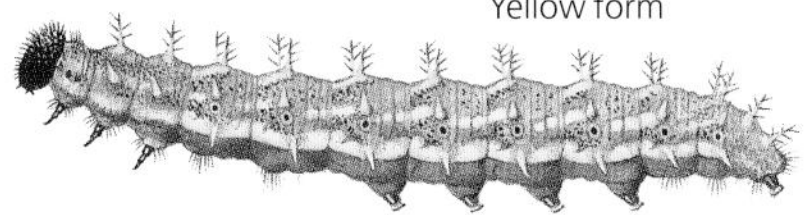

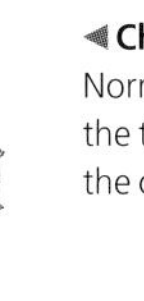

◀ **Chrysalis** [× 1½]
Normally formed within the tent constructed by the caterpillar.

▲ **Feeding adult**
In late summer, adults are attracted into gardens to feed on flowers and rotting fruit.

	Jan	Feb	Mar	Apr	May	Jun	Jul	Aug	Sep	Oct	Nov	Dec
Egg												
Caterpillar												
Chrysalis												
Adult												

In Scandinavia, where the return migration is well studied, the trigger to fly south is a fall in temperatures. Most Red Admirals wait for clear, still conditions, and then fly within a metre of the ground to escape the wind. Most take a straight line south, using the sun as a compass, or follow the sheltered lee of local topography in that general direction. On reaching the sea in south Sweden, they gain some altitude and change course to fly by the shortest possible route to Denmark, about 24km away to the west. On a clear day the Danish coastline is just visible to human eyes, but how the Red Admirals detect it is unknown: they perhaps perceive the difference between sea and distant land through the 'water sky effect', an optical phenomenon, detectable by insects, that results from light becoming linearly polarised after reflectance from the water surface.

There are reliable reports, recently confirmed by radar, that many other Red Admirals fly at high – but not mid-range – altitudes on their migrations south (and later north), gaining height by soaring on thermals before taking advantage of strong following winds. It now seems certain that the adult Red Admirals of north Europe, including those that breed in the British Isles, fly to central or southern Europe each autumn in a single journey that probably lasts two to three weeks. Thus, near the Catalan coast, weekly counts of Red Admirals are a mirror image of those in Britain: the butterfly is non-existent during summer, but increases rapidly through October and early November.

Hibernation myth

It was once thought that adult Red Admirals hibernated in the warmer south, but this is another myth. When the adults return to the Mediterranean, nettle beds throughout the lowlands are recovering from the parched conditions of summer, and have developed lush, fresh leaves ideal for breeding. As in Britain in spring, the male Red Admirals establish territories on every hillock, where they perch and court the virgin female immigrants that fly there. The females then lay eggs in late October and November, after which both sexes die. The resulting caterpillars develop slowly in the cool weather of winter, and produce fresh adults in March and April. As far as is known, these virgins desert their lowland breeding sites immediately, and migrate north again across Europe. In fact, the pattern is a little more complicated, for some adults make an altitudinal instead of a latitudinal migration. They fly to cooler altitudes high in the Pyrenees, where a parallel sequence of breeding occurs, followed by a mini-migration in autumn down to the lowlands.

Although a few Red Admirals may seem to hibernate in the British Isles during mild winters, these generally settle in exposed places, such as on tree-trunks or under branches, and usually perish. It is likely that these are late-emerging adults that became trapped after the sudden onset of conditions too cold for flight. The few adults that are seen on sunny days from December to February are believed to result from late caterpillars that were able to develop in warm spots.

The main northwards migration of the Red Admiral generally lasts from April to June, and brings varying numbers to Britain every year. The males establish territories on any elevated mound, and the females start to lay almost immediately after mating, gradually spreading northwards through the country. There is some evidence that we receive fresh influxes from the Continent through the summer. Some adults reach Ireland and Scotland every year, and are often quite common there. It is usual to see a few individuals even in Shetland; some occasionally reach Lapland. Then the Red Admirals again start to fly south, and by mid-September the traffic is more or less one-way.

Magnificent flying machine

As befits one of the true European migrants, this butterfly is a supreme flying machine. Its aerodynamics were neatly studied by Adrian Thomas and R B Srygley, who trained adults to fly to and from artificial flowers in a wind tunnel, and took high-speed photographs of trails of smoke disturbed by different wing beats. Unable to explain how such a heavy insect could fly so fast, or with such manoeuvrability, using conventional steady-state aerodynamics, they found that the Red Admiral employs an array of unconventional aerodynamic mechanisms that enable it to take off, generate uplift and propulsion, and manoeuvre and land. These include two types of leading-edge vortex that create turbulent flows of spinning air over each wing, active and passive upstrokes, an ability to rotate the wings during wing beats, wake capture (whereby the butterfly dips its wings into the swirling air generated by the previous wing beat to recapture some of its energy) and the fondly named 'Weis-Fogh clap-and-fling'. The last mechanism involves clapping the wings together above the butterfly's body and then flinging them apart. As they fly open, air is sucked in and creates a vortex above each wing, but during the clap the vortex is created on the other side, resulting in a rapid circulation of air that is capable of lifting much heavier bodies than is achievable from conventional leading-edge vortex effects.

Fluctuating numbers

With the uncertainties of a double migration, and a generation bred under Mediterranean climates, it is not surprising that the number of Red Admirals in the British Isles fluctuates markedly from one year to the next. The only discernable pattern is that numbers have gradually increased since accurate monitoring began, 33 years ago. This is a long-term trend: there is no correlation between the butterfly's populations in consecutive years, and very large numbers may be followed by exceedingly low counts the following year, as occurred in 2003-2004. Ernie Pollard argues plausibly that the overall increase in 'our' Red Admirals results, in fact, from improved survival on their Mediterranean breeding grounds, leading to greater numbers of immigrants arriving from southern Europe in most years.

Attracting Red Admirals into gardens

In Britain, the Red Admiral is especially conspicuous in autumn, not only because numbers are higher but also because it visits garden flowers to fuel up for the journey south. As with all summer butterflies, buddleia is a favourite, and it is worth pruning a bush hard in spring to ensure a late flush of flowers to attract this magnificent butterfly. Windfall apples and split plums are still more attractive; the butterfly will flutter for long hours in the early autumn sunshine to drink with wasps from fallen fruit.

Growing nettles rather than flowers increases the likelihood of attracting Red Admirals to gardens earlier in the year. While the Red Admiral is less selective than its relatives, and will lay eggs on semi-shaded shoots, the nettles will be ignored if they are relegated to a shady corner. It is essential that they are vigorous, with plenty of fresh green tips, rich in nitrogen. Many nettles are in ideal condition in May, but most Red Admiral eggs are laid in August, when plants tend to be elderly and grey. It is worth cutting patches back in late June to encourage fresh growth, as well as watering them during dry weather. The fresh shoots will also attract other nymphalids.

Eggs, caterpillar and chrysalis

An egg-laying Red Admiral is quite easy to spot owing to its rambling, fluttery flight, when it fussily investigates every nook and cranny, giving the impression of a broody hen. When a suitable nettle has been chosen, she quickly lays a green egg on the growing tip, on the upper surface of a tender leaf. One egg is laid per tip, placed in exposed spots if the nettle is half shaded, or sheltered within a patch if the clump is in full sun. Moses Harris, writing in 1766, had his own interpretation:

> 'I have often perceived her, when about to lay an Egg, creep in among the Nettles; which I imagine is not only to place the Egg from the heat of the Sun, but likewise to see if those Nettles are frequented by Ants, these Creatures being very destructive to caterpillars'.

A nettle-leaved tent
In its final stage, the caterpillar fells a nettle top, spins the leaves together, and lives inside this tent, feasting on the tender crown.

Attractive though this may be, there is not a shred of evidence to support Harris's theory of ant avoidance.

Not every egg is laid on Common Nettle. Small Nettle and its close relative Pellitory-of-the-wall are also used. But whatever the plant, the behaviour is the same. The egg hatches after about a week and the tiny caterpillar spins a tent around itself by fastening a tender young leaf double with silk. Over the next four weeks it lives and feeds in a succession of nettle-leaf tents – or 'Places of Security', as Moses Harris called them – each progressively larger and distinctively shaped. Tents are extremely easy to find, and often several occur up a single stem, looking rather like rows of stuffed vine leaves. When opened, the plump black-and-yellow caterpillar will be found on a silk pad, its body curled in the shape of a figure 6.

The caterpillar frequently abandons its one-leaf tent for the final bout of feeding, and constructs another that is equally conspicuous in its own way. It selects a vigorous-growing spike, which it fells by chewing two-thirds of the way through the stem, about 15cm below the tip. The upper section topples, but remains attached to the plant, and the caterpillar crawls into the downward-pointing terminal leaves, spinning them together in a cocoon. It may also form its chrysalis in this contraption if sufficient room exists. More often, it spins two or three large nettle leaves together and pupates inside the shelter. No British butterfly has a chrysalis that is easier to find in the wild than this species, and they are easy targets for parasites. Watch, in particular, for the black ichneumon wasps that buzz around the nettles, injecting eggs via a long ovipositor through the wall of the tent deep into soft, young chrysalises. The wasps are as interesting as they are beautiful, and few will begrudge the loss of some butterflies for the sake of this elegant parasitoid.

Painted Lady

Vanessa cardui

The Painted Lady is one of the world's most successful butterflies. A cosmopolitan species, it breeds mainly in warm, well-drained habitats, has no hibernating stage in its life cycle, and its caterpillar perishes at temperatures lower than about 5°C. As a result, there is probably no place in Europe where this species is permanently resident, yet in most summers it is one of the commonest butterflies across the Continent.

Like the Monarch and the Silver Y moth, this butterfly is a true trans-continental migrant. In a typical year, three generations of Painted Ladies breed across Europe every spring to early autumn, before migrating *en masse* to live for the next two to three generations on winter breeding grounds around the desert edges of North, West and even sub-Saharan Africa, as well as in Arabia. Then, come spring, a fresh brood of adults teems north again across Europe as their winter habitats desiccate, reaching some part of the British Isles every year and Orkney two or three times a decade. In exceptional summers they penetrate far above the Arctic Circle, making a round trip of 13,000km that takes six generations to complete.

Patterns of migration

Our knowledge of this remarkable migration has been transformed in recent years thanks to the pioneering work of Constantí Stefanescu, to the ground records of 10,000 European volunteers, and to Rothamsted's vertical-looking radars, which can identify not only the species but also the flight directions of migrating insects at altitudes of up to 1,200m. It was once thought that the Painted Lady resided permanently in northern Africa, with offshoots temporarily colonising Europe in most years only for their offspring to perish with the onset of winter. This concept, known as 'The Pied Piper Effect', made no evolutionary sense, yet nonetheless was widely believed. Today, we know not only that the entire summer population of Painted Ladies migrates southwards in autumn, but also that – like the better documented Silver Y moth – those generations of Painted Lady that breed in Europe can reach much higher numbers than those emanating from the south. For example, in the glorious migration of 2009, Jason Chapman's radar counts show that an estimated 11 million Painted Ladies immigrated to the British Isles in the spring of that year and no fewer than

Distribution
A regular immigrant. Usually common throughout lowland England and Wales; found almost anywhere in its best years.

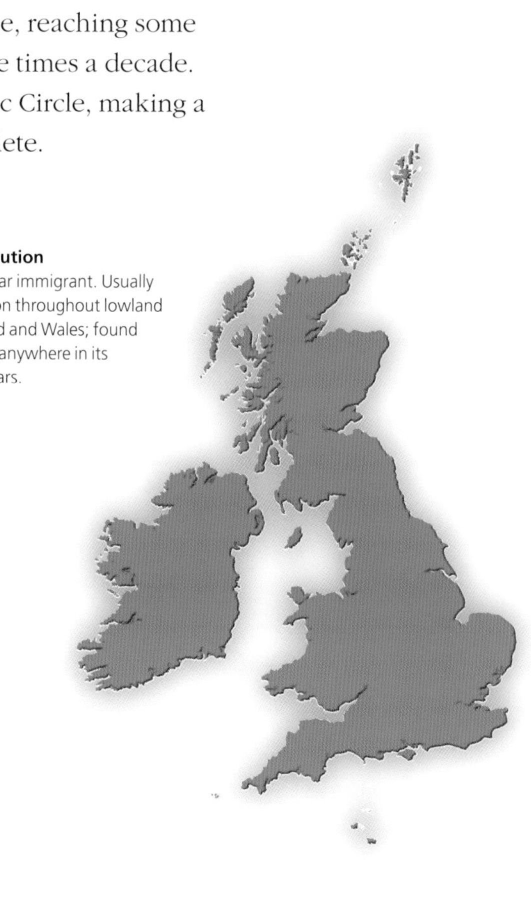

▲ **Male upperside**
Background coloration fades to dull orange-brown as the butterfly ages.

▲ **Male underside**
Sexes are similarly marked on both upper- and underwings.

▲ **Colour variant**
Rare *rogeri* form, an extreme example of aberration.

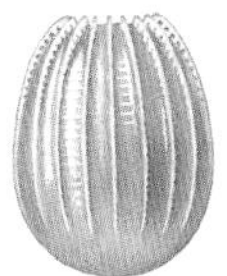

▲ **Egg** [× 22]
Laid singly on various species of thistle; other foodplants occasionally used.

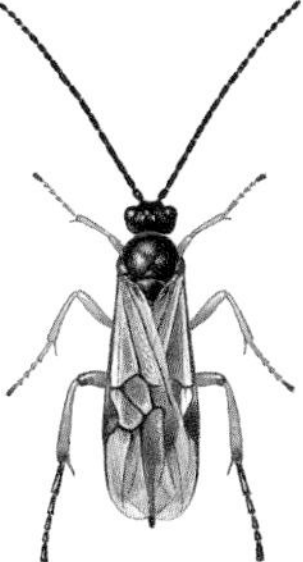

▲ **Parasitic wasp**
Small wasps, such as this braconid *Microgaster subcompletus*, cause the death of many caterpillars.

▲ **Caterpillar** [× 1½]
Solitary, living inside a succession of tents made of folded leaves.

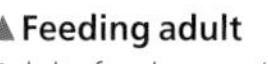

▲ **Feeding adult**
Adults feed on a wide variety of flowers, including scabiouses.

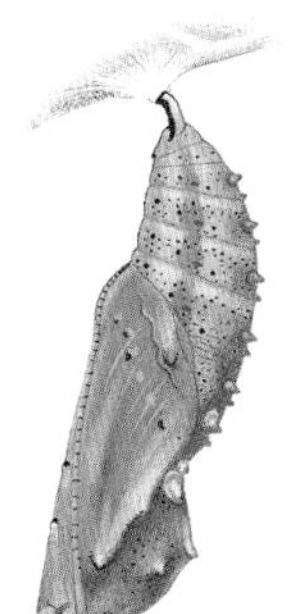

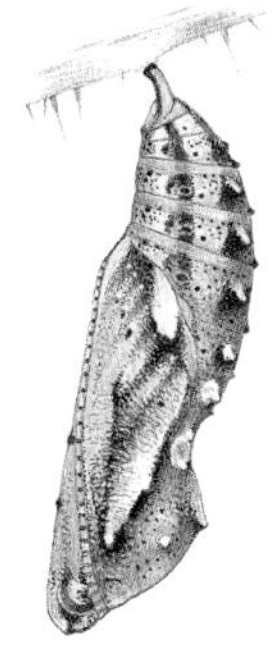

◀ **Chrysalis** [× 1½]
Suspended under vegetation; there are two main colour forms.

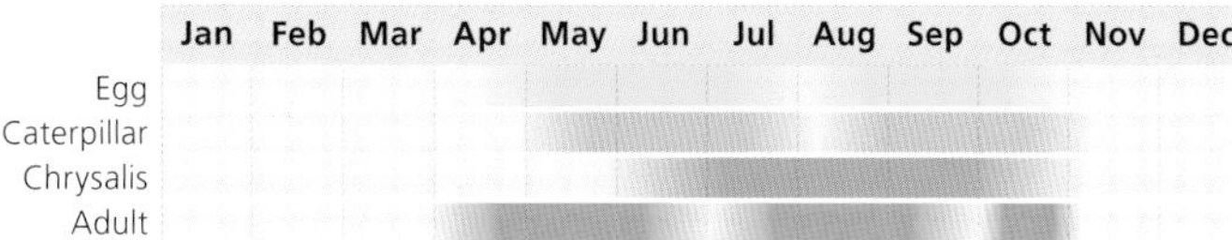

21 million butterflies left for Africa the following autumn. In the case of the Silver Y, a trivial proportion of adults is lost during migration: the same is likely to be true of the Painted Lady, making Europe a net exporter of this butterfly in high-density years.

Prior to Stefenescu, very few observations have been made of the start of the flight to Europe. The best account, by a Mr Skertchly, occurred in the desert, beside the Dead Sea, in 1869:

> 'From my camel I noticed that the whole mass of the grass seemed violently agitated, although there was no wind. On dismounting I found that the motion was caused by the contortions of pupae of *V. cardui*, which were so numerous that almost every blade of grass seemed to bear one. The effect of these wrigglings was most peculiar, as if each grass stem was shaken separately – as indeed was the case – instead of bending before a breeze… Presently the pupae began to burst, and the red fluid that escaped sprinkled the ground like a rain of blood. Myriads of butterflies limp and helpless crawled about. Presently the sun shone forth, and the insects began to dry their wings; and about half-an-hour after the birth of the first, the whole swarm rose as a dense cloud and flew away eastwards towards the sea. I do not know how long the swarm was, but it was certainly more than a mile, and its breadth exceeded a quarter of a mile.'

It will be noted that these butterflies swarmed before they had mated, and were all of exactly the same age. What triggers such migrations is not clear, but they could be responding to day-length, rising temperatures, or a deterioration in larval food. It is equally unclear why the Painted Lady swarms stop flying to settle, mate and breed in a particular district.

European migrations of the Painted Lady vary enormously in their size and exact flight-paths. In many years, vast numbers are seen streaming north across the Mediterranean, flying by day or night, flitting their wings powerfully and then gliding, so that they can rest on the wing. There have been many descriptions of these flights. Most are of low-flying butterflies that typically cruise at 20-25km per hour in the calmest available air, a metre or so above ground level, flying incessantly in a straight line over every hill or obstacle as it is encountered. This is not, however, the preferred option. Radar traces recently revealed that, should the wind be in the right direction, the majority of individuals ascend to high-altitude, typically 500m and up to 1,000m above the ground, to exploit fast-flowing tailwinds. This enables them to travel at 50kph on their desired course, allowing each butterfly to cover several hundred kilometres a day. High-altitude cruising appears to be especially common during the autumn migration from Europe to Africa, which no doubt accounts for the belief that no significant return migration occurred.

Most migrations of Painted Lady occur at low densities on fronts that may be hundreds of kilometres wide; a few fly in compact swarms no more than 3-4.5m across, making a continuous procession that can take several hours to pass. Although individual Painted Ladies fly at least a metre apart, the greatest flights cast a shadow over the ground and contain millions of adults. High-altitude flyers are, perhaps, less able than the Red Admiral or Silver-Y to compensate precisely for wind direction, for they arrive less predictably in different parts of Britain at any time in spring and summer, depending on their point of origin and the prevailing wind direction.

Landfall in Britain

British immigrations of Painted Ladies generally contain smaller numbers than the great continental swarms. They arrive as early as January or February in some years, when regular sightings are made along the south coast. It is more usual, however, for the main swarms to arrive in late May or June, often followed by further immigrations later in the summer that mingle with the offspring of the early arrivals. One well-documented series of immigrations occurred in 1980, when westerly winds first carried them on an anticyclone from Spain or North Africa, depositing large numbers in the Western Isles and west Wales. This was followed by another large immigration in late July, this time arriving on easterly winds along the east coast of England and Scotland. These were probably the offspring of earlier flights to central and eastern Europe.

In typical years, however, the migrations are more uniformly south-north, and it is southern England that receives the most butterflies. This was the case with both of the great Painted Lady years of recent times: the mass flight of 1996, then considered to be the largest arrival for 150 years, and the even greater swarms of 2009, when more than 10 million butterflies crossed our southern coastline during the weekend of 23rd-24th May. Happily, large immigrations have been commonplace in recent decades, rather more so than in 1766, when Moses Harris wrote that 'These Flies are not very common ... yet there are particular Seasons when they are very plentiful, which happens once in about ten or twelve Years.'

A tented caterpillar

Having made land, the Painted Ladies fan out through the country and can be seen in any flowery habitat, although by far the greatest concentrations occur on dry, open terrain such as heaths, dunes and downland. Each male soon establishes some form of territory, choosing a warm patch of bare ground, sand or rock on which to settle. In weak sunshine he basks with wings spread wide, pressed firm against the dusty ground, flying up to investigate any passing object. Females are courted with a zest that marks most of the Painted Lady's activities.

Both sexes feed frequently on flowers. The females then disperse throughout the countryside in search of thistles on which to lay eggs. They use a wide range of species, but nevertheless have favourites. Martin Warren found over 200 caterpillars on Musk Thistle on the south Dorset downs in 1985, but only one on Spear Thistle and none on Creeping Thistle, both of which were abundant on the same sites. I suspect the freshness of the thistle leaves, as well as the prominence of the plant, are important, for others have found the latter two species used frequently on other occasions. That was certainly my experience in June 2009 when, in my rather weedy fields on the central Dorset downs, I counted 498 well-grown Painted Lady caterpillars in two hours (and estimated about 50,000 to be present on the 12ha of land), all on Spear and Creeping Thistles. The large majority were on tall specimens: none occurred on thistles shorter than 15cm in height, but from five to 20 caterpillars fed on every growth that was more than 1m tall. Regardless of height, I also found nearly twice as many caterpillars on Spear as on Creeping Thistles. A similar situation existed in the Cotswolds in 2009, where the large Woolly Thistles were scarcely used at all. Female Painted Ladies occasionally lay on other plants, including mallows, Globe Artichoke and nettles, although in much smaller numbers, and the caterpillars fare poorly.

The Painted Lady is thus one of the few butterflies that can breed in intensive farmland in Britain, for even the best-kept farms have the occasional thistle patch. The small green egg is easy to find in Painted Lady years. Although laid singly, there may be several eggs per plant, mainly on the uppersides of the thistle leaves. Each hatches after about a week, and the small caterpillar crawls to the undersurface of its leaf. Here, it spins a pad of silk and feeds on the lower cuticle, leaving a silvery patch of epidermis above, which is also quite easy to spot. It becomes much more conspicuous as it grows, for it constructs a tent out of one or more folded leaves, fastened firmly by silk, and lives within this, eating all but the sturdiest ribs and spines. Successive tents are spun as it grows older, leaving very obvious abandoned quarters consisting of skeletons of spines, silk and desiccated droppings. Finally, after its last skin moult, the caterpillar lives openly on its thistle before forming a beautiful burnished chrysalis. This exists in two main colour forms, and is suspended in a larger tent of vegetation, rather like that of a Red Admiral.

The egg, caterpillar and chrysalis together last four to six weeks in warm weather, so several generations of Painted Lady can occur in the Britain Isles during a hot summer. This, together with the erratic immigrations, means that it is impossible to predict where or when Painted Ladies will be seen. As a general rule, they are less common than the Red Admiral but more common than the Clouded Yellow, although this is by no means invariably so. Peak numbers are usually seen in July to September, but may be much later in some seasons.

Junkbug and other names

The Painted Lady is by no means restricted to its European and African populations. Similar migrations occur in all other continents except South America, leading to the unattractive alternative names in the USA of the 'Cosmopolitan Butterfly' and 'Junkbug', the latter a derisory term used by collectors to reflect the extraordinary abundance of this butterfly in some years. It was estimated, for example, that one Californian swarm contained 3,000 million adults.

European names are more agreeable. Apart from an abortive experiment with 'The Thistle' in the 18th century, the butterfly has been known as the Painted Lady or 'Papilio Bella Donna' in Britain since 1699, 'Belladonna' in Scandinavia, 'La Belle Dame' in France, and 'Bella Dama o Cardero' (Beautiful Lady of the Thistles) in Spain.

Small Tortoiseshell

Aglais urticae

There are few more common butterflies than the Small Tortoiseshell in the British Isles, and few that are held in greater affection. Except, that is, in Scotland, where it was once known as the 'Devil's Butterfly' or, scarcely more flatteringly, as the 'Witch's Butterfly'. These unworthy names have long disappeared, but some Scots continued to plough an independent furrow by calling it the 'Red Admiral'. Welsh, English and Irish names have been more constant, the 'Tortoise-shell Fly' and 'Nettle Tortoiseshell' being the only old variants. In Germany it is known as 'Kleiner Fuchs' (Little Fox).

This is one of our loveliest butterflies, familiar to all who have gardens. It can vary slightly in size and appearance, depending on the temperature. Cool conditions result in unusually dusky individuals, very like the beautiful forms of Scandinavia; these are especially common in the north of Scotland after a cold summer. On the other hand, those emerging after hot summers tend to be brighter, with larger orange marks. These and more bizarre forms can easily be produced in captivity by rearing the chrysalises at different temperatures. Extremes are sometimes also found in the wild.

Broods and migrations

Temperature and day-length affect the number of adult emergences each year. There are usually two distinct broods in the south, one emerging in June and July, with their more numerous offspring appearing at any time from August to mid-October. The second brood goes into hibernation quite quickly, reappearing in the first warm days of spring to mate and breed, and often surviving well into May. However, after a cold, late spring, many first-brood (mid-summer) adults also enter hibernation, saving their eggs for the following year. The trigger for this switch is the changing day-length: early-emerging adult butterflies respond to lengthening hours of daylight by producing a second brood, but only in the south, for there is also geographical variation in the response. Small Tortoiseshells from further north are less apt to lay eggs under a given day-length, and in central

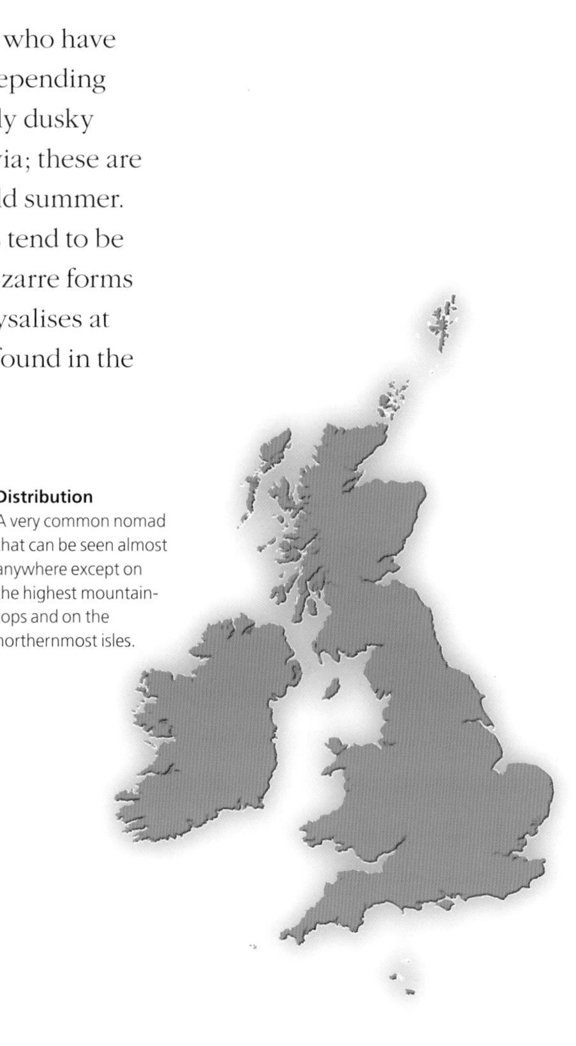

Distribution
A very common nomad that can be seen almost anywhere except on the highest mountain-tops and on the northernmost isles.

▲ Male
The markings of both sexes are identical. Cool conditions produce duskier butterflies, which are seen particularly in northern Scotland.

▲ Female
Aberrant form *semi-ichnusoides* produced by high temperatures during the pupal stage.

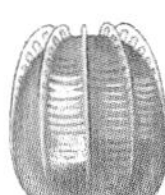

▲ Egg [× 15]
Laid in large clusters; more than one female may lay eggs on the same leaf.

▲ Male
Aberrant *lutea* form.

▲ Hibernating adult
Camouflaged underwings conceal the butterfly during hibernation.

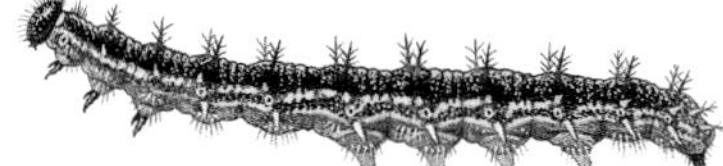

▲ Caterpillar [× 1½]
Bright colours warn that the body contains poisons. The amount of yellow and black is variable.

◄ Feeding adult
Michaelmas-daisies are a favourite nectar source in gardens in late summer.

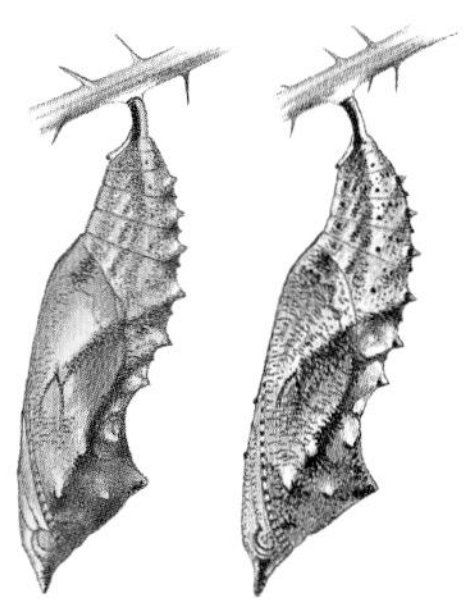

▲ Chrysalis [× 1½]
The colour varies from golden to brown, depending on the pupation site.

	Jan	Feb	Mar	Apr	May	Jun	Jul	Aug	Sep	Oct	Nov	Dec
Egg												
Caterpillar												
Chrysalis												
Adult												

Single brooded in Scotland (emerging in August).

Scotland, where summer days are long, no amount of exposure to light will induce a second brood. Thus there is just one generation of adults a year in Scotland, which emerges in July, hibernates, and reappears in March.

The adult Small Tortoiseshell does not live in distinct colonies, but travels through the countryside laying eggs wherever suitable conditions are encountered. Typical tortoiseshells move a kilometre or two a day, but some fly much further, and even migrate across the English Channel. Migratory flights have been recorded in both directions, and tired butterflies often rest on lighthouses up to 50km out to sea. F W Frohawk described one influx, coming from the south-east towards Swanage, Dorset, in August 1929. Although adults flew in singly, this went on for a week, and the eventual number of immigrants was considerable. Other large immigrations have been reported, such as when hundreds of dead specimens were found off Flamborough Head, in Yorkshire. But this is unusual: there is little doubt that the vast majority of Small Tortoiseshells are resident, even on the remoter Scottish isles, although breeding has yet to be confirmed in Shetland. The fact that tortoiseshells from different regions respond in different ways to day-length provides further evidence that there is no mass mixing of adult butterflies across the British Isles.

Feeding and hibernation

The Small Tortoiseshell's behaviour varies much with the time of year. Those destined to hibernate are preoccupied with feeding, and large numbers gather in gardens from midsummer onwards to gorge on buddleias, valerians, hebes, Ice Plant, Michaelmas-daisies and other nectar-rich flowers. Much feeding also occurs on wild flowers, and adults gather in any habitat where these abound.

Having fed for many days, adults next search for places to hibernate. More than any other butterfly that overwinters in this stage, the Small Tortoiseshell seeks roofs, outhouses and other human habitations, and many are thwarted by well meaning householders who 'rescue' and release them outside. This is a mistake, except from centrally heated rooms where the butterflies will awake when the heating comes on in the autumn, and then waste essential fat reserves by fluttering around windows.

Early autumn is always a dangerous time for the adult Small Tortoiseshell. Individuals that continue to visit flowers through September and into October become increasingly cumbersome and slow-flying as their weight increases and the temperature cools, making them ill-equipped to evade tits and other insectivorous birds. Like the Peacock, the Small Tortoiseshell has three lines of defence. Camouflaged underwings help to conceal the butterfly at rest, but when a bird approaches it tries shock tactics by flashing open the bright upperwings and producing a hissing noise. This species, however, lacks the Peacock's startling eyes, and by falling midway between possessing a really effective alarm system or instead relying on near-perfect camouflage, as does the Comma, the Small Tortoiseshell appears to have the worst of both worlds (see page 191). Tits, in Sweden, kill many more Small Tortoiseshells than either of the other vanessids.

So much for the dangers of staying outside in late autumn. Equally vulnerable are the butterflies that hibernate early, in mid-September, for many are killed when comatose by mice. Christer Wiklund found that about 60% of Small Tortoiseshells that hibernated early were discovered and carried to slaughter places, where their disembodied wings were left in neat piles. In contrast, only a third of Small Tortoiseshells entering roofs in October were killed, perhaps because the mice themselves were slowing down or had hibernated.

Courtship territories

The individuals that survive awake in the first warm days of spring. Their behaviour is exactly the same as the adults in the summer brood that go on to breed. At night, both sexes roost deep in nettle patches, where their dark underwings merge into the earthy background. Males become active around 10am, and spend the next two hours flying through the countryside, with frequent pauses to feed and bask. This ceases around midday, after which they concentrate on mating. Each first establishes a territory, generally near a large nettle bed and almost always in a sheltered, sunny spot on the south-facing side of a hedge or wall. He waits there, basking on the ground, soaring up to investigate any passing Small Tortoiseshell. Most turn out to be males, for there is considerable competition for perching places. Both spiral high into the air, then dive and ascend again, each striving to fly just behind and above his opponent. This is repeated until the intruding male flies off, leaving the victor to return to the exact spot he had left.

If, after about 90 minutes, no female has flown past, the male abandons his post and sets up another territory which he occupies until about 4pm. Any passing female is pursued at high speed, although it can take up to three hours before she succumbs. Mating finally occurs deep inside a nettle patch.

The search for egg-sites

Fertilised females spend the rest of their lives searching for egg-sites. As with most butterflies – and especially those that lay large batches of eggs – great care is taken to choose the ideal plant. Common and Small Nettle are both used, but the Small Tortoiseshell has a strong preference for laying eggs on young, tender plants that are growing in full sunshine on the edges of large nettle beds. There is generally no shortage of suitable sites in spring, but the summer emergence depends on the re-growth of young shoots, where patches have been disturbed, cut, grazed back or trampled earlier in the year.

Even in spring, the vast majority of nettles are rejected as unsuitable. This includes most that are deliberately left or planted for this butterfly in gardens, for they typically consist of small growths that are tolerated in a shady corner of the vegetable garden. To be of use to the Small Tortoiseshell, large clumps of nettle should be planted along the sunniest edges of garden beds, and some should be cut back and watered in early June to produce tender re-growth in July.

Female Small Tortoiseshells congregate on suitable nettle beds, each laying 60-100 eggs at a time under the topmost leaf of an edge plant. They often choose leaves that have already been laid upon, and it is not unusual to find two females egg-laying simultaneously on the same leaf. Great clusters of up to a thousand eggs can sometimes be found, piled several deep and looking like green caviare. Once suitable nettles are known, it is quite easy to find these batches, but remember that the sting of young nettles is fierce.

A new parasite from Europe

The egg-batches hatch after 10-14 days, and the tiny caterpillars spin a dense web of silk over the growing leaf tips. Large batches split into groups of about 200 each. These mass together within the web and, like other species with dusky caterpillars that live gregariously, can regulate their body temperatures with remarkable precision by basking in the sun (see page 222). As each plant is defoliated, they spin new nests on adjoining nettles. Feeding damage is highly conspicuous, as are the nests of caterpillars themselves. After the penultimate skin change, the groups split up and each caterpillar lives alone for the remainder of its life; it is still easy to find, basking on a nettle leaf or devouring the younger growth.

The caterpillars vary in colour from almost uniformly black to yellow with black spots, and are paler when young. All have spines in the later stages, and the curious habit of jerking their heads and rippling their spines in unison when disturbed is probably a warning to predators. They are by no means immune to attack by parasites, however, and large numbers are killed, either by braconid and ichneumon wasps or by tachinid flies. To most entomologists, the latter are the less appealing parasitoid, and are responsible for the greater number of larval deaths. Several species infest the Small Tortoiseshell. Once inside, the tachinid egg hatches to produce a minute grub which feeds on the internal tissues of the caterpillar, literally eating it alive.

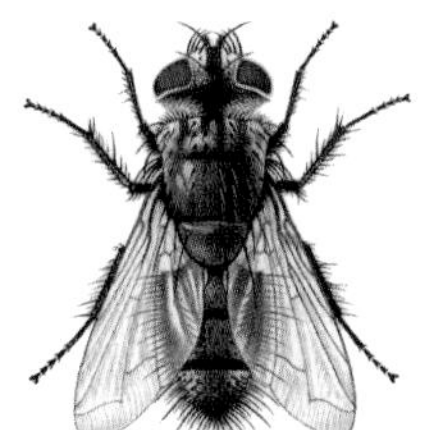

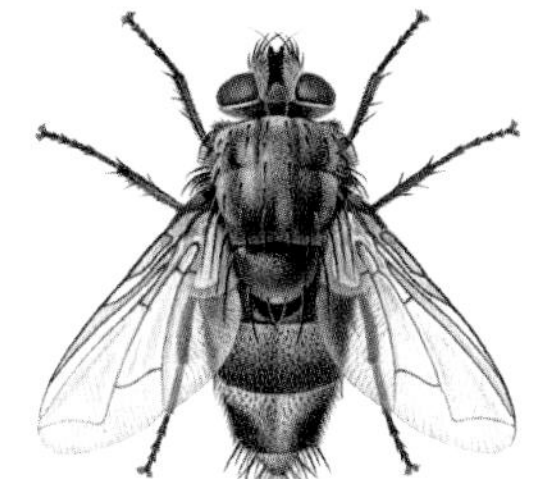

Parasitic flies [× 3]
The grub of the tachinid fly *Phryxe vulgaris* (right) lives within the caterpillar, killing it before pupation. *Sturmia bella* (below) has recently colonised Britain. The adult emerges from the chrysalis.

Until recently, the main parasitoids of the Small Tortoiseshell were the ichneumonid wasp *Phobocampe confusa* and a variety of tachinid flies, including *Phryxe vulgaris*. In the past decade, a new tachinid, *Sturmia bella*, has invaded from the Continent and spread rapidly from Hampshire, reaching Merseyside and Durham by 2009. Unlike the other tachinids, which are also generalists that infest many butterflies and moths, *S. bella* specialises mainly on nymphalids and has been bred from British Commas and Peacocks, as well as the Small Tortoiseshell. All these parasitoids lay their eggs on or into their hosts, apart from *S. bella*, which lays its eggs near to caterpillars on the nettle leaves; in due course these are ingested whole. A rough and ready means of distinguishing *S. bella* is that it is the only tachinid to emerge from the Small Tortoiseshell's chrysalis, its white maggot descending to the ground on a characteristic silk-like thread, where it forms a brown pupa. *Phryxe vulgaris* and related tachinids emerge from their host's caterpillar.

In the 15 years since *Sturmia bella* was recorded in Britain, Small Tortoiseshell numbers have declined by roughly half. There has been much popular speculation that this is cause and effect. In a comprehensive study of the incidence of *S. bella* up to 2009, Sofia Gripenberg, Owen Lewis and Nia Hamer found that the parasitoid was indeed more widespread and was killing more caterpillars than had been suspected; in 2008 and 2009 respectively, about 18% and 11% of the caterpillars collected produced this fly. Moreover, the butterfly's decline was considerably more severe in parts of the country that had been invaded by *S. bella* than those that had not. On the other hand, synchronised lesser declines by the Small Tortoiseshell occurred in localities and northern regions where the parasitoid is unknown, as well as across north Europe as far east as Poland, where the tachinid is not a newcomer. Gripenberg and colleagues conclude that the jury is still out on *S. bella*, which at the very least may have exacerbated our declines, especially in south-east England.

It is also possible that the unusual weather patterns of the past decade may have driven numbers down. Fortunately, not every recent year has seen low emergences; 2002 saw the fifth highest counts of Small Tortoiseshell in the 33 years since monitoring began.

Fluctuating numbers

Surviving caterpillars disperse to pupate, and chrysalises have been found more than 50m from the nearest nettles, generally suspended in a hedge or on a wall. They vary in colour from lilac-pink washed with copper to dull brown, and all are beautiful. Unfortunately, they too are vulnerable, and many are killed. Birds are believed to be the main culprits.

Small Tortoiseshell numbers have fluctuated greatly from one year to the next since 1976, when systematic monitoring started, but the overall trend has been strongly downwards. The five great years for Small Tortoiseshells were 1982, 1984, 1992, 1997 and 2002. As described, numbers were exceptionally low in 2005-2008, but those for 2009 and 2010 suggested a modest recovery in some parts of the country. Alas, they were followed by further declines in 2010 and 2011; at the time of writing, it is unclear whether the large numbers apparent during the glorious summer of 2013 represent more than a temporary gain. Irrespective of whether the new tachinid has influenced recent falls, Ernie Pollard and I found that the butterfly's numbers generally crashed after a hot, dry summer, presumably because few nettles contained the lush nitrogen-rich leaves preferred by egg-laying females. I suspect the droughts of 2005-2007 played a part in the recent declines.

Notwithstanding these declines, the Small Tortoiseshell remains one of the commonest butterflies in town and countryside. The highest densities occur on rich soils where nettles are abundant, but the butterfly is extremely well distributed and can be expected almost anywhere in the British Isles. It breeds on the smallest and remotest islands and has been found at altitudes of 1,200m on Scottish mountains; the highest recorded breeding site was around 330m high.

Large Tortoiseshell

Nymphalis polychloros

This splendid tortoiseshell mysteriously declined six decades ago, and – despite recent sightings – has probably been extinct as a resident butterfly for approaching 50 years. Only about 150 sightings were reported in Britain between 1950 and 2005, and the large majority were almost certainly specimens of the butterfly imported from the Continent, or misidentifications of other vanessids. In contrast, more than ten times as many Camberwell Beauties and Monarchs have been seen in Britain during the past 60 years.

Such figures would have dismayed Victorian collectors, who regarded the Large Tortoiseshell as scarcely worth recording in wooded regions of the south. Edward Hulme, writing a century ago, is typically offhand: '[one] may sometimes come across a patch of spear-plume thistles in an open clearing in the wood, and here we shall probably find in plenty the Red Admiral, the Peacock and the Large Tortoiseshell'. The really noteworthy vanessid in those days was the Comma, a close relative of the Large Tortoiseshell whose increase during the 20th century was as remarkable as the latter's decline. A similar pattern of change has occurred over the same period in Belgium, The Netherlands and, more recently, northern Germany.

Colonies and dispersal

Like its close relatives, the Large Tortoiseshell is a mobile butterfly that flies through the countryside laying eggs wherever suitable conditions are encountered. There seem to have been more or less permanent colonies in the New Forest, Essex and East Suffolk, but in general it would appear in a particular district or wood, breed for a few years and perhaps become abundant, then dwindle and disappear. This made it hard to be certain in the latter years of the last century whether or not the species had entirely vanished as a resident, for there were no fixed sites to check for its occurrence.

Former distribution
An extreme rarity, once widely distributed in wooded regions of England and lowland Wales. Extinct as a resident.

The life of the Large Tortoiseshell most closely resembles that of the Peacock, with one generation of adults a year, emerging in July and August. Fresh adults are active in the morning and late afternoon, when they sup voraciously to build up reserves for a seven-month hibernation period. Like all vanessids, they will visit garden flowers. They also share with the Red Admiral and Camberwell Beauty a passion for sap bleeding from a wounded tree, and for the droplets of honeydew at the bases of leaves. At other times they are hard to approach, and it takes a mere flit of their great tawny wings to send the adults soaring and gliding at high speed around the treetops. This elusiveness was well known to early collectors, who observed that 'it requires nimbleness to take'.

Courtship and egg-laying

Large Tortoiseshells hibernate a week or two after emergence, settling in dry piles of wood, in garden sheds and, almost certainly, in rotten holes in tree-trunks. They emerge the following spring to feed on the flowers of Goat Willow *Salix caprea*, and to bask on warm tree-trunks with wings held as wide apart as possible, reflexed backwards and pressing hard on the warm bark, as if to encircle the trunk. Females spend long periods in this pose, sitting head downwards, about one or two metres up the trunks on the sunny side of a wood.

The Danish lepidopterist, H J Henriksen, gives a fine description of courtship. While the female basks on the trunk, the male at first hunts close to the ground, skimming above the surface with small flits of his wings. He glides between tree-trunks until he notices a female, whereupon he flutters up to within a few centimetres of her. She, however, flies off, with the male following perhaps 30cm behind in a curiously sinuous flight. This can last several hours, but eventually mating occurs on a tree-trunk or under dead leaves on the forest floor.

Most Large Tortoiseshells mate in early April, and few survive the month. Having paired, females fly among the treetops searching for places to lay their large batches of eggs. In Britain, most were laid on the terminal twigs of elms, particularly Wych Elm, sometimes along tall hedges or on suckering scrub, but more usually from 3m up, and occasionally as high as the crown of a large tree. Sallows and willows were also often used as, more rarely, were Aspen, birches, Wild Cherry and Pear. The eggs, which I have seen only on the Continent, are most attractive. They are laid in batches of 100-200 in an encircling band on a young twig, and are always on the sunny side of the tree.

An array of defences

Large Tortoiseshell eggs hatch after two to three weeks, and the little caterpillars quickly spin a web of silk over the tender leaf-tips, which they chew in bouts of group feeding. They remain together until fully grown, basking much of the time on their web. They are prickly even when small, and the full-grown caterpillar has a fearsome array of spines, each capable of piercing human skin. The large groups of caterpillars must be an unappetising prospect for insect-eating birds, and they deter predators further by jerking their bodies in synchrony, making the whole web look like one twitching mass of spines.

The webs of Large Tortoiseshells were once as easily found as the adults in many parts of southern England, and were one of the main ways in which collectors obtained this species. It is ominous that there has not, to my knowledge, been a record of a nest of these caterpillars for 60 years in Britain, despite many efforts recently to find them. This strengthens the belief that the rather few genuine records of adults in recent decades emanate from migrants or from released or escaped captive stock. It is still easy to collect webs in central Europe, and I know several people who have brought them back, and then tried to photograph the resultant adults in natural situations, only for a few to escape. I have also, with some dismay, witnessed the unauthorised release of a large number of French adults onto a west-country nature reserve, and have periodically heard of similar releases.

Large Tortoiseshell caterpillars are generally fully grown by late June, when they leave their trees to form solitary chrysalises. These are situated on dead wood, such as on fences, wooden sheds, and probably among dry sticks. F W Frohawk describes how he watched the caterpillars drop off 'the topmost branches of a lofty Elm', falling one after another to the ground, where they dispersed to search for pupation sites. I have seen the chrysalis only in captivity: it is unspectacular and must be hard to distinguish from a dead leaf in the wild.

Abundance and scarcity

On the Continent, as once in Britain, the Large Tortoiseshell is chiefly found in sheltered wooded districts, especially where elms, willows, poplars or cherries are common. Although most frequent in woods, nests of caterpillars are also seen along avenues and tree-lined lanes, while adults can be found almost anywhere in years of abundance. The species has, however, always been subject to great fluctuations, which have yet to be satisfactorily explained. In Britain, long periods of absence or rarity were punctuated by shorter spells when the butterfly was locally common, and in occasional years it built up to an extraordinary abundance. This last happened in Britain in 1947-48.

In view of these fluctuations, and the semi-migratory behaviour of the adults, it is difficult to tell just how widespread and common the Large Tortoiseshell once was. It has been recorded over the whole of southern Britain and the Midlands, and there are a few records extending up to Aberdeen, but none from Ireland. Although known to the earliest collectors, it was not thought particularly common, yet there were periods in the 19th century when it was com-

▲ **Male**
Markings of both sexes are identical. Colours are less bright than those of the Small Tortoiseshell, and there are no white spots on the forewing.

▲ **Male**
Aberrant form *testudo*, with dusky upper hindwings.

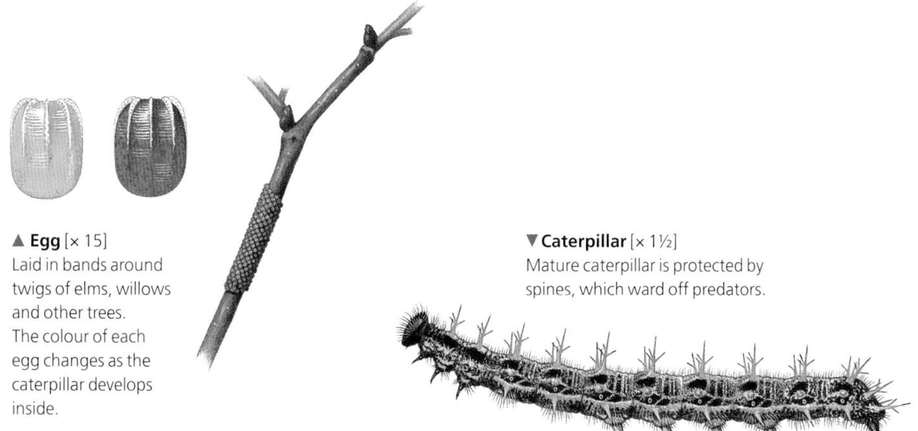

▲ **Egg** [× 15]
Laid in bands around twigs of elms, willows and other trees. The colour of each egg changes as the caterpillar develops inside.

▼ **Caterpillar** [× 1½]
Mature caterpillar is protected by spines, which ward off predators.

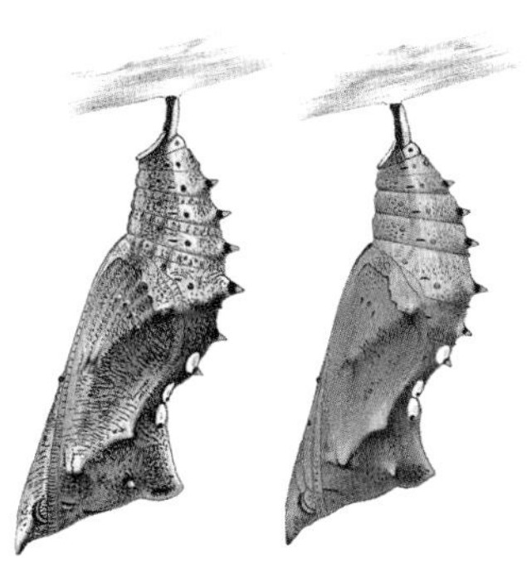

◀ **Chrysalis** [× 1½]
Usually formed on dead wood. Colour of chrysalis may vary.

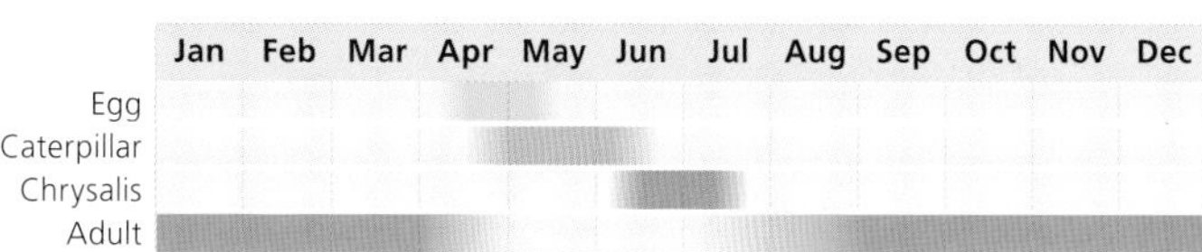

paratively abundant. Its strongholds were south Devon, the New Forest, Essex, east Suffolk and Sussex, where it could be extremely numerous in the outbreak years. W H Harwood wrote, in 1901, that caterpillars were 'so excessively abundant in north Essex and on the southern side of the River Stour that I could have taken hundreds of broods had I required them'.

There was then a period of about 40 years when the Large Tortoiseshell was distinctly scarce over most of southern England, although still common enough around Ipswich. Then, during the warm summers of the mid-1940s, there was one final build-up and Large Tortoiseshells were again locally common, especially in southern East Anglia, Kent, Sussex and the Isle of Wight. This revival lasted five years, before numbers suddenly dipped in 1949. The butterfly had virtually disappeared by the early years of the following decade.

Occasional genuine sightings continued to be made in the next 50 years over wide areas of southern Britain, but there was no wood where one could expect, let alone guarantee, to see this butterfly. Most records were of solitary adults in a single year, although repeat sightings were made over two or more years in north Wales, Buckinghamshire, Hampshire, West Sussex, Wiltshire and the Isle of Wight. Then suddenly, from 2005 onwards, a cluster of sightings was made in south Dorset, Hampshire and, especially, from the Isle of Wight, peaking in the 'unprecedented influx' of 37 genuine sightings in the two latter regions in 2007. Numbers were again high (12 records of 18 individuals) in 2009, but have been lower since, most centred on Woodhouse and Walter's copses on the Isle of Wight

At the time of writing, it is difficult to interpret these – undoubtedly genuine – records. To date, no nest of caterpillars has been found despite assiduous searching, and the consensus is that they are adult immigrants from France. The fact that several records are of battered individuals in spring indicates successful hibernation, even if breeding has yet to be confirmed. Recent records are too many and widely distributed to make the release of captives their likely origin, yet the migration theory is not wholly convincing, either. For example, this is not a species that is picked up with other migrants coming ashore or flying around lightships. Furthermore, if the old British race were an international migrant, one would have expected at least a few individuals to have been recorded in Ireland during the periods of its historical greatest abundance. In fact, there has never been an Irish record, despite the fact that strong colonies once bred in west Wales. In contrast, the Camberwell Beauty, which reaches our shores from Scandinavia, has been reported from 11 different 10km squares in east Ireland. It could be – as with some other butterflies – that different races of the Large Tortoiseshell vary in their powers of dispersal, and that those in mainland Europe are more migratory. If so, the possibility that the British Isles may once again be colonised by this, the most beautiful of the vanessids, remains a realistic hope.

Peacock

Aglais io

This familiar insect has been known as the Peacock, or 'Peacock's Eye', since 1699, but the 'Owl' would be more appropriate, as can be seen if you rotate the illustration and view the adult upside down. Now the twin hindwings assume the image of a Tengmalm's, Scops or Little Owl peering angrily from behind the forewings. The more one looks, the more realistic this appears: the fur and scales between the eyes sweep back towards a dark, rounded profile that is topped by two small ear tufts. The butterfly's cigar-shaped body looks very like a beak.

Shock tactics

This is, of course, no chance resemblance but another example of the remarkable mimicry and defensive markings that have evolved in butterflies. The underside offers a different image, but excellent camouflage nonetheless. By resembling bark, the Peacock is practically invisible when settled on a tree-trunk. This is its first line of defence. The second is to flash the wings open when an enemy comes near, simultaneously rubbing the forewings and hindwings together to produce a warning hiss from the friction between elevated veins at the base of each surface. This is easy to induce if you lunge at a resting Peacock, and the sound is startling enough to human ears. The shock to a mouse must be severe when, after sniffing to investigate an appetising insect odour on the bark, a glaring owl suddenly looms above it, rustling and hissing dangerously.

These shock tactics are also an effective defence against birds. In a fascinating experiment, Adrian Vallin and Christer Wiklund compared the impact of Blue Tits confined for 40 minutes in a large cage containing settled Commas, Small Tortoiseshells and Peacocks, all three of which are highly palatable to insectivorous birds. The Comma was the most perfect leaf mimic of them all, but relied wholly on its ragged undersides for defence; it was invariably the last species to be discovered, but ultimately suffered high mortalities, with nearly three-quarters of adults being found and consumed. The Peacock fared best – indeed, every single Peacock survived – for although it was usually the first butterfly to be discovered, it was also the most intimidating, flicking its wings at a greater distance from the attacking bird than the Small Tortoiseshell. The Small Tortoiseshell was a halfway house, more conspicuous than the Comma

Distribution
A common nomadic species found throughout the lowlands of England, Wales and Ireland. Less common in Scotland.

but less intimidating than the Peacock, and more than 90% of Small Tortoiseshells were eaten.

As if these defences were not enough, the Peacock hiss, created by rubbing together its wing membranes, is overscored with ultrasonic clicks. These startle and deter any bat that encounters the adult butterfly in late summer, perhaps if they enter the same tree hole at the start of the hibernation period.

A familiar sight

We are fortunate that this glamorous butterfly is one of the commonest garden visitors in England, Wales, Ireland and, most recently, Scotland, a species that has not merely held its own during the past 40 years but which has expanded many kilometres northwards in range. It has one brood a year, but the adults live for up to 11 months – hence the need for a sophisticated defence – and can be seen on any sunny day, even occasionally in winter. The emergence of fresh adults generally starts in late July, and continues for another month or two. They then hibernate and reappear the following spring, when they mate and breed, mainly from March to early May. A few individuals linger on until late June, almost overlapping with the next generation. There may be a very small second generation in the warmest localities after hot summers, with caterpillars feeding in September to produce adults in early October.

The behaviour of this butterfly is well understood, thanks to excellent studies by Robin Baker and Christer Wiklund. It is a nomad rather than a true migrant, for while the adults have a tendency to fly north in spring and south in late summer, the furthest any individual had been shown to travel is about 95km. Only seldom do they cross the sea.

Hibernation

A Peacock's requirements alter according to the season. The freshly emerged adults of late summer have two priorities: to feed up for winter and to find a hibernation site. They begin in August, often visiting gardens, hedgerows and any flowery habitat, although the greatest concentrations occur in woods. At first they are quite mobile, but once an overwintering site has been found and autumn progresses, they tend to remain nearby. They will roost in their chosen site by night and feed every day until entering hibernation, usually in early September. By then, huge concentrations may build up, especially if it has been a good summer. I have counted up to 185 adults feeding from Wild Teasels along a single woodland ride.

Peacocks choose dark crevices, sheds or holes in trees in which to hibernate, and also enter houses, although less often than their close relative, the Small Tortoiseshell. Large aggregations have occasionally been found in hollow trees, and this could well be how most Peacocks hibernate. The Victorian lepidopterist, Edward Newman, found 40 together inside an oak, while A B Farn found 'a large assemblage' and noted, as others have since, that the group produced a loud, hissing, snake-like sound when disturbed.

Despite their ability to hide from predators and, if found, to startle them, the long winter period is a dangerous time for adult vanessid butterflies. Wiklund describes a 'slaughter place' in a Swedish attic that contained the wings of 57 Small Tortoiseshells and eight Peacocks. All had apparently been killed by Yellow-necked Mice, which caught the stationary adults, carried them to a favourite spot, and neatly gnawed off their wings before devouring the bodies. The main pulse of winter killing occurs during the first fortnight of hibernation, when roughly half the adult Peacocks studied by Wiklund were eaten. Thereafter, 98% of adults survived, probably because the mice themselves were hibernating, but perhaps also because the most accessible Peacocks had already been found.

The daily routine

Adult Peacocks awake on the first warm days of spring to feed on a wide variety of wild flowers. The males sup mainly in the morning, but soon start wandering and travel, on average, about 500m a day. Their flights stop from around 11.30am onwards, when they begin to establish solitary territories on the ground. The process is complete by about 1pm. Typical territories are situated beneath the sunny edges of a wood or tall hedge, and are usually near a roost site. Corners are especially favoured, for it is here, above all, that females are most likely to fly.

The male swoops up to investigate any dark object that flies over his territory. Even birds may be challenged, and Robin Baker found that the sexes could be distinguished by throwing a clod of soil a metre or so above a Peacock. Females ignore the missile, but males fly up to intercept. Many intruders prove to be other males that are still searching for a territory. Most are seen off with ease, although the chase may continue for up to 200m. Some stay their ground, leading to spectacular dog-fights as each male tries to outmanoeuvre and fly above his opponent. This causes both to spiral upwards to a great height before plunging to the ground. The ascent may be repeated two or three times, but eventually one male achieves dominance and the other speeds off in search of another territory.

Courtship and egg-laying

Females that fly through male territories are pursued with a zeal and persistence that is scarcely deserved. Those that have already mated are especially contrary, and try to lose their suitors on and off during the afternoon, flying across

▲ **Male**
Markings of both sexes are similar. Camouflaged undersides conceal the butterfly while it hibernates in crevices and holes in trees.

▲ **'Blind' Peacock**
A rare aberration that lacks the blue pupils on the hindwings.

◄ **Egg** [× 15]

▲ **Egg batch**
Large batches of eggs are laid on the undersides of Common Nettle leaves, usually in sheltered situations.

▲ **Caterpillar** [× 1½]
The caterpillars are gregarious, but separate before pupation.

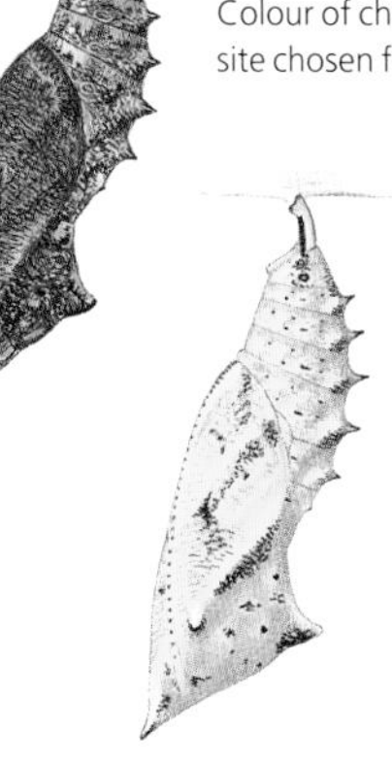

◄ **Chrysalis** [× 1½]
Colour of chrysalis depends on the site chosen for pupation.

▲ **Feeding adult**
In gardens, buddleia flowers may attract many adults, which feed on their nectar.

	Jan	Feb	Mar	Apr	May	Jun	Jul	Aug	Sep	Oct	Nov	Dec
Egg												
Caterpillar												
Chrysalis												
Adult												

open country, then suddenly folding their wings to drop from the sky into vegetation. Another escape strategy is to fly around the back of a tree and hide by settling on bark. Females may even lead an amorous male into another male's territory, escaping during the inevitable battle. A virgin is scarcely more solicitous, and it may be several hours before she succumbs.

Having mated, the next aim in a female's life is to lay eggs. Like most close relatives, the Peacock lays in massive batches, taking considerable care over where they are placed. Common Nettle is nearly always used, despite occasional reports of caterpillars being found on Hop. Egg-laying generally occurs around midday, and is restricted to nettles that are in full sunshine at that time. Large, vigorous plants are usually chosen, especially those growing in sheltered situations such as in a woodland glade, or along a wood edge or a tall hedgerow.

Once she has selected a plant, the female hangs with closed wings beneath a tender young leaf, curves her abdomen onto the undersurface, and slowly pumps out 300-500 sticky green eggs. They remain for one to two weeks, piled several layers deep. On hatching, the mass of caterpillars immediately spins a communal silk web over the nutritious growing tip of the nettle plant. They live and feed within this web, until the leaf is a skeleton and the silk a grubby cobweb of shed skins and droppings. Fresh webs are spun whenever the food supply is exhausted, but as they grow larger the caterpillars live and bask increasingly in the open. Like other black caterpillars that bask communally (see page 222), they elevate their body temperatures to at least 32°C, even though the air temperature may be 20°C cooler.

Caterpillar and chrysalis

Peacock caterpillars are easy to find at all stages of growth, and can be distinguished from those of the Small Tortoiseshell by their blacker, white-specked bodies and their much longer spines. They are presumably as conspicuous to birds and small mammals as they are to us, but evidently obtain sufficient protection from their awesome spikes. These appear all the more formidable when the caterpillars are disturbed, for they jerk their heads and bodies in unison, making the spines sway in waves over the black mass. None of this deters parasitic wasps, which are responsible for the deaths of countless Peacock caterpillars of all ages.

The caterpillars that survive are ready to pupate in successive batches from early June to mid-July. They invariably desert their nettles at this stage, and disperse to form solitary chrysalises. F W Frohawk gives an excellent account of how a band of fully grown caterpillars walked 3m from their foodplant before climbing 4.5m up an oak, still in one group, and then separated, each to pupate apart on the tree. I have yet to find a wild Peacock chrysalis, but have reared many in captivity. Its shape and colours are captivating. It comes in two main types: one is blackish or grey, and this tends to be formed on dark tree-trunks or fences; the other is a much prettier yellow or golden shade, which gives good camouflage when suspended beneath leaves.

The Peacock caterpillar has less specialised requirements than many species described in this book, and this may explain why it is still a common butterfly throughout most of the British Isles. Open sunny woodlands, rich in flowers, are its favoured sites, but the butterfly can be seen almost anywhere and on any soil. Peacock numbers fluctuate considerably from one year to the next, and it is often abundant after a cool, wet summer, possibly because Common Nettles are lush and nutritious in such years.

Today, the Peacock is a very common species throughout lowland England and Wales, and is well distributed across Ireland, where it has increased. Its status in Scotland has changed dramatically for the better in recent decades. From being widely distributed in the south in the 19th century, the butterfly declined and was often scarce until an expansion began in the 1930s. This accelerated in recent years, and the Peacock is now a common mainland resident in all but the Highlands, with breeding reported from Ardnamurchan and Moidart in the west, and from Grantown-on-Spey, near Aviemore, in the east. Regular sightings are now made on most Scottish isles, including much of the Shetlands.

Comma

Polygonia c-album

The Comma is a familiar visitor to gardens throughout England and lowland Wales, where it feasts alongside Peacocks, Red Admirals and Small Tortoiseshells, building up reserves to last the winter. Although seldom as numerous as its close relatives, this represents an astonishing change in fortunes compared to a century ago, when the Comma was on the verge of extinction and more or less confined to the Welsh Borders. There were fewer than half a dozen records of the butterfly from 1830 to 1929 in counties such as Suffolk, Surrey, Sussex and Dorset, whereas today one expects to find it in almost every wood or scrubland. In the present decade it has colonised Scotland, after an absence of 140 years.

Camouflage and seasonal colour forms

This is a particularly interesting and attractive species. The adult, with wings closed, offers one of the finest examples of camouflage found among British butterflies. Its ragged outline and the marbled pattern of grey, tan and brown, with occasional green speckles, combine perfectly to resemble a dead oak leaf that has just begun to moulder. There even appears to be a crack in the centre where the 'leaf' is breaking up – the white, comma-like mark that has given this species its name.

The normal male Comma is slightly darker than the female, but both sexes have a brighter golden form called *hutchinsoni*, which is produced by the earliest caterpillars to develop in spring. These lovely Commas may constitute up to two-fifths of the midsummer emergence, but are not on the wing for long. They quickly pair, lay eggs and then die, in contrast to their darker siblings which live for ten months but do not breed until the following spring. The offspring of *hutchinsoni* Commas develop quickly to produce normal-looking adults in August and September. These feed for a few weeks before joining their uncles and aunts in hibernation.

Like other tortoiseshells, the Comma lives in loose, open populations, and wanders through the countryside selecting spots in which to feed, hibernate and breed. It is not, however, a true migrant, although it may recently have colonised the Isle of Man, and a few vagrants have been reported from the east of Ireland. The adults undoubtedly roam widely at a local scale, and the species managed to re-

Distribution
Common in woods, hedgerows and gardens throughout much of England and Wales. Has expanded its range considerably in recent decades.

occupy southern England in the early 20th century within the space of 25 years. Similarly, under the warmer climates of recent years it has spread north at an average rate of 20km a year, recolonising more than a third of Great Britain, in latitudinal terms, since the mid-1970s.

Spring and summer peaks

Adult Commas can be seen at almost any time from March to late October, but there are two main peaks, one in early April, the other from July to mid-September. The butterfly's requirements differ through the year, and this takes them into different types of habitat. For hibernation, adults rely mainly on the dry parts of woods, settling low down on elevated tree roots and other exposed surfaces, choosing places where drifts of dead leaves will later accumulate to complement their remarkable camouflage. They awake quite early in March, and feed in the weak sunshine on the catkins of sallows and other spring flowers. Males, in particular, remain near their woods, feeding first thing in the morning and again in late afternoon, but otherwise spend most of the day looking for mates. Each establishes a separate territory in a sunny nook on the wood edge, beside a glade or where two rides cross. The male sits at this vantage point for long periods, poised a metre or two up on a prominent branch, leaning forwards with wings half-open. He intercepts any passer-by in a swift but economical flight, achieving high speed with no more than a rapid whirring of wings, punctuated by swooping glides. He soon overtakes and loops around the intruder. After a thorough investigation, the male either returns to his perch or, if it is a female, gives chase.

Courtship, egg-laying and mimetic caterpillars

Pairing takes place high on a shrub or tree, and the female later wanders away to seek suitable places for her eggs. Many of these sites are along woodland edges, but she will also investigate hedgerows, scrub and especially hop-gardens. Hop was undoubtedly the main foodplant in the days when every village or farm had its own brew-house. The butterfly was once so plentiful in the hop-gardens of south-east England that pickers had their own names for it, calling the caterpillars 'hop-cats' and the chrysalids 'silver-grubs'. Hop remains a favourite foodplant, but is unavailable nowadays over large tracts of the countryside, forcing the majority of eggs to be laid on Common Nettle. They are also laid on elms and, occasionally, currants.

A female ready to lay will flutter slowly around shrubs, periodically alighting to test leaves by tapping their surfaces with her legs, tasting each to assess its suitability. The foodplant need not be in full sunshine – I have found caterpillars deep in the shade of a north-facing wall – but most are laid in sunny, sheltered nooks on the tenderest sprouting leaves of Hop, Common Nettle or trimmed hedgerow elm. Although laid singly, the glassy green egg is quite easy to find in good years. It is laid on the upper edge of the leaf, often on the extreme tip. Prominent nettles beside a wood are preferred; exposed plants growing in open fields are almost invariably avoided.

The *hutchinsoni* adults of midsummer behave in much the same way as the normal form in spring, and are therefore seen perching or egg-laying mainly around hedges and woods, where they are frequently mistaken for fritillaries. However, the darker Commas that emerge in midsummer must prepare for hibernation, as do the offspring of *hutchinsoni* later on. Both broods spend long periods feasting at flowers or on rotten fruit and oozing tree-trunks, bringing a succession of dark adults into the garden from mid-July to late autumn.

Comma eggs hatch after two to three weeks. The little caterpillar immediately crawls under its leaf, where it spins a fine silk web and feeds on the tender tissue, making distinctive perforations. As it grows larger, the skin is shed four times and the caterpillar increasingly comes to resemble a bird-dropping. At first it is small and crusty, a mere sparrow's drop of black and white sticking to the underside of a leaf. But the full-grown caterpillar is tan, with a gleaming splash of white, looking like a large deposit from a thrush lying curled in the open over a leaf. The chrysalis is strangely beautiful, resembling a withered leaf as it hangs on a thread beneath a Hop or Common Nettle. Chrysalises can be found in the wild, but the caterpillar is easier to spot and the egg is the simplest of all.

A history of captive breeding

This is an easy species to rear, and of exceptional interest on account of the changing camouflage from young to old caterpillar, followed by that of the chrysalis and adult. There is the added fascination, starting with springtime eggs, of producing normal and *hutchinsoni* adults.

The trick of breeding golden Commas was apparently first mastered by Emma Hutchinson, who also established that the species could be double-brooded. Mrs Hutchinson's name was inseparable from that of the Comma in the late 19th century, and very properly lives on through its golden form. She lived near the market town of Leominster, in the heart of Comma country, at a time when the butterfly was exceedingly rare, and was famed for her generosity. Large numbers of Commas were collected from the hop-gardens there, or were bred artificially, enabling her to supply entomologists throughout the country.

F W Frohawk also bred *hutchinsoni* from a female sent by Mrs Hutchinson in 1894. It says much for Frohawk's

▲▶ **Male**
Typical male. Camouflaged underwings conceal the butterfly during hibernation.

▲▶ **Female**
The outline of the female's wings is less ragged than the male's, and the colour is lighter.

▲ **Male**
Hutchinsoni form, produced by early spring caterpillars.

▲ **Male**
Aberrant form *suffusa*, found in Oxfordshire, 1940.

◀ **Egg** [× 15]
Laid singly on the upperside of leaves of the foodplant.

▶ **Feeding adult**
Hutchinsoni adult on Bramble, a nectar source in summer months.

▼ **Caterpillar** [× 1½]
White splash on the caterpillar's back mimics a bird-dropping.

◀ **Chrysalis** [× 1½]
Formed on low vegetation, and camouflaged to resemble a withered leaf.

expertise that he kept this creature alive for 47 days, and obtained 275 eggs, the caterpillars of which he reared on a variety of foodplants to produce 200 perfect adults. Forty-one were *hutchinsoni* and, with few exceptions, came from the first adults to emerge, while virtually all that were of normal appearance derived from the later bouts of egg-laying. And there the story remained for nearly a century, giving rise to the myth that 20% of springtime caterpillars produce *hutchinsoni* adults. In fact, there is a complex switching mechanism at work, which responds primarily to the amount of daylight to which the caterpillar is exposed during development.

Colour forms and day-length

In Stockholm, Sören Nylin reared Commas from both Sweden and Oxford. He confirmed that, in early summer, English caterpillars can alter their development either into golden *hutchinsoni* adults that will breed that same summer or into normal (dark) specimens that will hibernate. Nylin found that this is influenced by two main factors: the length of day experienced by developing caterpillars, and whether the days become shorter or longer as they grow. When English caterpillars are given just 12 hours of light each day, all turn into dark hibernating adults. But with 18 or 20 hours of light a day, 90% turn into golden *hutchinsoni* Commas. The way the day-length changes during a caterpillar's life can tip the balance. In general, more *hutchinsoni* adult butterflies are produced if the days are lengthening and fewer if they are becoming shorter. Indeed, under shortening days, only caterpillars exposed to more than 18 hours of light per day develop into *hutchinsoni*.

This switching mechanism is not quite so simple because, when the day-length is marginal, caterpillars are more apt to develop into the golden form if the temperature is warm or if they have fed on especially nutritious leaves. On the whole, however, the system enables the butterfly to capitalise on early springs and warm seasons, allowing caterpillars that have completed their development before the summer solstice (21st June) to give rise to a second generation in the same year. Interestingly, Swedish Commas never form *hutchinsoni* in the wild, but can do so if reared in artificial conditions that favour the golden form. They are, however, less prone to do this than our British race.

Nylin and Christer Wiklund have answered many fundamental questions in evolutionary biology through a painstaking series of studies of this lovely butterfly. Of interest to us here is that the Comma can use additional foodplants to those eaten in Britain, although Common Nettle and Hop are favourites, and result in faster growing caterpillars that survive better and produce adults that give rise to more offspring than those feeding on other plants. A key ingredient is the high concentration of nitrogen found in Hop and nettle leaves. This element tends to be in short supply, and is reserved for making melanin and other polymers that strengthen the thorax and darken and camouflage the wings in the overwintering form of the adult: in Wiklund's words, 'Animals that are going to live a long time need to be built to last.' Certainly, the dark form of adult is found less often by birds and mice compared with *hutchinsoni*. But the latter does not need to live long. The onus for *hutchinsoni* Commas is to mate and lay as many eggs as possible, and in their case the vital nitrogen resources are concentrated not in their wings (hence their light colour) but in the abdomen, to fortify the sperm and eggs.

19th-century decline

As more is discovered about this fascinating insect, the more fortunate seems its recovery in Britain. There is perhaps no butterfly that has experienced such dramatic fluctuations. To the earliest English naturalists it was not a particular rarity, and was among the 36 butterflies illustrated by Moses Harris, who found it 'very swift in Flight, and [so] timorous when settled that it is difficult to... lay the Net over it; they fly in Lanes, by Bank Sides, often settling in dry clayey Places, and against the Bodies of Trees'.

The heyday of the Comma was perhaps the early 19th century, when it was widely distributed over England and Wales, and reached Fife, Alloa and Clackmannanshire. There then followed an extraordinary decline, beginning around 1818-20 in most southern counties. By 1830, it had more or less disappeared from a large triangular segment of south-east England, from Dorset to Kent and north up to Lincolnshire. It also declined further north, where it had never been especially common, yet somehow survived there and experienced a minor increase in the mid-19th century. Thus the Comma was seen in Derbyshire in 1855, and was present and occasionally common in Lancashire, east Yorkshire and north Lincolnshire around that period, only to disappear again in the second half of that century. Despite one or two sightings in Kent and Epping Forest, the butterfly remained excessively rare and its demise was much lamented. 'It is one of the receding species which we so greatly regret,' wrote Barrett in 1893, while a Scottish entomologist lamented, 'It is very unpleasant to think that *P. c-album*, the butterfly of our youth, has left us for good and all.'

An unexpected recovery

The nadir of the Comma occurred between the end of the 19th century and about 1914. It is impossible to say whether it completely disappeared from all northern, central and south-eastern counties, for although there were one or two isolated records, introductions and frauds prevailed during this period. For example, one collector noted that 'Before 1881 hundreds of larvae and pupae were released in Surrey hoping to introduce the species, but without success.' Many were supplied by Emma Hutchinson.

What is certain is that the butterfly all but disappeared, and could be guaranteed only in the Welsh Borders, especially Gloucestershire, Herefordshire and Monmouth, and in a few adjoining areas. Even here, there were enormous fluctuations. In 1875, Mrs Hutchinson wrote that 'Pupae must have been very plentiful on the Hop plants cultivated here. I have had two hundred brought me, and thousands must have been thrown on the heap with the hop-bine, and burnt. Last year we had none.'

Then, suddenly and inexplicably, came the recovery. This was first noticed in 1914, with single adults seen over most of the southern counties during the next 10 years, and regular sightings thereafter. In Sussex, for example, one was spotted at Eastbourne in 1915, and although none was seen in the early 1920s, it was 'abundant' at Chichester in 1929, reaching 'thousands' on the South Downs by 1935. Dorset had been colonised slightly earlier, so that Commas were 'everywhere' by 1929, but it was not until 1935-36 that Suffolk was fully re-occupied.

The triggers of population change

By the 1950s, the Comma had expanded to become well established over the whole of lowland Wales and southern England, extending up to Manchester in the west and Scarborough in the east. Numbers were then quite stable for 25 years, before the current expansion began. Its most striking feature, as we have seen, has been the rapid recolonisation of northern breeding grounds, so much so that it is close to reaching its early 19th century maximum. There have been equally welcome local expansions elsewhere, for example into unpopulated parts of East Anglia, the West Country and western Wales. And on sites where the Comma was already well established, numbers have increased by nearly three-fold during the last 30 years, while in the last decade the butterfly has colonised the Isle of Man and five stretches of the eastern coastline of Ireland.

Quite why these fluctuations have occurred is a mystery. There is no doubt that hop-gardens declined greatly during the 19th century, especially after 1870, and that the new techniques of bine-burning and washing with insecticides can only have exacerbated the situation during the 1880s. It is also true that Common Nettles have increased and become more luxuriant in the present century – particularly since the last war – as a result of nitrates applied on farms. This, however, neither explains the sudden decline of this butterfly in the 1820s nor its recovery early this century, before artificial fertilisers became common.

The weather is a more plausible driver of the Comma's fluctuations. There is little doubt that its rapid northward spread has been mediated by a warming climate. It is possible that British races of the Comma were ill-suited to the curious cold weather patterns of the 19th century, and perhaps it took several generations for a better-suited race to evolve. But this is to speculate beyond current knowledge. For the present, we can be thankful that this beautiful ragged butterfly is again common in the British countryside, and hope that it will remain so for another century.

Queen of Spain Fritillary

Issoria lathonia

The Queen of Spain Fritillary was first recorded at Gamlingay, Cambridgeshire, about 300 years ago. Except in the Channel Isles, it is ill-equipped to survive our winters, and apart from three short-lived possible colonies at Minsmere, Chichester and on Jersey, all British sightings are of immigrants or their immediate offspring. Although an attractive, even spectacular, species, it has been reported fewer than 500 times since its discovery, with most records dating from the 19th century. It was, indeed, sufficiently well known to early English collectors to have acquired a range of names, including the 'Scalloped Winged Fritillary' and the 'Lesser Silver-spotted Fritillary'.

There is, unfortunately, some doubt about the authenticity of many 19th-century records, and even those from Gamlingay are not devoid of suspicion, on account of the unrealistic number of early rarities, including Chequered Skipper and Bath White, reported from the one wood. But the main period for fraudulent specimens was in the second half of the 19th century, when the demand for 'British' specimens of rarities led to a thriving business in counterfeits. Chief among the perpetrators were George Parry and the 'Kentish Buccaneers', whose exploits are wittily chronicled by P B M Allen. Their ploy was to rear Continental specimens of the Queen of Spain Fritillary in captivity, and then to release numbers on the Kentish Downs for colleagues to net and sell. These activities died down after the 1870s, and perhaps partly for this reason the Queen of Spain Fritillary was one of the scarcest of all recorded immigrants during the 20th century except, that is, for the *annus mirabilis* of 1945 and the exciting upsurge since the 1990s.

The Queen of Spain Fritillary is not particularly rare on the Continent, nor is it especially associated with Spain. It is a beautiful, medium-sized fritillary that lives in a patchwork of apparently residential populations from which migrants spread every spring, and there is some evidence that their offspring return south in the autumn. Despite this, some permanent populations occur quite far north, for example on dry limestone pavement, beaches and heaths in southern Scandinavia, and among the coastal dunes of The Netherlands, Belgium and northern France. Here, it breeds only on Field Pansy and the beautiful Wild Pansy, although violets are eaten in southern Europe. This may partly explain the absence of colonies in Britain, for neither pansy is particularly common in our dry southern grasslands, where summer temperatures are warm enough for this butterfly.

In northern Europe, the sites that support this fritillary are generally hot, arid localities, where the adults fly in strong, zigzagging flights between flowers. They frequently alight to bask on bare patches of ground, sitting with wings held open in a V towards the sun, so that the beautiful silver patches on the undersides are partly visible. The caterpillars, too, spend much time basking, allowing their dark, velvet bodies to absorb maximum heat from the sun.

Although the Queen of Spain Fritillary is a powerful flier, and makes genuine migrations every year, the distances covered are short compared with those achieved by its relatives, the Painted Lady and Red Admiral. Thus, having arrived in Britain it is seldom seen north of the southern coastal counties, and there are just three records from Ireland. There was a strong predominance of 19th-century records from the extreme south-east, mainly on the cliffs around Dover. F W Frohawk attributed this to the presence of permanent colonies just across the Channel on the coast near Calais. These had largely been destroyed by the 1930s, and Frohawk believed that the marked drop in immigrants in the first third of the 20th century was due to the loss of this source. A dramatic increase in sightings during the past two decades has been concentrated along the coastline of East Anglia, extending north into Lincolnshire and south into Kent. These perhaps reflect the known recent increases of this fritillary on the dunes of The Netherlands and Belgium.

Most sightings of this arresting fritillary are made in September and early October, with some in July and a very few in June and May. Some late records are probably the offspring of single early immigrants, for the butterfly has two or three broods on the Continent, and egg-laying

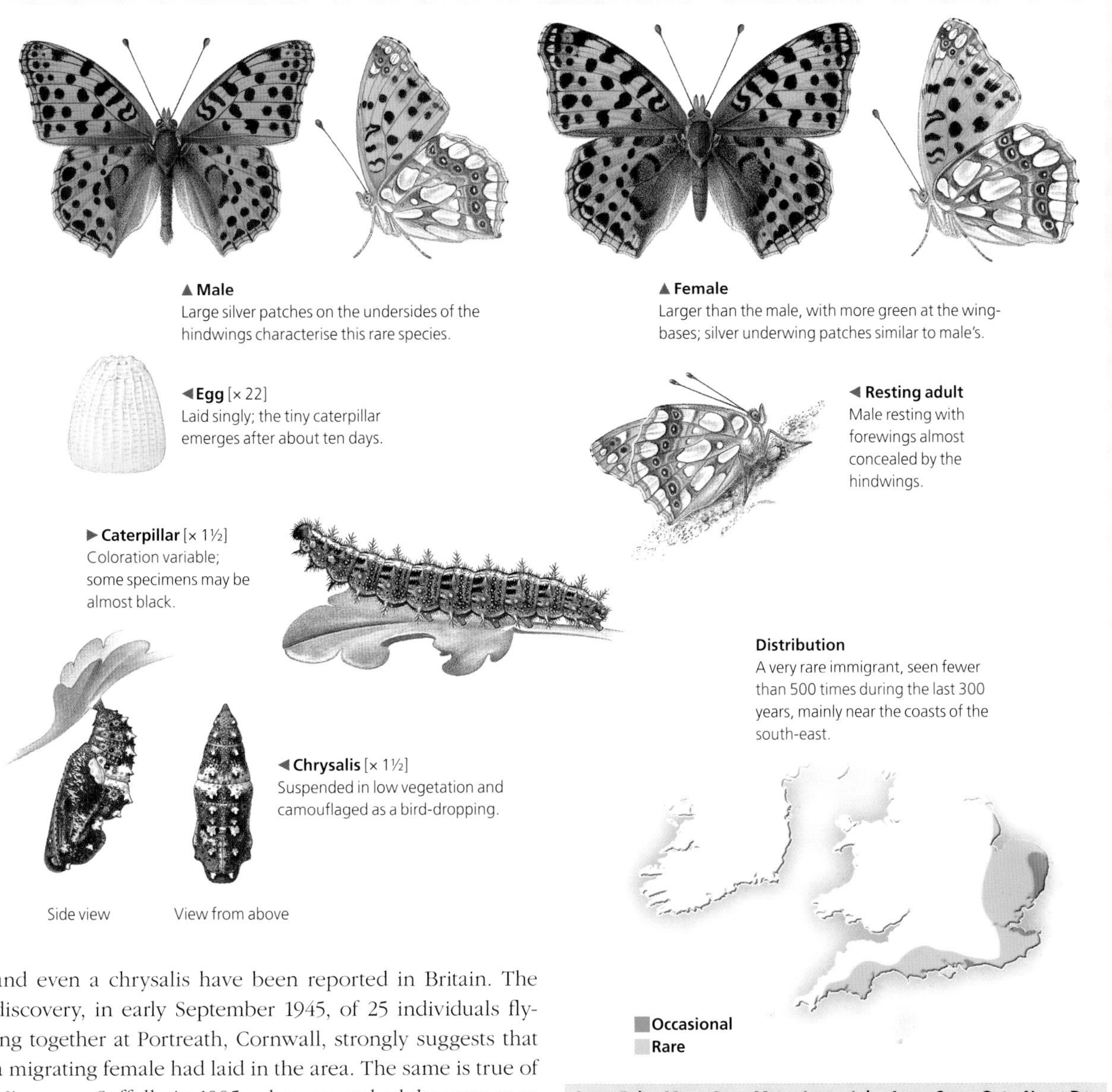

▲ **Male**
Large silver patches on the undersides of the hindwings characterise this rare species.

▲ **Female**
Larger than the male, with more green at the wing-bases; silver underwing patches similar to male's.

◀ **Egg** [× 22]
Laid singly; the tiny caterpillar emerges after about ten days.

◀ **Resting adult**
Male resting with forewings almost concealed by the hindwings.

▶ **Caterpillar** [× 1½]
Coloration variable; some specimens may be almost black.

Distribution
A very rare immigrant, seen fewer than 500 times during the last 300 years, mainly near the coasts of the south-east.

◀ **Chrysalis** [× 1½]
Suspended in low vegetation and camouflaged as a bird-dropping.

and even a chrysalis have been reported in Britain. The discovery, in early September 1945, of 25 individuals flying together at Portreath, Cornwall, strongly suggests that a migrating female had laid in the area. The same is true of Minsmere, Suffolk, in 1995, when several adults were seen in two distinct peaks, and again in the following two years. Similarly, in Sussex an adult was seen at Brandy Hole Copse, near Chichester, in September 2008, followed by another on 14th July 2009 and by further sightings of about seven individuals on the same site in September 2009, followed by a mating pair seen in October. These, alas, failed to produce offspring, and it was not seen in this area in 2010 or since.

The greatest early years for the Queen of Spain Fritillary were 1868 (46 records), 1872 (50) and 1882 (25). There was then a long gap in which none was seen at all until the great immigration year of 1945, when 37 were reported, including the group at Portreath.

Queen of Spain Fritillaries were then extremely scarce until the 1990s, when matters changed. This began with five sightings on one site in the Channel Isles, in 1991, followed by several reported from Spurn Head, Lincolnshire, in 1993. In addition to the records from Minsmere and Chichester, 'many' Queen of Spain Fritillaries were recorded from the Channel Isles and south-east England during the five survey years of 1995-99 for *The Millennium Atlas*.

There is little evidence that a permanent population has become established in England or the Channel Isles, although there have been several sightings of the adult butterfly in each year of the past decade. However, it is a species that we may hope to welcome as a resident when our climate becomes warmer, especially if conditions are also drier.

Small Pearl-bordered Fritillary

Boloria selene

The caterpillar of this species, and of the four fritillaries that follow, feeds exclusively on the leaves of violets. All five butterflies were fairly common until the 1950s, but then an extraordinary decline began which continues to this day. By far the heaviest losses have been in eastern and central southern England, where the Small Pearl-bordered Fritillary is reduced to a handful of populations in Sussex, Hampshire and Dorset. In contrast, it remains widespread and locally common in Wales, south-west and north-west England, and especially in Scotland, where it can still be seen in abundance during June and July, gliding across woodland clearings, rough grassland and moorland.

Breeding sites for this attractive little fritillary invariably contain violets in abundance, growing in a warm, damp, grassy, sheltered sward. Colonies occur in four main types of habitat: woodland glades and clearings; grasslands interspersed with scrub or Bracken; northern moorlands and damp grasslands; and, in Scotland, open woodland pasture that is lightly grazed by deer or sheep. A few large populations also breed in dune slacks.

A colonial life

In many of the northern colonies, Small Pearl-bordered Fritillaries breed at low densities over extensive tracts of land, with larger numbers occurring in moist, sheltered pockets where the Marsh Violet grows. The adults can be abundant in good years, and seem to be fairly mobile, with individuals flying from one concentration to another. In south-east and central England their lifestyle is very different. Typical colonies consist of dozens rather than hundreds of adults, and many are restricted to small patches of isolated land. Adults in these relic populations are highly sedentary, and their slowness in colonising new breeding sites can be remarkable. Like the Pearl-bordered Fritillary in these regions, just 100m of tall shrubs may represent an insurmountable obstacle to the colonisation of a new clearing in a wood.

The males behave very similarly to Pearl-bordered Fritillaries, with which they fly on several sites, the smaller species appearing much brighter on the wing due both to the natural difference in markings and to its two- to three-week

Distribution
A declining species; colonies survive in woodland clearings throughout its range and on wild, sheltered grassland in the west and north.

▲ **Male**
The underside has a more contrasting coloration than that of the similar Pearl-bordered Fritillary.

▲ **Female**
The Scottish form *insularum*, shown here, is more brightly marked than butterflies further south.

◄ **Resting adult**
Adult of form *insularum* resting on the leaf of a violet, the foodplant of the caterpillar.

▲ **Colour variant underside**
Specimen caught in the New Forest in 1920.

▼ **Basking male**
Males fly close to the ground, perching from time to time to bask in the sun.

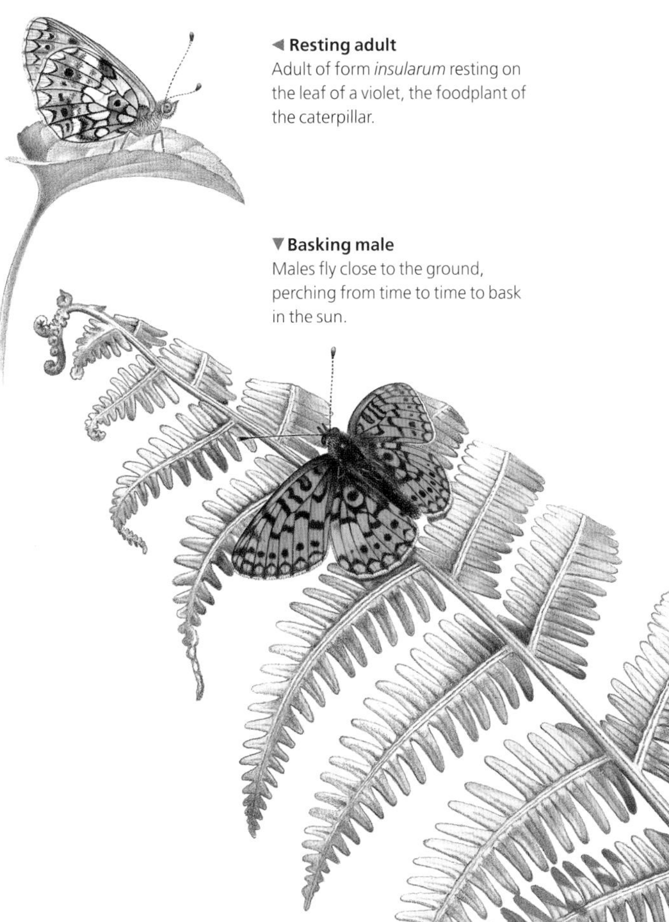

▲ **Egg** [× 22]
Eggs are scattered near Marsh Violet or Common Dog-violet.

▼ **Caterpillar** [× 1½]
Identifiable by two long, forward-facing spines.

◄ **Chrysalis** [× 1½]
Pupation takes place low down in vegetation near the foodplant.

	Jan	Feb	Mar	Apr	May	Jun	Jul	Aug	Sep	Oct	Nov	Dec
Egg												
Caterpillar												
Chrysalis												
Adult												

Principal breeding sites of violet-feeding fritillaries

▼ **Dark Green and Small Pearl-bordered Fritillaries**
Patchy grassland, where violets grow in clumps.

▼ **Pearl-bordered and High Brown Fritillaries**
Violets growing in warm ground, within a year or two of clearance.

▼ **Small Pearl-bordered Fritillary**
Lush violets in woodland regrowth, a few years after clearance.

▼ **Fritillaries absent**
Mature coppice or shady woodland.

▼ **Silver-washed Fritillary**
Sunny woodland with an open canopy.

later emergence dates. They skim rapidly across their breeding areas and beyond, about 1m above ground level, with rapid whirring of the wings punctuated by longer, graceful glides. They have an uncanny knack of detecting females half hidden in a grass tussock or shrub, and court them assiduously until eventual rejection or acceptance.

Early development

Having mated, the females hide until their eggs are ripe. They then flutter slowly above the ground, sifting the air for the scent of violet leaves, and frequently alighting to tap the vegetation with their sensitive feet. All violets cause some excitement, but most are rejected for egg-laying. The main species chosen are Marsh Violet and Common Dog-violet. Eggs are generally laid on medium-sized plants growing in open, sunny situations. In English woodlands, the preferred growth-form of violet occurs when the ground vegetation is beginning to close over, about two to five years after a clearance, or along sunny but moist ditches, ride edges and boundary banks, where the soil is sufficiently crumbly for successive growths of violets to be maintained. In grassland and moorland, warm, sparse areas are chosen, especially humid patches recovering from a fire or from heavy trampling by cattle. I have often also watched females dotting their eggs on Marsh Violets growing proud above small flushes of water.

Once a suitable place has been found, the female scatters her eggs with little regard as to whether they land on violets or not. Some are neatly placed beneath their leaves as she scuttles crab-like over the vegetation, but many are simply dropped in flight. She hovers a few centimetres above the ground – usually over a moist depression – holding her abdomen still, pointed downwards as the white, cone-shaped eggs are squirted out. The caterpillar hatches after about two weeks, and begins to feed on lush violet leaves. After shedding its skin for the third time, it settles down to hibernate within a withered twist of vegetation. Feeding is resumed the following spring but, unlike most other fritillaries, the caterpillar continues to be elusive. It hides in humid pockets beneath its foodplants, stretching upwards when hungry to nibble moon-shaped bites in the lobes.

The challenge of a changing habitat

There is no doubt that the Small Pearl-bordered Fritillary has been badly affected by the virtual cessation of coppicing in the 20th century. Central and eastern England have been worst affected. The butterfly disappeared altogether from the Midlands and East Anglia during the second half of the century, and in the last decade from Kent and Surrey; in Sussex, it was reduced to one tiny colony in 2013. It is also extremely rare in Hampshire and Dorset, counties where it was once ubiquitous in woods on damper soils.

The decline has been less severe in the West Country, Wales and the Lake District, where the butterfly can still be seen on marshland, moorland, in coastal valleys and in moist grassland throughout these regions, and in some woodland clearings, too. Its stronghold, however, is south-west and most of central Scotland, where it is widespread and locally common at altitudes of up to 800m on Highland moors. Although present on several of the inner isles, the Small Pearl-bordered Fritillary apparently never recolonised the outer ones following the last Ice Age. It is absent, too, from Ireland.

Pearl-bordered Fritillary

Boloria euphrosyne

The Pearl-bordered Fritillary is one of our most rapidly declining butterflies, and has been lost from many counties where once it was too common to register. It is the earliest British fritillary to appear, emerging in April after a warm spring. Its first English name was, indeed, 'The April Fritillary', but that was before the Gregorian calendar was adopted and dates moved back by 11 days. During the 20th century it was unusual to see adults before the second week in May, but with recent climate warming it has become a true butterfly of April; I have even seen males flying in March in Devon.

The identification of medium-sized fritillaries can be confusing. This species is distinguished from all but the Small Pearl-bordered Fritillary by the seven silver 'pearls' along the borders of each lower underwing. The main difference between the two pearl-bordered species is found on this same wing: the Pearl-bordered Fritillary is more uniformly golden, and has two bright patches of silver on either side of a central pentagonal cell that houses a black dot. The Small Pearl-bordered Fritillary has a larger black spot in this position, surrounded by seven or eight patches of white or silver. It also has darker, brighter uppersides.

Colonies, courtship and breeding

Except in Scotland, this is a butterfly that lives in close-knit colonies. These vary enormously in size on different sites and from one year to the next. Many contain fewer than 20 adults, breeding in a minute portion of the wood or scrubby grassland that they inhabit. But numbers can increase to thousands in the two to four years following a clearance, before falling just as rapidly as the breeding area becomes overgrown.

Each sex occurs in roughly equal numbers, yet females are seen less often because they huddle in the vegetation, emerging only to flutter slowly above the ground to look for egg-sites or to feed on Bugle and other spring flowers. Males, by contrast, are much in evidence. When not drinking nectar, they spend most of every warm day flitting and gliding across their sites, flying about a metre above ground level, scanning every nook for an emerging female. They

Distribution
A rapidly declining species, confined to dry, sheltered grassland in the west and to woodland clearings throughout its range.

dip to investigate any coloured object, and hover around females until rejected or accepted.

The female hides for a day or two after mating, while her eggs ripen. She then embarks on short bouts of egg-laying. Eggs are laid singly on or near violet leaves. As with related fritillaries, any species is used, although most English, Welsh and Irish colonies are supported by the Common Dog-violet, whereas many in Scotland depend on the Marsh Violet. On the heathy grasslands around Dartmoor, I have seen large numbers of eggs laid on the beautiful Pale Dog-violet.

When laying eggs, the female Pearl-bordered Fritillary has a strong preference for small, young violets that sprout in abundance from warm, bare ground or rock, or which surface through an equally warm layer of dry leaf litter, such as crisp oak leaves or the golden fronds of dead Bracken. On heavy soils and in northern Britain, egg-laying is often confined to south-facing slopes, or even to the sunny side of a ditch or boundary bank, where narrow strips of warmed foodplants grow among the myriads of cooler violets growing elsewhere in the wood.

The egg hatches after about a fortnight. The caterpillar feeds intermittently from late June onwards, concentrating on the freshest leaves of established violets and, when available, on dense flushes of seedlings growing in crumbly, warm soil. By September it has completed the fourth of five skin moults, and settles down to hibernate in a twisted dead leaf. It usually reappears in early March, to embark on a month of serious feeding. Violet seedlings are again its preferred food, but almost as acceptable are the leaf lobes of one-year-old plants. The caterpillar at this stage has a beautiful velvet-black body, with yellow, black-tipped spines, and can be found quite easily in early April in strong colonies, basking in sunshine in the curl of a dead leaf.

The role of coppicing

Violets in a suitable growth-form for this fritillary occur mainly on dry soils or on banks in the first years after a perturbation. Until recently, most breeding sites were in woods or scrubland, but the colonies that survive in the west mainly use rough grassland that is regularly burned back in winter or has been trampled by cattle, then left to recover. Some Scottish populations breed on warm, dry moorland, but most are in the shelter of woodland pasture and clearings, typically along the south-facing edges of birch or oak woodland, interspersed with patches of Bracken that are broken down by deer or sheep.

In the past, ideal habitat was created annually in most British woods through the traditional practice of coppicing. As every collector knew, the Pearl-bordered Fritillary bred 'where the undergrowth has been cut down for two or three years and the ground is carpeted with wild flowers such as wild hyacinth, bugle, violet and primrose,' (F W Frohawk 1914). That is an ephemeral habitat, and this butterfly is unable to thrive in the slightly older clearings inhabited by the Small Pearl-bordered Fritillary. With David Simcox, I have had the pleasure of studying several hundred Pearl-bordered Fritillary breeding spots in three types of habitat during the past 30 years, including the beautiful lungwort coppices at Bloxworth, in Dorset, the dry scrubby grassland at Brackett's Coppice, also in Dorset, and the grassy heaths surrounding Dartmoor. On every site, the females laid most of their eggs on violets growing in spots where the shrubs had been cut back or the gorse burnt during the previous 18 months; few were deposited among the larger, more abundant violets growing in two- to three-year-old re-growth, and virtually none among older vegetation.

On Dartmoor, the burning (swaling) of gorse traditionally occurred on a 5-7-year cycle at any one spot, resulting in a patchwork of annual clearings interspersed with regenerating vegetation. Under this regime, the females need shift no more than a few metres each year to reach fresh patches, although they also traversed up to 400m of unsuitable ground to colonise more isolated clearings. I suspect, however, that there was selection for more sedentary races in the many regions where the butterfly depended wholly or largely on coppicing. For example, we found in Dorset that the females readily shifted into fresh coppice clearings, so long as these were immediately adjacent to panels that had been cut and colonised the previous year. This, of course, was how rotational coppicing was traditionally carried out: it was convenient for the woodmen to stand in last year's clearing while they cut next year's crop. But in nature reserves, where the location of new clearings has been more haphazard, we found that even a 100m stand of ten-year-old coppice growth was a complete barrier to the low-flying females. At the other end of the spectrum, the populations in Scotland appear to be much more dispersive, and may again be adapted to the historical dynamics of their habitat.

Decline from the north

Until late into the 19th century, the Pearl-bordered Fritillary was a widespread and common butterfly across much of England, Wales and Scotland. It was almost ubiquitous in the woods of lowland Wales and southern England, except on the most waterlogged soils. The decline began towards the end of the century, coinciding with a period when many coppices were being abandoned. At first, this was serious only in north-east England and south-east Scotland, where suitable sites were probably always few and far between. Thus, in 1934 F W Frohawk was still able to describe the

▲ Male
Markings are fairly uniform over its range in the British Isles. Seven silver 'pearls' edge the hindwings of both sexes.

▲ Female
Larger than male, often with paler, yellowish marginal spots. Two bright silver patches mark the inner hindwings of both sexes.

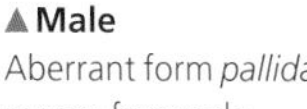

▲ Male
Aberrant form *pallida* ranges from pale orange to white.

▲ Male
Aberrant form *pittionii* has darker wing-bases.

► Feeding adult
Both yellow and purple flowers are visited by the adults. Here, a male feeds on a buttercup.

◄ Egg [× 22]
Laid singly on or near various species of violet.

▼ Female egg-laying
The female has a strong preference for small, young violets in warm micro-climates.

▲ Caterpillar [× 1½]
Between feeds, the caterpillar often basks in the sunshine.

▲ Chrysalis [× 1½]
Suspended in a nest of withered vegetation, just above the ground.

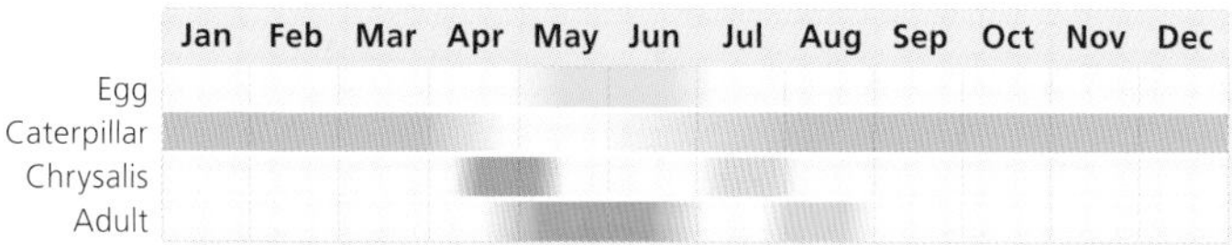

butterfly as being 'one of the commonest of our woodland butterflies in the spring and early summer months... [and] sure to be met with... in sunlit openings'.

The situation changed dramatically after the 1950s, when the aim of most forestry was to generate crops of mature trees. Plantations and other clearings were often suitable during the first years after a planting, but became far too shaded for breeding during the ensuing 50-70 years before the trees were felled. And as the butterfly became rarer, few isolated woods were colonised during the brief period when the habitat became suitable again, if indeed any violets appeared after the decades of heavy shade. Among the fritillaries, the Pearl-bordered and the High Brown were especially vulnerable to local extinction owing to their dependence on the newest clearings. Small Pearl-bordered and Dark Green Fritillaries lasted several years longer, and often also persisted in sheltered grassland nearby; the Silver-washed Fritillary breeds among trees, and often survived for many years in maturing broadleaved woods (see page 204).

Management for survival

Twenty years ago, it was still possible to write that 'because it was originally so common, the Pearl-bordered Fritillary has yet to become a national rarity', although I added a warning: 'Most surviving colonies occur among the youngest trees in large conifer plantations. Nearly all face extinction in the near future.' That moment quickly arrived. Already in the early 1980s the butterfly had disappeared from 198 (60%) of the 10km squares from which it had recently been known, although against this another 130 'squares' were located in under-recorded parts of Scotland. Things then became critical. During Butterfly Conservation's national surveys of 1997-2004, a third of the remaining English colonies became extinct, with more than half of those in the south-east disappearing in the seven-year period, and a similar rate of loss was reported from Wales. Unbelievably for entomologists of my generation, the butterfly is today extinct in Dorset, Kent and Somerset, and is reduced to single sites in Surrey and Gloucestershire, and to six small locations in Sussex.

At the time of writing, perhaps 170 colonies survive in England and Wales, many of which are small, with most located in central southern Devon, the New Forest and on the scrubby limestone pavements around Morecambe Bay, in Cumbria. Populations have been considerably more stable in central Scotland, where the butterfly may still be under-recorded and where – with perhaps 150-200 occupied sites – the majority of British populations now occur. The Pearl-bordered Fritillary has also persisted in Ireland, where it has a curiously restricted distribution. It is found only in the Burren, where it remains a comparatively common butterfly, mingling with the Wood Whites that also breed among the Hazel scrub that grows sparsely on the warm limestone pavements in this beautiful region of County Clare.

In England and Wales, the best chance for the survival of this charming fritillary is on nature reserves and in Forest Enterprise woodlands, where the management is tailored to ensure continuity of its ephemeral breeding habitat. This is perfectly feasible, and isolated successes have recently been reported from Wyre Forest, the New Forest, West Sussex, Morecambe Bay and Denbighshire.

Large Blues and fritillaries

The longest-running example of recovery to date occurred fortuitously as a by-product of targeted conservation management, not for this butterfly but for the Large Blue, on the southern edge of Dartmoor. These are the grassy heathland sites, owned by the National Trust, where annual burning of gorse results in a flush of violets among a regenerating sward that is cropped by ponies and winter cattle. This type of pastoral farming was becoming obsolete by the early 1970s, and Pearl-bordered Fritillaries were then restricted to two small breeding patches. However, a massive extension of this management for the Large Blue, and its application to an abandoned hillside nearby in 1975, and to a fourth site in 2000, generated extraordinary populations of Common and Pale Dog-violets: up to a million plants of Pale Dog-violet, itself a national rarity, flower there today. Interestingly, the spread of both violets was hastened, in part, by the same *Myrmica* ants required by the Large Blue. For, as Zoë Randle has demonstrated, attached to each seed of both violet species is a small fat-body, called an eliasome, which is irresistible to red ants. The ants gather the seeds, bite off the eliasome, and scatter the remainder around their nests in spots where seedlings thrive.

The Pearl-bordered Fritillary was quick to colonise the new 'Large Blue sites', and increased by up to 20-fold on the old ones, during a 35-year period when four-fifths of English colonies were lost and the surviving ones declined, on average, to one-tenth of their former densities. At the time of writing, in 2013, they exist as a network of four interlinked populations in this part of Dartmoor, each supporting several hundreds, and sometimes thousands, of this fritillary, and also lower numbers of Small Pearl-bordered, High Brown and Dark Green Fritillaries, as well as thousands of Grayling and other scarce insects, each utilising slightly different patches or stages of regeneration in the scrubby grasslands. It is a reason for hope, but is not a chance occurrence: its continued success relies on the annual efforts of conservationists.

High Brown Fritillary

Argynnis adippe

This handsome fritillary once bred in most large woods in Wales and southern England, and on a few scrubby grasslands in the west. But an extraordinary decline afflicted all types of colony, beginning in the late 1950s in eastern woodlands and continuing to this day. As a result, this has become the most endangered butterfly in the British Isles, now reduced to about 70, mainly small, colonies that occupy about 5% of its former range. Thanks, however, to targeted conservation programmes, populations have now more or less stabilised. At present, the High Brown Fritillary survives primarily in wooded valleys and acid grasslands on the southern edge of Dartmoor, in one valley system on Exmoor and, especially, among the limestone hills of the Lake District, at the extreme north of its traditional range.

It would be unfortunate if this butterfly were lost, for it is an impressive and beautiful insect, first known in the early 1700s as 'The greater silver-spotted Fritillary'. Although arguably less fine than its close relatives – the Silver-washed and Dark Green Fritillaries – it was once the commonest large woodland fritillary in several southern counties. Today, it has virtually disappeared from woods, making a nonsense of the assumption that any large fritillary with silver patches seen flying in woodland is a High Brown. This rule of thumb held true up to 60 years ago, but nowadays thousands of Dark Green Fritillaries must be seen in woods for every one of this species.

Distinguishing the High Brown and Dark Green Fritillaries

It is worth dwelling on the distinguishing features of these similar-looking fritillaries, for they still fly together on a few sites. The most obvious characteristics are on the underside of the wings. The High Brown Fritillary has an extra row of small silver spots, each ringed by a reddish-brown halo, between the two bands of silver patches that extend down the outer edge, halfway in on the hindwing. The Dark Green Fritillary has no markings in this gap, but usually has additional silver patches near the tip of the forewing, extending a short distance down the outer edge. In addition, as their names imply, the High Brown has brownish crescents around the outer silver patches on the underside of the hindwing, and the inner silver patches are often encircled by brown. The Dark Green Fritillary instead has green crescents, and the whole of the background to the inner

Distribution
One of our rarest and most rapidly declining species. Strong populations occur only in the southern Lake District.

two-thirds of the hindwing is tinged with the same beautiful olive-green. It should be noted, however, that this green tinge varies in intensity, and that the presence or absence of the little rust-ringed spots is the only infallible guide.

These two fritillaries are much harder to distinguish with their wings open. One trick is to look at the outer edge of the forewing, which is straight or slightly concave in the High Brown Fritillary, whereas it curves outwards on the Dark Green Fritillary. Basking males can also be distinguished by the sex-brands or scent marks near the middle of the forewings. These are elevated, black and shiny, on the second and third veins up, on the High Brown Fritillary, rather like a smaller version of the streaks on a male Silver-washed Fritillary. On the Dark Green Fritillary they are flat, insignificant and confined to the first two veins.

Finding foodplants

In flight, the High Brown Fritillary is swift and powerful, and is hard to approach except while feeding on flowers, when it becomes oblivious to everything around it. Both sexes soar high among the treetops, roosting on the uppermost branches in poor weather and at night. They float down when the sun shines, the males to search for virgin females, and the mated females to lay eggs. Male High Brown Fritillaries pound back and forth over the breeding areas, dipping rapidly to examine anything resembling a female.

Although a colonial butterfly, both sexes may wander 1-2km from their breeding sites, especially in dry summers when the flowers wither or die. I shall never forget seeing them during the drought of 1976, when scores of males congregated in a few marshy meadows below the wooded fringes of Dartmoor. Here they jostled, up to four to a flowerhead, on Marsh Thistles, gorging on the rich nectar. They were so abundant that I gave up bothering to photograph single specimens.

When laying her eggs, the female High Brown Fritillary flutters just above ground level through warm, sunny areas that are sheltered by a sparse growth of Bracken or shrubs. She responds to the scent of violets, the caterpillar's foodplant, but seldom homes in on individual clumps. Instead, she lands nearby and crawls around until a sheltered spot with broken plant cover is found.

The eggs are usually laid on twigs, dead leaves or even stones, and always near the ground, with three or four often around one patch. In the acid grasslands of the Malvern Hills and on the edge of Dartmoor, entomologists have found eggs on the crisp fronds of dead Bracken, in spots trampled down by cattle. Yet on the Devon sites known to me, the females largely ignore this habitat. Instead, they lay eggs on the sunny side of, or deep within, a large gorse bush, depositing them by the score amongst parched leaf litter through which vigorous clumps of large-leaved Common Dog-violets and Pale Dog-violets sprout.

The eggs are attractive, cone-shaped and apricot-pink when laid, but soon turn a dull slate-grey as the little caterpillars develop inside. Although fully formed after three weeks, the caterpillar does not hatch until the following spring, when it bores out of the toughened shell before searching for violets. Having located a source of food, the young caterpillar feeds on the tender growing leaves, biting increasingly large holes in their lobes. It feeds through April and May, and can be found quite easily on the best sites, for it basks openly in sunshine. It is a most attractive creature that exists in two main colour forms when fully grown. The usual lighter form is shown opposite, but do not be surprised to find others that are dark brown or grey.

The importance of coppicing

In the past, the High Brown Fritillary bred in abundance in many of the larger coppiced woodlands of England and Wales, probably in comparable situations to the spots used for egg-laying in scrub and rough grassland. I suspect they depended on short-lived sites, using the largest clumps of violets that had survived in the shade of mature coppices, and which suddenly took on a new lease of life in the spring after a block was cut. After one or two seasons, the site would become overgrown, and it was only under a continuous cycle of coppicing (see page 204) or annual clearings that the butterfly appears to have prospered. This would explain why the High Brown was usually the first of the five violet-feeding fritillaries to disappear from a district following the abandonment of coppicing.

While there is no doubt that the decline of coppiced woodland is largely responsible for this butterfly's disappearance during the second half of the 20th century, it is surprising that the onset was so sudden. I suspect that rabbits may have played a part, for it is often forgotten that they were once the major herbivore on many woodland floors, and were not just confined to heaths and grassland. At the very least, the loss of rabbit grazing when myxomatosis struck in the 1950s must have added to the shading effect in woodland clearings, allowing tall grasses, sedges and coarser vegetation to swamp violets and the warm patches of leaf litter or bare ground.

Today's survivors

As late as the 1950s, the High Brown Fritillary occurred widely in the woods of East Anglia and the south-east, and was considered to be the commonest of the woodland fritillaries in Sussex, being found 'in local abundance in suitable spots throughout the county'. Today, it is extinct in the eastern two-thirds of England, having disappeared

▲ **Male upperside**
Outer edge of the forewing is less rounded, and sex-brands on the 2nd and 3rd veins up from base are more obvious, than in the similar Dark Green Fritillary.

▲ **Male underside**
Red-ringed spots on hindwings are a characteristic feature.

▲ **Female**
Females are more heavily marked with black than males.

▲ **Egg** [× 15]
Egg turns grey as the caterpillar develops inside. Hatching is delayed until spring.

▶ **Feeding adult**
Flowers of Bramble are a favourite nectar source.

▼ **Caterpillar** [× 1½]
Light form; darker brown or grey specimens also occur.

▼ **Chrysalis** [× 1½]
Formed in a loose web of silk. The leaf-like, brown coloration varies in hue.

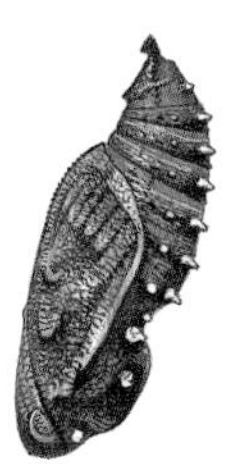

Side view

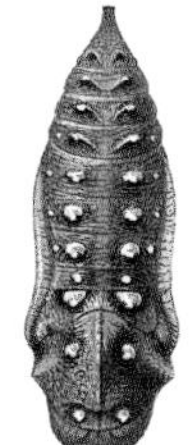

View from above

	Jan	Feb	Mar	Apr	May	Jun	Jul	Aug	Sep	Oct	Nov	Dec
Egg												
Caterpillar												
Chrysalis												
Adult												

in recent decades from Dorset, Wiltshire, Hampshire and the last localities in the Midlands. It also declined greatly in Devon. On Exmoor, a former stronghold, it is probably now extinct from all but the Heddon Valley where, however, numbers have stabilised or recently increased on at least six sites thanks to the 'Two Moors Threatened Butterfly Project' led by Butterfly Conservation's Jenny Plackett and Nigel Bourn. Working mainly with volunteers, Jenny's team has greatly improved and extended the habitat available for High Brown Fritillaries on National Trust and other land within colonisation distance of surviving colonies, and already there is evidence of an expansion: so long as this excellent initiative can be maintained, here at least there is hope for the future.

The populations along the southern edge of Dartmoor have fared similarly, and even increased, on a few sites following conservation management, some again thanks to the Two Moors project. Elsewhere on Dartmoor there is an extended colony that breeds on four of the National Trust's restoration sites for the Large Blue. There, a regime of annual rotational gorse- and Bracken-burning, coupled with rough winter grazing by cattle and ponies, generates an adequate habitat for the butterfly. Although by no means large in most years, these populations of High Brown Fritillary have not only survived during 40 years of management for Large Blues, but have increased by more than tenfold, albeit from a very low baseline, and have colonised two new areas. As elsewhere on the moor, the 2013 emergence was exceptionally high in this species, and it was a huge pleasure to see multiple High Browns, amid a few Dark Greens, gliding around the recently cleared ramparts of Hembury Castle. Today, these sites are among a handful in Britain where, in season, the naturalist can watch all five species of violet-feeding fritillary.

The Dartmoor and Exmoor sites, together with a few woods to their west, are possibly the only places where the High Brown Fritillary survives in its former stronghold of the West Country; they currently support about 15 and six populations respectively. The situation in Wales and the Welsh Borders is dire. At present, one population is known from Wales and the butterfly has now been pronounced extinct in the Malvern Hills, where until recently it bred in many sheltered patches of rough grassland. Indeed, 25 years ago I wrote that the Malverns was 'one of the few places where High Brown Fritillaries can be seen in any numbers, sweeping up the hillsides or fluttering around gaps in the bracken. It is heartening to know that the management of this area is specifically designed to encourage this and other Fritillaries.' It was evidently a much harder task than was envisaged in those optimistic days.

The only other region where this lovely fritillary survives is on the carboniferous limestone surrounding Morecambe Bay. This remains the British stronghold for the species, and – with 53 confirmed breeding sites – supports two-thirds of our current populations. Here, the adults fly, but do not necessarily breed, over wide areas. Amid the many small colonies still found here are a few very large ones, breeding mainly among patchy scrub on violets that flourish in the cracks between the natural limestone pavements, for example on Arnside Moss, Warton Crag and Gait Barrows. It also breeds on more acidic, Bracken-dominated grassland in this region. As on Dartmoor and Exmoor, the High Brown Fritillary has responded positively to conservation management on several nature reserves in the Lake District, where sheltered clumps of violets sprout annually following scrub-cutting, among the limestone rubble and pavements.

Dark Green Fritillary

Argynnis aglaja

This spectacular fritillary inhabits dunes and flower-rich wild grasslands from Orkney to the Channel Islands. Although seldom particularly numerous, it is not a species likely to be overlooked. A single airborne male will quickly catch the eye, particularly when it battles alone against the breeze or, in calmer weather, cruises among the browns, blues and skippers that abound on most of its sites.

Flight and concealment

For all its powers of flight, the Dark Green Fritillary is not a particularly mobile species. Most adults remain within the boundaries of their clearly defined breeding grounds, although the occasional individual will fly 2-3km to a neighbouring site. In lowland England, sites may be quite small in area and often well separated from the next. Others are more extensive, but even along the coast there is little evidence that many adults fly between sites. Although Dark Green Fritillaries are generally seen in ones and twos, the better sites support a few hundred adult butterflies, and the best occasionally contain several thousands. On one memorable day in the 1990s, at Porton Down, Wiltshire, the adults were so abundant that they 'were swarming like flies'.

Male Dark Green Fritillaries spend much of every warm day on the wing, scanning the hillsides for hidden females. These sit in tussocks and are probably located by scent. Mating quickly follows, low down in the grass. There is elegance as well as power in these patrolling flights: the wings whirr swiftly for a second or two, causing the butterfly to surge forward and then glide, before another burst of wingbeats becomes necessary. They seldom pause to rest, and the only realistic chance of seeing or photographing either sex when settled is to wait beside thistles and other purple flowers in the early morning or late afternoon.

Females remain hidden for the rest of the day, waiting for their burden of eggs to ripen. Each then embarks on a brief investigative flight, fluttering slowly above the grass and often alighting to tap vegetation as she sifts carefully through the scents rising from below. Her reaction to the smell of violets is spectacular. She spins around and makes brisk, jerky movements back and forth in their vicinity, squeezing herself deep into the tangled vegetation. Any violet species can prompt her to lay, but clumps of robust, large-leaved species are favoured. Thus, in the north and on wet sites, Marsh Violet is preferred, as is Hairy Violet on dry calcareous soils. Elsewhere, Common Dog-violet suffices.

Although catholic in her choice of violet species, the female is extremely choosy about where she lays. More than any violet-feeding fritillary, except perhaps the Small Pearl-bordered, the Dark Green prefers violets growing in lush, cool or humid spots, and selects large clumps that protrude

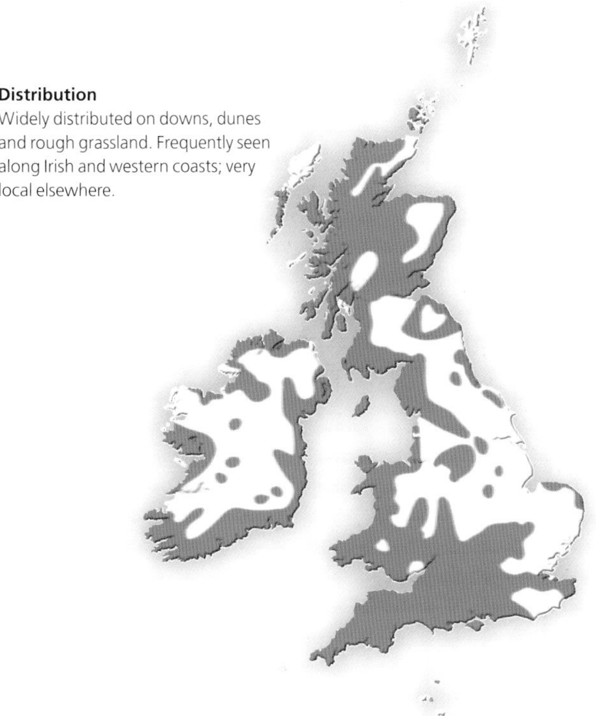

Distribution
Widely distributed on downs, dunes and rough grassland. Frequently seen along Irish and western coasts; very local elsewhere.

above a fairly dense sward, typically 8-15cm in height, or which grow sideways through a wall of dense grass into a sheltered, sunny spot. There is no holding a female once a suitable patch is found. Her abdomen curves around, stabbing left and right, and a series of eggs is ejected in rapid succession.

The eggs hatch two to three weeks later, and the caterpillars immediately hibernate among the leaf litter. They re-emerge in spring to take hurried bites out of large but tender violet leaves. They are nervous, rather wasteful feeders, especially when older, and they hide for long periods between scuttling from one plant to another to snatch a few bites from the lobe of a leaf. In cooler weather or localities, the lovely inky blue-black caterpillars bask quite openly on vegetation.

A survivor among fritillaries

This splendid butterfly is the commonest fritillary over much of the British Isles, and the only species present in the far north and on many islands. Unfertilised, flower-rich open habitats are its main abode, particularly where the turf is regularly disturbed and then left to recover. Such conditions occur naturally on dunes and undercliffs along much of the coast, and have been created for centuries by farmers on rough, scrubby grazing land and moorlands, where it also breeds on Marsh Violets in wet flushes. Many other colonies occur on unfertilised chalk and limestone downs, but only where grazing is light or erratic. A few breed in woodland glades and on wide, sunny rides where the soil has been churned up by forestry machinery.

There is no doubt that countless inland colonies of the 'Darkned Green Fritillary', as it was once known, have been lost since the early days of entomology, when it was sufficiently widespread to be included in Thomas Moffet's first list of 18 British butterflies (1589). But despite many losses in eastern England, this remains the least threatened of our eight fritillaries. Its strongholds have changed little over the years. In England, look for it particularly on the chalk downs of the south, especially on Salisbury Plain, on the limestones of the Cotswolds, on Dartmoor and Exmoor grassland where the Bracken has been trampled by cattle or ponies, in the Lake District and the Peak District, and anywhere along the coast where there are extensive dunes, undercliffs or grassy valleys rich in violets. It is also almost ubiquitous – although often in low numbers – along the coastlines of Wales and Scotland, and occurs on many rough inland grasslands and moorlands in both countries, too.

The Scottish form *scotica* is often considered to be a distinct subspecies. It is undoubtedly the largest and finest form of this butterfly, with dusky large-spotted wings that look magnificent when battling against coastal breezes or gliding in still air among dune slacks. This beautiful form is found throughout the Highlands and Western Isles, including the Outer Hebrides, as well as southern Orkney. It also breeds more or less continuously along the north Irish coast. Elsewhere in Ireland, the Dark Green Fritillary occurs sporadically along the coast and throughout the Burren, with occasional populations found further inland on eskers and bogs.

▲ **Male upperside**
Wings are more rounded than those of the similar High Brown Fritillary.

▲ **Male underside**
Underwings are flushed with green; those of the female are similar.

▲ **Female**
Ground-colour of the wings is paler than in the male, especially towards the margins.

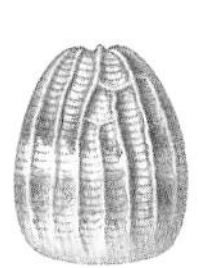

▲ **Egg** [× 15]
Laid singly, close to any of several species of violet.

◀ **Female, Scottish form**
The Scottish form *scotica* is more heavily marked than southern specimens.

▼ **Feeding adult**
Purple flowers, such as Dwarf Thistle, are a favourite source of nectar.

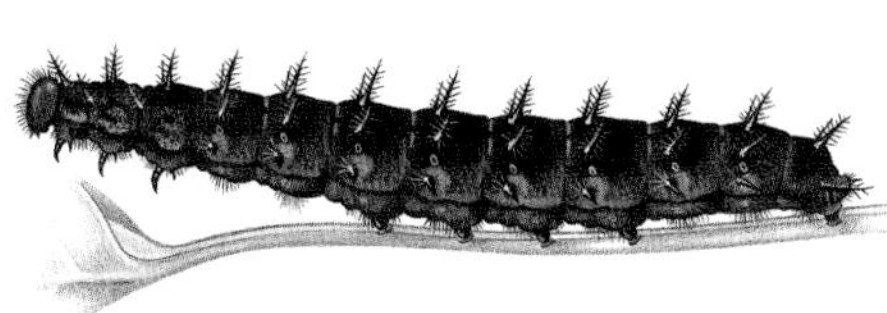

▲ **Caterpillar** [× 1½]
Feeds on the leaves of violets, cutting out characteristic moon-shaped patches.

▲ **Chrysalis** [× 1½]
The distinctively curved chrysalis is formed inside a loose cocoon of silk and grass leaves.

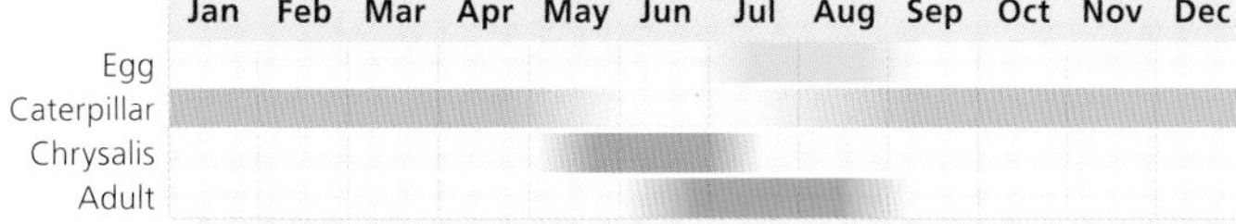

Silver-washed Fritillary

Argynnis paphia

Only one of the 50 fritillaries of Europe exceeds this butterfly in size, and none is more beautiful or more magnificent. It is, by some margin, the largest British fritillary, with a male wingspan that is nearly 1cm wider than that of the Dark Green. Fortunately, the Silver-washed Fritillary is less rare than most of its near relatives, although it too experienced worrying declines during much of the 20th century. Since then it has regained some – but by no means all – of its former range, and remains a relatively common inhabitant of most large woods, and many small ones, in Ireland, lowland Wales and the West Country. It also flies and breeds along hedged lanes in these regions.

Silver-washed Fritillary colonies are strictly confined to woodland everywhere else in Britain. They were once found as far north as southern Scotland, but in recent years have been virtually unknown beyond a line between the River Mersey and the Wash. South of this, the species became extinct or reduced to a scattering of small populations in all the eastern counties of England at its nadir in the 1970s. This even included the New Forest, once its most famous stronghold. Although still locally common within some New Forest woods, its abundance a century ago was legendary. According to F W Frohawk, 'It used to be in such profusion that it was common to see forty or more assembled on the blossoms of a large bramble bush.'

Colour and scent

Adult Silver-washed Fritillaries can be distinguished from other large fritillaries by the beautiful markings on the underwings. These lack the large, clear-cut silver patches of the High Brown and Dark Green Fritillaries, and look rather as if they have been given a watercolour wash of delicate greens and silver streaks, hence both the current name and its predecessor, the 'Greater Silver-streakt Fritillary'. A similar line is taken on the Continent, where in Germany it is called 'Perlmutterfalter' (the Mother-of-Pearl butterfly), and in France 'La Grande Nacre' (Great Mother-of-Pearl).

The upperside is equally distinctive in the male, owing

Distribution
Locally common in woods and scrubby lanes in south Wales, the West Country and Ireland; present in most large woods in the rest of its range.

to four thick, black ridges along the veins of the forewing. These are sex-brands, also known as androconial organs, that are formed from two types of modified scales on the wings. They play an essential role during courtship and mating, the behaviour of which is well understood, thanks to the classic research of Dietrich Magnus and others in Germany.

The Silver-washed Fritillary has one generation a year, with adults flying throughout July and August in Britain, reaching peak numbers around 1st August. Most modern colonies contain no more than a few dozen adults in an average year, although populations of hundreds, or even thousands, fly on the best sites. Adult numbers tend to be fairly stable in a particular wood from one season to the next, so long as no major felling occurs. But every so often there is a spectacular increase. This generally takes place after there has been unusually warm weather in early summer. Many naturalists in south-west England will remember the staggering numbers that emerged in 1976, when Silver-washed Fritillaries could be seen from the car along every wooded lane in large parts of Devon; I saw more than 100 at a time in some woods.

An aerobatic courtship

This fine fritillary is considerably more mobile than its smaller relatives. The adults glide at high speed above the treetops, flying from one end of their wood to the other in a matter of seconds, before descending into any sheltered opening in search of flowers, mates or places to lay eggs. Although in the West Country they make frequent sorties along wooded lanes, little mixing has been detected between colonies in neighbouring woods.

Adults of both sexes spend long periods on treetops drinking aphid honeydew, but they also frequently visit flowers. The male matures about two to three days after hatching, and is preoccupied for the rest of his life with finding a mate. He seeks his females on the wing, flying along a zigzag path, generally 1-2m above ground level, and concentrating on rides, wood edges and glades. The female is first recognised by her colour. He glides rapidly to investigate any golden

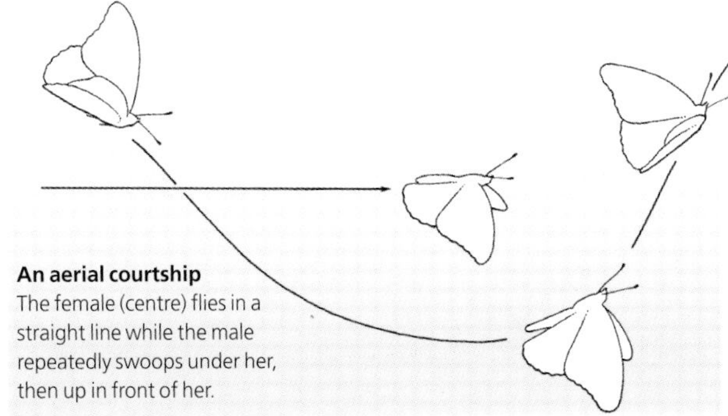

An aerial courtship
The female (centre) flies in a straight line while the male repeatedly swoops under her, then up in front of her.

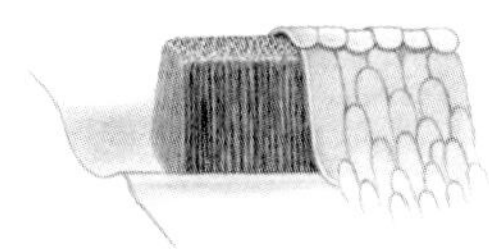

Scent scales
During courtship the outer scales of the male's sex-brands burst open, showering scent scales over the female.

object but is especially attracted by movement. This was well known to the old Continental collectors, who would make orange paper models and wave these on the end of fishing lines, netting the males that approached.

Even more irresistible to the male is the flickering pattern that is produced when a female flaps her wings. She, if willing, solicits him further by emitting a scent from the tip of her abdomen. This gives rise to a beautiful looping courtship flight which can be seen throughout July in the glades and widest rides of all our larger western woods.

Egg-laying

Except during courtship, female Silver-washed Fritillaries are comparatively inconspicuous until late July, when they begin egg-laying. They are then much in evidence as they flutter slowly through their woods, weaving between the tree-trunks in search of suitable sites. They have very particular preferences, choosing much shadier places than any of the other fritillaries whose caterpillars feed on violets. Thus they never lay in fresh clearings, although I have often seen females flutter around the edges of these before crawling a short distance into the neighbouring older coppice growth to lay in the base of a dense stool. But most eggs are laid under trees, or rather on the trees themselves, for the pale, cone-shaped egg is usually deposited singly in chinks in the bark, nearly always on the mossy north or west side of the trunk. Having found a suitable area, they then fly slowly above the ground, seeking the scent of violets. Any species attracts them, although the commonest in these situations is Common Dog-violet. They hover over the clumps, sometimes brushing the violets with their wings before fluttering on to the nearest shaft of sunlight.

It is fascinating to watch the egg-laying female as she crawls crab-like with closed wings up a tree-trunk, probing every crack with her plump, curved abdomen, then pausing to press firmly into a chink for a second or two, before the tip slowly recoils. This is a sure sign that an egg has been laid. She often lays two or three more on the same trunk before moving on, usually to the next tree.

Although most eggs are laid between 1m and 2m above ground level, I have frequently watched females through binoculars and seen them 6m up, probing and laying along

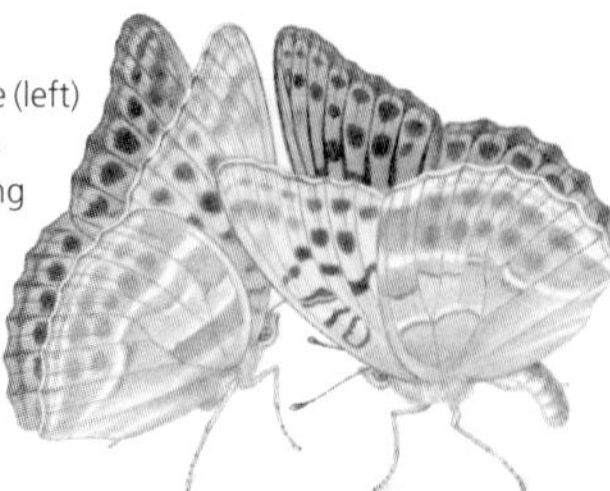

Courtship concluded
After their courtship flight, the female (left) lands on the ground. The male stands opposite with quivering wings, wafting scent scales over her. He then draws her antennae over the sex-brands on his upperwings, and the butterflies soon pair.

the mossy upper branches of ancient oaks. Indeed, any rough niche appears to be suitable. Eggs are sometimes laid on dead Bracken, and I have even had some on my clothes on a few occasions. I simply stood still in an ideal spot, and the passing females alighted one by one on my trouser leg, probing the material as they crawled up. They would eventually reach a seam, invariably laying an egg in it. Five eggs were laid within a few centimetres on one afternoon, all so firmly affixed that it was impossible to detach them: they perished in the wash.

Caterpillar and chrysalis

The egg hatches after about a fortnight, but the little caterpillar does no more than eat the eggshell before spinning a tiny pad of silk on which it hibernates, still on the tree-trunk. It descends to ground level the following spring, and immediately starts searching for violets on which to feed.

Temperature regulation

Making adjustments (1)
The butterfly keeps its body temperature near 32°C by opening or closing its wings.

Absorbing heat (2)
The upper surfaces of the wings are opened wide when the butterfly needs to absorb heat.

Keeping cool (3)
As it gets hotter, the butterfly raises its body above the ground, and begins to close its wings.

Reflecting heat (4)
When it is too hot, the undersides are exposed in order to reflect the sun's heat.

The caterpillar's feeding damage is easy to detect, for it leaves large, curved bites in the sides of the violet leaves. The caterpillar is much harder to spot. It spends most of the day basking in shafts of sunlight, often on dry, dead leaves up to 30cm away from the violet clump. Although it lies quite openly, the spiny brown body and yellow stripes blend perfectly with the sunlit background. By early June, the first caterpillars are fully grown, and wander in search of pupation sites. The few chrysalises found in the wild have generally been a few metres up in a tree or shrub, suspended beneath a leaf or twig, looking very like a curled up dead leaf with silvery patches.

Valezina *females*

Entomologists eulogise over all stages of this fritillary's life cycle, but there is one form that causes particular excitement. This is the lovely *valezina* type of female, in which the upperwings have a dusky greenish sheen and the underwings are distinctly pink. It looks strikingly different on the wing, and is quite common in some central southern populations. *Valezina* females were once thought to be more or less confined to the New Forest and neighbouring woods, but I have found them just as frequently in certain woods in the west Weald of Surrey, and I expect to see them in any population in north Dorset. Generally between 5% and 15% of females are of the *valezina* type in the larger colonies within the species' main range. There is, however, an abrupt cut-off to the west, north and east of the central south.

Just why *valezina* females are restricted to central southern England is a mystery, although the same patchy pattern occurs across the Continent. A gene that finds expression only in the female controls this form of the butterfly. It is dominant over the normal type when fritillaries of mixed parentage are bred in captivity, being produced in greater numbers despite being rarer in the wild. This indicates that *valezina* females must be at some disadvantage in the field, which keeps their numbers down. They are, for one thing, considerably less attractive to males than normal females. They also behave rather differently, avoiding sunny clearings and rides and instead skulking in shady woodland. This is possibly because their dusky bodies are prone to overheating. They are certainly capable of flying in much cloudier weather.

Changing fortunes

The increasing shade in many woodlands created by the decline in coppicing, together with the increased maturity of woods, caused the extinction of this fine fritillary in large areas of its former range. Its need for fairly open woodland is strikingly demonstrated whenever trees are thinned, for

▲ **Male upperside**
Four black sex-brands lie on the veins of each forewing.

▲ **Male underside**
Markings are 'washed' rather than clear-cut.

▲ **Female upperside**
Females have more extensive black spots and fewer linear markings.

◀ **Female**
Valezina form, found principally from Dorset to the west Weald.

▲ **Egg** [× 15]
Laid singly on tree-trunks, usually in bark crevices.

▼ **Caterpillar** [× 1½]
Back is striped with characteristic yellow lines.

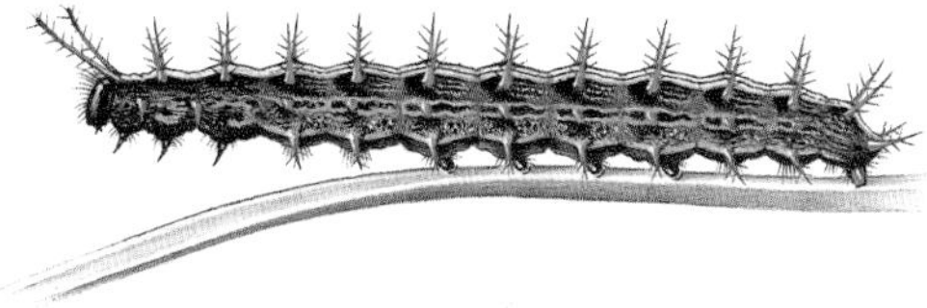

▶ **Feeding adult**
Adult *valezina* form, feeding at the flowers of Bramble.

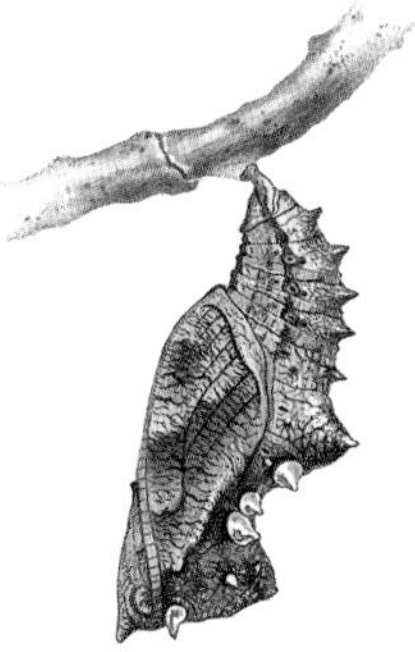

◀ **Chrysalis** [× 1½]
Camouflaged as a dead leaf and suspended among vegetation.

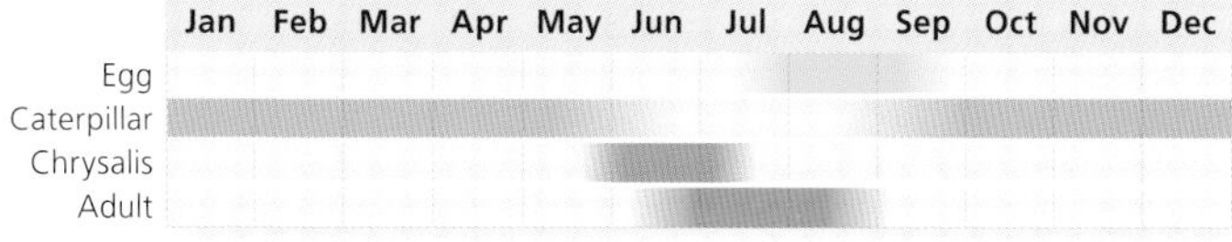

the butterfly's numbers shoot up from a few individuals to many scores during the next few years. Although generally tolerant of much shadier conditions than the other fritillaries that feed on violets, and often the last of the group to disappear, it cannot switch to breeding in rough, sheltered grassland. This means that it has been lost from many shady woods in which the Small Pearl-bordered Fritillary still lingers on in glades and along rides. A great many populations of Silver-washed Fritillary have disappeared from both mature conifer plantations and deciduous woods in which the canopy has more or less closed. This has been particularly severe on the heavy, flat soils of eastern England, where the butterfly was always much more localised.

The Silver-washed Fritillary has made a minor recovery during the past 25 years. This largely reflects a natural expansion into woodlands to the immediate north and east of its breeding sites in the Midlands and south-east England, although records are confused by the deliberate – and generally unreported – release of adults in some former localities. Although still extinct in Lincolnshire and Cambridgeshire, there are a few recent records from East Anglia, and it is more widespread than for many a year in the western Weald of Surrey and Sussex. Twenty years ago, its decline in Hampshire was so great that I wrote, 'Its status is parlous – Victorian entomologists who once flocked to the county for the *valezina* form would be shocked by the decline, and by learning that the Purple Emperor now occurs in nearly 50% more of the county's woods than this species.' Fortunately, it has now recovered much of this lost ground.

The butterfly's strongholds remain, as ever, in south-west England, Wales and Ireland. Colonies still breed in most woods of Dorset and Wiltshire, and in considerably larger numbers in Gloucestershire, Somerset, Devon and east Cornwall, where it is so widespread and common in some years that it is hard to imagine that it could ever decline; ominously, the same was said of the burgeoning New Forest populations 75 years ago. Many Welsh colonies seem to have disappeared, but it remains locally common and widely distributed at low altitudes over the southern two-thirds of the country.

It is difficult to assess any change in status of the Silver-washed Fritillary in Ireland, for until recently large areas were under-recorded. For example, Butterfly Conservation's survey of 2000-04 revealed its presence in 123 'new' 10km grid squares, in addition to the 166 squares in which it was recorded in 1995-99. This probably reflects little more than an increased recording effort. It is, nevertheless, welcome confirmation that this spectacular fritillary remains a widespread and locally common butterfly in many parts of the country.

Marsh Fritillary

Euphydryas aurinia

This lovely fritillary is one of the most rapidly declining butterflies in Europe. It is primarily a wetland species, and the chief cause of its decline has been the relentless drainage of meadows for agriculture. This process has been going on for centuries, but the combination of modern farming methods and subsidies has resulted in a devastation that would have seemed inconceivable to Victorian entomologists. It is, indeed, little more than a century since Marsh Fritillary caterpillars became so abundant in County Fermanagh that Irish villagers were forced to barricade their homes with peat bricks against the onslaught, and farmers raked up huge piles for burning. Questions were asked in the House of Commons about the problem.

Such outbreaks were a regular, but infrequent, occurrence throughout the 19th century. Another example occurred in County Clare, when the Revd S L Brakey drove to see a reported 'shower of worms', but instead found Marsh Fritillary caterpillars 'so multitudinous in some fields that a black layer of insects seemed to roll in corrugations as the migrating hosts swarmed over one another in search of food'.

Changing wing patterns

It would be wrong, however, to conclude that the Marsh Fritillary was ever common in the British Isles as a whole. Colonies were distinctly local and, more often than not, numbers were low. A typical population was monitored by the distinguished geneticist E B Ford and his father in Cumberland, from 1881 to 1935. They watched it through two periods when the butterfly was 'excessively common', which were separated by 30 years when it was very rare indeed.

The Fords noticed that the adults varied enormously in size and wing pattern during the period when numbers were increasing, but that they then settled down to a uniform pattern. This, however, was recognisably different from that of typical adults during the previous period of abundance. They concluded that this species possessed great inherent variety in its genetic coding, but that this was seldom expressed because most variants were comparatively unsuited to the prevailing conditions, a conclusion supported by recent genetic analysis. The majority of the variants survived only in the most favourable seasons, and it was in these years too that the colony expanded.

Distribution
A declining species on boggy and unfertilised grassland. Locally distributed in Ireland, but rare in the rest of its range.

The Marsh Fritillary can, indeed, be a most variable butterfly in appearance. The smallest males are a fraction of the size of the largest females, and although the chequered pattern is constant, the wing colours of both sexes can vary from bright to dull and from pale to dark. It is the latter specimens that were responsible for its early English name of 'Mr Dandridge's Midling-black Fritillary', later altered to 'The Dishclout' or 'Greasy Fritillary' on account of its waxy sheen, before finally settling down to 'Marsh'.

Emergence and mating

Marsh Fritillaries breed year after year in the same patch of grassland, which may be less than 0.5ha but occasionally is more than 20ha in size. They emerge once a year, starting in mid-May to early June, depending on locality and the warmth of the spring. The males emerge first, and are by far the more conspicuous sex. By the time the females emerge, most males are quartering the ground in low, zigzagging flights, searching for a freshly hatched mate. A typical female emerges early in the morning, waits an hour or two for her wings to dry, then crawls out into the open to sit with wings held wide open. This soon attracts a male, who alights beside her. He walks around to the front and flutters his wings, probably showering her with scent. Pairing quickly follows and lasts for about two hours. The male then disengages, but not before he has sealed her genitals with a 'chastity belt' of foam, which prevents rival males from fertilising her later.

The female spends the remainder of the day searching for a suitable plant of Devil's-bit Scabious on which to lay her eggs. Although just a few hours old, her swollen body may already contain 300 mature eggs – so many that flight is often impossible unless the weather is warm. She usually just crawls among the vegetation until she encounters a prominent, medium-sized plant growing in a warm sunny situation, with its leaves standing proud of the surrounding vegetation. She grasps the edge of a leaf between her legs, and curls her abdomen around to lay a batch of about 150 eggs, glued in neat rows to the undersurface. When this is complete, a second layer, and often a third, is cemented onto the first until, three hours later, the entire load has been laid. By then, her body is a slack and flabby bag, but she is now light enough to fly properly, and the last hours of the day are spent flitting between flowers to drink nectar.

Many females are caught by predators on the day of emergence, but those that survive quickly develop more eggs, enabling some to lay a smaller batch on their second day, and possibly a third after that. As in the Small Tortoiseshell and Glanville Fritillary, these are often placed on plants that already contain eggs, and huge clusters can be found. Keith Porter, on whose work much of this account is based, once found 1,500 eggs under a single scabious leaf in Oxfordshire, probably originating from five or more females.

Keeping warm
Marsh Fritillary caterpillars can become 20°C warmer than their surroundings by basking communally on their web. Individual caterpillars commute down to the cool scabious leaves to feed, then return to the warm huddle to digest their meal.

Temperature regulation

The eggs hatch after about three weeks, and the tiny caterpillars immediately spin a dense and conspicuous silk web, which is perforated with tunnels that lead to the centre of the plant. The original scabious is usually devoured within a week, causing the caterpillars to crawl, *en masse*, to a nearby plant, where the process begins again. This continues until late August. The caterpillars then moult for a third time, change colour from brown to black, and spin a new, denser nest deep among the grass. They spend the winter inside it, huddled together in small groups, each suspended in a pocket of silk. The silk insulates the caterpillars in a warm blanket of trapped air, and on swampy sites the whole nest can remain submerged for several weeks without the caterpillars coming to harm.

The bristly little caterpillars reappear on the first warm days of spring to spin a fresh web over another communal scabious plant. The air is often cold at this time of year, and the caterpillars must raise their body temperature to around 35°C if they are to digest their food. They achieve this by clustering together above the vegetation to form a black mass that absorbs the warmth of the sun.

Nests of basking fourth-stage caterpillars are easily found in early spring, particularly on cool but sunny days. Caterpillars are also conspicuous in their fifth and sixth stages, even though they separate into small groups, and eventually live singly. By now, they need to eat considerable quantities of scabious. The hungry caterpillars are surprisingly agile, and can spread out over the entire site when numbers are high;

▲ **Male**
Both sexes have a row of spots on the hindwing upper- and undersides. Underwing does not have silver patches.

▲ **Female**
Females are often larger and slightly paler than males.

▲ **Female, Irish form**
The Irish form *hibernica* has a brighter and more contrasting coloration.

▲ **Egg** [× 22]
Laid in large batches, sometimes three layers deep. The eggs gradually darken and hatch after about three weeks.

▲ **Feeding adult**
Adult on the flowerhead of Meadow Thistle, a favourite nectar source.

▼ **Caterpillar** [× 1½]
Gregarious when young, dispersing and becoming solitary when fully grown.

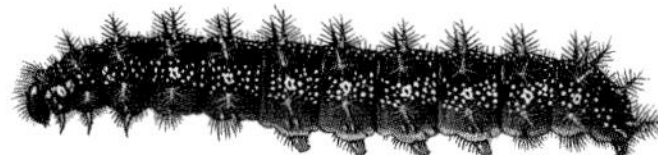

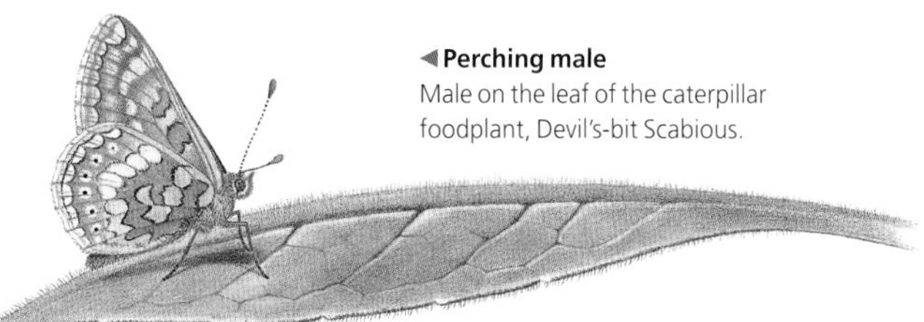

◀ **Perching male**
Male on the leaf of the caterpillar foodplant, Devil's-bit Scabious.

◀ **Chrysalis** [× 1½]
Suspended from a leaf or stem in low vegetation.

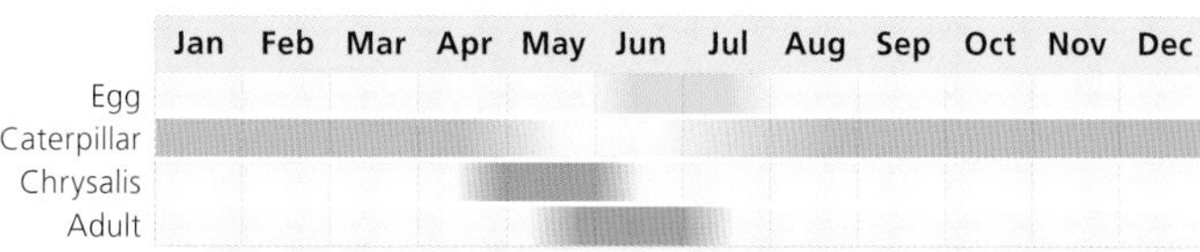

it is not unusual for every scabious to be chewed down to an unrecognisable stump. If there is still insufficient food, the caterpillars wander far from their fields, and will tackle other plants, especially Honeysuckle, in the surrounding hedgerows. Although Honeysuckle is the main foodplant eaten in Mediterranean habitats, few British caterpillars survive once the Devil's-bit Scabious has been exhausted, and in outbreak years starvation is a major cause of death.

Attack by parasitic wasps

Apart from a lack of food, the other main threats to Marsh Fritillary caterpillars are from ground beetles, spiders, bugs and tiny parasitic wasps. Birds are a lesser problem due to the protection afforded by the spiny bristles. Parasitic wasps are the greatest cause of death. Two species are involved: in southern England, Wales and southern Ireland, caterpillars are mainly infected by the wasp *Cotesia bignelli*, whereas in the north *C. melitaearum* is the major parasite. There is, however, a slight overlap; both parasites are found attacking colonies on Dartmoor. *C. melitaearum* is the same parasite that infests Glanville Fritillary caterpillars on the Isle of Wight, but so far as is known, *C. bignelli* has no other host. Both parasites are rather beautiful when examined through a hand lens, and are an integral part of this butterfly's natural history. Both merit conservation as much as their better protected host.

The life history of *Cotesia* wasps is an excellent example of how parasites are adapted to exploit their hosts. Each generation of caterpillars plays host to up to three generations of wasps, with individual caterpillars being attacked at any stage in their lives. Newly hatched caterpillars are afflicted in July, with the wasps injecting minute eggs into any caterpillar that has emerged from the safety of its nest. The eggs soon hatch into wasp grubs within the caterpillars' bodies. These first-generation grubs feed on the caterpillars' tissues, but do not cause their death until August, when the parasites are ready to depart to spin their own cocoons.

By the time that the wasps hatch from these cocoons and are ready to infect the next batch of caterpillars, their prospective hosts are already spinning hibernation nests. So the second generation of *Cotesia* grubs spend the winter inside the medium-sized, fourth-stage caterpillars within their webs. Infected caterpillars are killed in spring, and adult wasps emerge from their cocoons a week or two later. This leaves time for them to lay a third generation of eggs, which develop in the caterpillars as they enter their final stage.

The parasites in this third generation have a problem, for there are now no caterpillars available, and the wasps are unable to infect chrysalises, eggs or adult butterflies. So they must wait for the next generation of caterpillars to hatch in July before they can continue their destructive breeding cycle. This apparent difficulty has been neatly solved in two ways. In the first place, the wasps slow down the development of the parasitised caterpillars, so that they are still present when all the uninfected Marsh Fritillaries have pupated. Thus the wasps do not themselves form cocoons until the adult butterflies of the next generation are emerging. At this point, the *Cotesia* spin especially dense cocoons encased in silk. Although the tiny adult wasps soon develop, they do not bite their way out for another four to six weeks, by which time a new generation of caterpillars is ready for infection.

These parasites can inflict immense short-term damage on a Marsh Fritillary colony: up to 70 wasps may emerge from a single caterpillar, and in some years three out of four caterpillars are killed. However, the parasite seldom gets out of hand, for it suffers setbacks in certain years, when the infection rate drops to less than 10% of the Marsh Fritillary population. This sometimes occurs when there is a cool but sunny spring. Marsh Fritillary caterpillars capitalise by basking and developing quickly, but the *Cotesia* cannot warm itself inside its cocoon. The wasps therefore grow more slowly, and many Marsh Fritillaries develop quickly enough to avoid being parasitised in their last stage as caterpillars.

A history of decline

It is impossible to say how many Marsh Fritillaries emerge on our best sites at the height of a population explosion, but it can certainly run into tens of thousands. However, typical British colonies seldom contain more than 200 adults, and it is clear that many decline to a few dozen individuals in poor years. Caroline Bulman has shown that these smaller sites often experience temporary extinctions, and depend on periodic recolonisations from neighbouring colonies for their persistence. In other words, the Marsh Fritillary survives in the long term as a meta-population of loosely connected colonies. The same is true of many of our butterflies, but the need for several linked populations to co-exist in a landscape appears to be especially critical in this fritillary.

Sadly, the number of Marsh Fritillary colonies now breeding in Britain is a minute fraction of the former total. Across Europe, the situation has become so serious that this is one of the few butterflies afforded protection on a continental scale, including the statutory preservation of all breeding sites.

Today, probably more colonies survive in Ireland than in any country north of the Mediterranean, with the possible exception of Poland. They have a beautiful bright form of the butterfly, called *hibernica*. But even in Ireland sites are disappearing at an alarming rate. In addition to the straightforward drainage and destruction of breeding grounds, many unfertilised swamps are no longer grazed, which causes the Devil's-bit Scabious to become smothered by tall vegetation and therefore unsuitable for the butterfly. Not that the Marsh Fritillary thrives under heavy grazing either: except on certain English downs, it is seldom, if ever, found in wet meadows where the sward is less than 15-25cm tall. Owing to its special conservation status, a major initiative to protect this butterfly was recently launched throughout the island, consisting of surveys to locate overlooked colonies, and the protection and appropriate management of known sites. Although probably overlooked still in some parts of Ireland, it is commonest in the northern two-thirds of the island, with strongholds in Counties Fermanagh, Sligo and Donegal, and in areas west of the river Shannon.

The Marsh Fritillary is much more localised in Scotland. All known colonies are in the central west, breeding among damp mossy moorland and on acid raised- and blanket-bogs, mainly in Argyll along both the coastal mainland of Argyll and, especially, on the isles of Mull, Jura, Colonsay and Bute. Scottish Natural Heritage, helped by Butterfly Conservation, have, commendably, launched a major initiative to ensure controlled light grazing by cattle or ponies across more than 100 sites, encompassing 20 meta-populations of Marsh Fritillary, in this region, with the aim of enhancing these internationally important colonies.

Scottish colonies are still outnumbered by the surviving populations in Wales and England. Again, all are in the west, and in lowland England agriculture has taken an especially heavy toll on this species. The butterfly is now extinct in the eastern half of the country, including Sussex, which was once considered a stronghold. It still maintains a toe-hold in the Lake District, and there are scattered relic populations in Wales and the West Country, where it is still worth searching any marshy piece of land where Devil's-bit Scabious flowers in profusion. Look, in particular, in the Culm grasslands of north Devon and Cornwall, and the Rhos pastures and heathy commons of south Wales. As in Ireland and Scotland, a major programme to maintain key sites is being spearheaded by Butterfly Conservation.

Colonies on chalk and limestone

Against this background of declines, there is one area where the Marsh Fritillary has colonised a different type of habitat, and has actually increased over the past 75 years, although nothing like enough to offset declines on traditional sites. In the Cotswolds, Wiltshire and Dorset, Devil's-bit Scabious frequently grows on unfertilised chalk and limestone downs. These were not colonised before the 1920s, probably because sheep and rabbits kept the sward too dry and short. However, as downland became overgrown, an increasing number of Marsh Fritillary colonies has been found on these hills, mainly on warm but moist western slopes that catch the prevailing rains. Some of these colonies are large, especially on the military ranges of Salisbury Plain, which today support some of the most extensive populations to be found in the British Isles. Although Marsh Fritillaries may look out of place among the Adonis Blues and other chalkland species, it is a delight to see this endangered butterfly enjoying a modest recovery somewhere within its range.

Glanville Fritillary

Melitaea cinxia

I have had the pleasure of studying the ecology of all eight British fritillaries and, to my mind, the Glanville is both the loveliest and the most interesting. It is probably also the rarest, being confined nowadays to perhaps a dozen sites on the Isle of Wight, to a few on Alderney and Guernsey, and to two introductions on the English mainland. Any of them is worth visiting in May. They are beautiful locations in their own right, most being warm, sheltered undercliffs carpeted with wild flowers such as Thrift and Common Bird's-foot-trefoil, which form the butterfly's principal nectar sources. The Glanville Fritillary's delicate underwings are especially attractive seen against these pink and yellow blooms, and it is a fine butterfly in flight, whether gliding swiftly through still air or battling with whirring wingbeats against a sea breeze.

A typical emergence begins in mid-May, reaching a peak in early June, but declining quite rapidly so that few butterflies outlive the month. This is variable, however, and is affected by the warmth of the season. The first males are seen in April during a warm spring, and in these years there may be a very small second brood in August if the weather holds.

A diet of plantains

Adult Glanville Fritillaries live in fairly close-knit colonies, with most remaining in the chines or undercliffs where they emerged. There is, however, some interchange. We found that one in 50 marked adults switched colonies if their sites were within 500m of each other. Longer flights occasionally occur, for one frequently finds a group of caterpillars on plantains along a cliff edge, resulting from an egg batch laid up to a kilometre from the nearest permanent centre. Strays are also picked up all over the Isle of Wight, and they periodically form temporary colonies on the chalk downs and in northern parts of the island.

This butterfly is certainly the most agile of fliers. Males are especially active and conspicuous as they patrol back and forth above the seashore. Like all small fritillaries, they constantly dip to investigate any orange object in the hope of finding a virgin mate. The females themselves are less conspicuous, preferring to hide for long periods in dense tussocks. At this stage, they are double the weight of a male,

Distribution

Apart from two introductions on mainland Britain, now confined to a few warm coastal cliffs and undercliffs on the Channel Islands and south coast of the Isle of Wight.

▲ **Male**
Males are slightly smaller than females, with angular wings.

▲ **Female**
Females are often darker than males; the wings are more rounded.

▲ **Male**
Aberrant *wittei* form, with less chequered markings, caught on the Isle of Wight in 1929.

◀ **Egg** [× 22]
The white eggs soon turn primrose-yellow. Shell has 16-20 vertical ridges, or keels.

▼ **Caterpillar** [× 1½]
Mature caterpillar feeds in the open on plantain leaves.

▼ **Chrysalis** [× 1½]
Concealed deep inside dense vegetation.

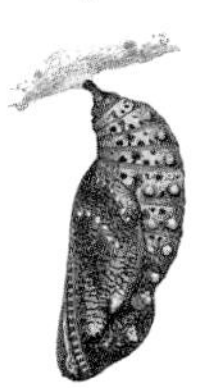

Side view

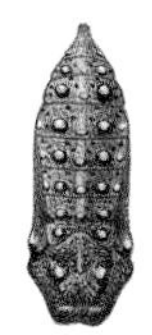

View from above

◀ **Feeding adult**
Common Bird's-foot-trefoil is the main adult nectar source on many sites.

▼ **Resting adult**
Adult on leaf of Ribwort Plantain, the caterpillar foodplant.

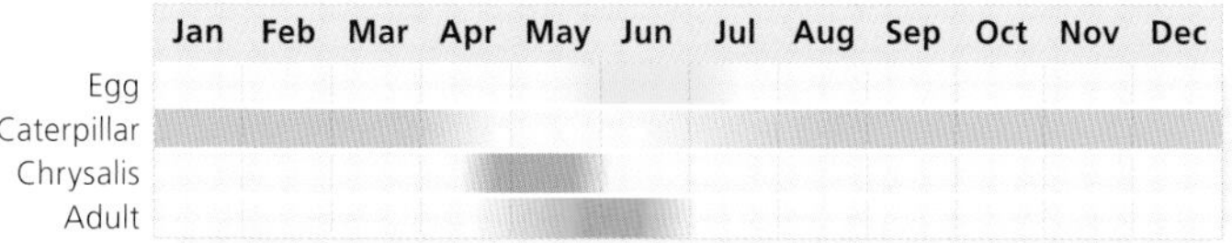

but for all their burden of eggs, they are far more agile than their close relative, the Marsh Fritillary, which can just about lumber off the ground when egg-laden, and then only in hot weather.

Mating usually occurs around midday, and is a conspicuous affair, for the female may continue to fly between flowers, dragging the male behind. She then starts searching for a site in which to lay her eggs. This takes a considerable time, for eggs are laid in batches of 50 to 100, and great care is taken in selecting a suitable spot. Most eggs are laid in sheltered nooks, where the ground is warm and well drained, and where abundant clumps of very young Ribwort Plantains grow vigorously in a short, sparse sward, with much bare soil around. The female avoids medium- and large-leaved plantains, even though these are often more abundant on her sites and, in captivity, provide equal nourishment for the caterpillars. In other words, she is more concerned with placing her eggs in a very warm environment than ensuring her offspring have the maximum supply of food. In good years, an ideal plant will often receive two or three egg batches from different females. The white eggs are deposited in layers on the underside of a young plantain leaf, where they soon turn an attractive primrose-yellow, and hatch two to three weeks later.

Conspicuous caterpillars

The Glanville Fritillary has the most conspicuous caterpillars of any British butterfly. They live gregariously from August until March, in dense webs of white silk spun over the clumps of young plantains. The caterpillars spend much of the day basking on top of these webs, where their black, bristly bodies absorb warmth from the sun in the same way as the Marsh Fritillary.

Using a miniature laser thermometer, David Simcox and I found that Glanville Fritillary caterpillars can be active in air temperatures as low as 7°C. This is partly because the egg-laying spots, so carefully chosen by their mothers, are invariably several degrees warmer than the ambient, but mainly because once the temperature in their niches exceeds a threshold of 13°C, they can warm their bodies by a further 20°C through huddling together in a black mass that absorbs heat from the sun. We frequently watched an individual caterpillar rush out from the warmth of its siblings to snatch a meal on fresh plantains, say 50cm from the nest. It generally had less than two minutes in the cool air to feed before the body temperature dropped below 20°C, at which stage it darted back in a wriggling sprint to warm up among its nestmates. Groups of caterpillars regulate their temperatures at 33-34°C with remarkable precision, and take care not to overheat. If this maximum is exceeded, they retreat into shade or extend their bodies to expose silvery connecting sections between each segment, which we believe reflect rather than absorb the sun's rays.

As the caterpillars grow they shed their skins several times, always within the safety of the nest. After the fourth moult, when the caterpillars are under 1cm long, they move to slightly taller vegetation where they spin much denser nests. These contain numerous little pockets of silk, suspended like miniature hammocks in a large communal chamber within the nest, and here the caterpillars gather to hibernate in small groups. Hibernation nests are easy to find in September and early October, until the first gales of autumn blow surrounding grass clumps down over them.

Escape underfoot

The caterpillars remain dormant until March, when they swarm out on sunny days to spin a fresh web over the young, newly sprouting plantains. They can be extraordinarily abundant, concentrating on warm, open patches such as footpaths, where it was once feared that many were trampled by ramblers. They have, in fact, an escape mechanism, admirably described over 200 years ago by Moses Harris: 'They are so remarkably timid that should you stir the Plant they are on, tho' never so little... they instantly roll themselves into the form of a Hedge-hog.' They also roll into the safety of the dense vegetation.

Glanville Fritillary caterpillars enter their final phase in April, and are extremely handsome when fully grown, with shining, russet-coloured heads and black, bristly bodies. At this stage the nests are abandoned, and the caterpillars swarm out over the plantains, greedily devouring the leaves. They still show a strong preference for young plants, but these are often exhausted, forcing the caterpillars to eat leathery old growths and, as a last resort, the strangely lobed leaves of Buck's-horn Plantain. Even these are sometimes used up, and although the caterpillars can cover considerable areas by running along paths and across short turf, many starve in certain years. Robin Curtis writes that in recent years 'the Promenade at Wheelers Bay, Ventnor, was often subject to swarms of caterpillars which so fascinate the public'.

As with other nymphalid butterflies, the spiny masses of Glanville Fritillary caterpillars are seldom attacked by birds or small mammals. However, this advantage is offset by their increased vulnerability to parasitic wasps, which find the conspicuous groups of basking hosts an easy target. Their main parasitoid is a small braconid called *Cotesia melitaearum*, which has no other known host on the Isle of Wight, and which kills anything from 5% to 75% of caterpillars on the sites where it occurs. Rather like the *Cotesia* wasps that parasitise the Marsh Fritillary (see page 224), *C. melitaearum* has two generations a year, and their grubs retard the development of infested caterpillars so that emerging wasps coincide with the presence of fresh batches of Glanville caterpillars to sting in mid summer, and again in spring.

The pale yellow cocoons of *Cotesia* are easy to find alongside the dead bodies and cast skins of caterpillars, in old silk nests that remain smeared across plantains in May and June on most Glanville Fritillary sites. Interestingly, in line with Swift's doggerel that '... *a flea has smaller fleas that on him prey; And these have smaller fleas to bite 'em, And so proceed* ad infinitum', many of these cocoons are themselves parasitised by a tiny ant-like wasp, *Gelis agilis*, which lays its eggs into the bodies of the pupating *Cotesia* grubs, where they develop to produce this hyper-parasite.

Breeding on the undercliffs

Although the Glanville Fritillary is a common and widespread insect in warmer parts of central southern Europe, it reaches the northern limit of its climatic tolerance in the British Isles, and so is confined to exceptionally warm places where there is also a regular supply of many thousands of young plantains. Such conditions occur along much of the south coast of the Isle of Wight, although they may not persist for long in any one place. This spectacular coastline is famed for its different geological formations: Glanville Fritillaries are found mainly on the well-drained sandstones, principally between Freshwater and Ventnor. They also breed on the south-facing slopes up several chines, but only in the first 100m or so in from the sea, where the ground is unstable and the vegetation sparse.

It is the instability of these cliffs and undercliffs that is responsible for the local prosperity of this rare butterfly. Some erosion occurs every winter, either as little slippages or major falls, and a flush of fresh plantains regenerates in the sandy exposures every spring. Many caterpillars are lost during the severest of landfalls, but enough survive to repopulate the new patches. It is far more serious if there has been little slippage for more than a year or two. Then a tall, dense sward develops that is utterly unsuitable for egg-laying.

For many years, David Simcox and I measured the changes in habitat and numbers of Glanville Fritillaries on the Isle of Wight, a task subsequently undertaken by Robin Curtis for more than a decade. At any one time, about 12 sites on the island typically support this butterfly, although some contain more than one population. However, only eight sites appear to have supported colonies more or less continuously since 1824, the year when the young Edward Newman, 'with a feeling of triumph', discovered the Glanville Fritillary on the island. The butterfly's numbers fluctuate wildly, and not necessarily in synchrony on neighbouring sites. We found a strong correlation between their density on a site in any year and the abundance of young plantains growing in warm conditions, an association subsequently confirmed by Robin Curtis, who also found a strong positive influence of warm summers on numbers. Crashes can also occur following years of high abundance, when the plantains are gnawed back to their roots. Smaller breeding patches regularly lose their Glanville Fritillaries and are recolonised a few years later. Again, the quality of the habitat, not the absolute number of plantains present, is the main factor that determines whether a potential breeding area is colonised or not. Its proximity to a neighbouring colony is also important. These dynamics are in accord with classic long-term studies of Glanville Fritillary meta-populations made by Illka Hanski in Finland.

Extinction on mainland Britain

Apart from a few generally short-lived introductions, of which one survives in Somerset and another in Hampshire, the Glanville Fritillary has long disappeared from the British mainland. It is difficult to gauge just how widespread the butterfly once was, for there is disagreement over the authenticity of some early records. The last, and most famous, mainland colonies were on the south-east Kent coast between Folkestone and Sandwich, where the species mysteriously disappeared in the late 1850s.

Other mainland colonies apparently bred on rough ground beside woods. These were almost certainly situated along warm, sheltered edges that were irregularly tilled. There are few contemporary accounts, and the butterfly was always a rarity, although known from as far north as Lincolnshire in the 1690s. It was here, indeed, that this fritillary was first discovered by Eleanor Glanville, after whom it was later named. Prior to this, the butterfly was christened the 'Lincolnshire Fritillary', later downgraded to the 'Dullidge Fritillary', after the more accessible populations around Tottenham and Dulwich; later still, it acquired the apt but unexciting name of 'Plantain Fritillary'. Finally, some 40 years after Eleanor's death, the name was changed again, as described by Moses Harris in one of the most famous passages in early entomology:

> 'This Fly took its Name from the ingenious Lady Glanvil, whose Memory had like to have suffered for her Curiosity. Some Relations that was disappointed by her Will, attempted to let it aside by Acts of Lunacy, for they suggested that none but those who were deprived of their Senses, would go in Pursuit of Butterflies.'

In fact, it was Eleanor Glanville's own son, Forest, who contested the will, and unfortunately for his mother's reputation, the challenge was upheld.

Heath Fritillary

Melitaea athalia

This attractive, small fritillary flies in a variety of warm habitats on the poorer soils of southern England. It breeds in patches where the vegetation has recently been cut, burned or cleared, but is exceedingly rare: colonies today are found only in a few sheltered valleys on Exmoor, in three woodland clearings in Devon and Cornwall, and in a handful of woods in Essex and east Kent. Yet the Heath Fritillary can still be abundant in optimum habitat. In good years, over 10,000 adults can emerge in a single colony, making a wonderful sight in June, with the males in particular evidence as they encircle small clearings, hunting for hidden females.

Rare though this fritillary may be, its prospects today represent a marked improvement on its plight in the late 1970s, when we expected it rapidly to follow the Large Blue to extinction. Although always a local species in Britain, the Heath Fritillary had been in continuous and serious decline for well over a century. By 1980, it was reduced to the large woods at Blean, near Canterbury, and to five woods in the West Country, although it must also have survived somewhere on Exmoor, despite the failure of surveys to find it. The outlook was equally depressing: two sites that had supported huge populations when they were set up as nature reserves had already lost their colonies, and the future management of existing sites made it unlikely that their Heath Fritillaries would survive.

Back from the brink

After the loss of the Large Blue, rescuing the Heath Fritillary became a matter of urgency in conservation circles. Fortunately, Martin Warren, now Chief Executive of Butterfly Conservation, provided the essential knowledge to save it. Warren's painstaking research revealed two unsuspected limitations to its persistence – low adult mobility and a specialised, yet ephemeral, breeding habitat for the caterpillars – constraints that are familiar enough to conservationists today, but which were novel concepts at the time.

It would have been sad if this fritillary had been lost, for it is an insect of charm and some beauty. It is on the wing from mid- to late May in the West Country, where it reaches a peak in the first third of June. Kentish colonies emerge a fortnight later, probably because spring temperatures are cooler there, and in warm years may produce a small second flight in August. Depending on the warmth of the season,

Distribution
A great rarity found in sheltered Exmoor valleys, and in woodland clearings in Devon, Cornwall, Essex and Kent.

▲ **Male**
The amount of black on the upperside varies between individual butterflies.

▲ **Female**
The ground-colour of the female is often lighter than that of the male.

▲ **Male**
Specimen showing unusually heavy markings.

▲ **Resting adult**
Male resting on the flowerhead of a plantain.

▲ **Egg** [× 22]
Laid in batches of up to 150, on or near the leaves of a variety of foodplants.

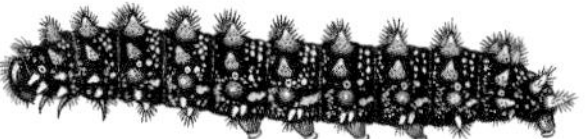

▲ **Caterpillar** [× 1½]
The caterpillars are gregarious at first, later dispersing.

▲ **Chrysalis** [× 1½]
Attached to a dead leaf or stem, usually near the ground.

▲ **Feeding adult**
Adults feed at a number of flowers, including Oxeye Daisy.

	Jan	Feb	Mar	Apr	May	Jun	Jul	Aug	Sep	Oct	Nov	Dec
Egg												
Caterpillar												
Chrysalis												
Adult												

most adults live for five to ten days, seldom straying from their breeding sites.

Female Heath Fritillaries spend most of their lives basking or hidden in the ground vegetation, but the males are extremely conspicuous. They are sun-loving insects, seldom flying when the temperature drops below 18°C, and then only if the sun shines. They quickly take to the wing when conditions are suitable, and are airborne for nearly half the day. Theirs is a smooth, flitting flight, powered by little flicks of the wings, which are then stretched flat as they glide just above ground level, traversing clearings in search of mates. Between flights, they often pause to feed on flat, open flowers.

Early development

Females mate within a few hours of emergence, and wait several more days while their eggs slowly ripen. Heavily laden, they then embark on ponderous flights, fluttering just above the ground vegetation before alighting to continue on foot. They often pause to bask in their wanderings, then burrow again in the undergrowth before settling to lay a batch of gleaming, pale green eggs. These are ejected at a rate of two a second, until a mass of perhaps 80 to 150 eggs has been laid, usually on the undersurface of a small Bramble or dead leaf, close to a foodplant. Smaller batches are occasionally found, but probably emanate from old females that have already made their first large drop. The eggs turn an attractive lemon-yellow, hatching after two to three weeks. The young caterpillars eat the eggshells and then move, *en masse*, to the nearby foodplant, where they spin a flimsy web. But soon they disperse into smaller bands, typically of ten to 20 caterpillars apiece, and again spin fine webs as they roam from plant to plant. These are not difficult to find if you search on warm, sunny days in August.

After feeding for about a month, the spiny young caterpillars enter hibernation, sometimes in small groups but usually singly, fastened to a silk pad spun in the curl of a dead leaf. They reappear on the first warm days of March or April, to bask and feed whenever conditions permit. Caterpillars are inactive when the air is colder than 12°C and, like those of the Marsh Fritillary, spend long periods basking to raise their body temperatures high enough to digest leaves. By the time they are fully grown in May, the curious blackand-amber caterpillars can be found with ease on any site where the vegetation is sparse, resting on dead leaves in ones and twos, or lying openly in the sunshine on any of their foodplants.

A wide-ranging appetite

It is always curious when a rarity such as the Heath Fritillary turns out to have several foodplants, some of which are common. The main food eaten by the caterpillars varies with the location. In woods they generally eat Common Cow-wheat, a lovely plant of mainly light acidic soils, which lives semi-parasitically on the roots of certain grasses, shrubs and trees. This is the only food of the eastern colonies, but on Exmoor, where Common Cow-wheat grows sparsely, older caterpillars sometimes switch to Foxglove and, on occasions, to other plants. In the abandoned hay meadows among woods and elsewhere in Devon and Cornwall, Ribwort Plantain and Germander Speedwell are the main food, although here again other plants may be tackled.

Given this lack of selectivity, a shortage of food is clearly not responsible for the Heath Fritillary's decline. Instead, a critical factor is that their foodplants must be lush, yet growing in a relatively hot environment. In woodland, the Heath Fritillary breeds only in the warmth and shelter of recently formed clearings, where its foodplants briefly flourish but then become too overgrown for the caterpillars only three to ten years later. This niche can be extraordinarily narrow: in the densest coppices at Blean, Heath Fritillaries will colonise an area in the first year after a clearing, increase to perhaps 10,000 adults by the second year, only to be completely shaded out two years later. A similar sequence occurs on freshly burnt heathland on Exmoor, and so, too, when a sparse hay meadow is abandoned, albeit on a slightly longer time-scale.

Poor adult dispersal

The remarkable speed with which events can change for this species was recognised by the Victorian collectors, who knew the Heath Fritillary as a butterfly that 'followed the woodman'. It is a pity that early conservationists did not heed this saying, for when the first two nature reserves were obtained for the butterfly, neither was cleared on a regular basis, and consequently their vast Heath Fritillary colonies disappeared. But although habitat change has been identified as the cause of most disappearances, poor dispersal is of equal importance.

Because adult Heath Fritillaries are so sedentary, they struggle to reach fresh clearings unless these arise within 200-300m of an existing colony. Even in the woods at Blean, many isolated clearings are never colonised in the four years that they remain suitable. At first sight, it seems surprising that any butterfly should have evolved to become dependent on a habitat that lasts for just a few years in one spot, while at the same time lacking the mobility to discover new clearances as and when these arise. The same problem is seen in several other species, including the small violet-feeding fritillaries, the Silver-spotted Skipper and some blues. There is a simple explanation, the key being that these butterflies became adapted to particular conditions that prevailed in the British countryside during the past 12,000 years, but suffered badly in the 20th century when radical changes occurred to the structure and dynamics of their habitats.

The effect of climate

To understand the constraints on these species, it is necessary to go back 13,000 years to the time when, at the height of the last Ice Age, all butterflies would have been eliminated from Britain. As the ice receded, our current species are believed to have reoccupied the country, spreading from southern Europe and beyond, roughly 10,000 years ago. There then followed a period of 5,000 or 6,000 years when average summer temperatures were 2-3°C warmer than today, rather similar, in fact, to those now occurring in central Europe.

Anyone who travels south to the Dordogne or Rhône Valley will be struck by the abundance of many species – and not just butterflies – that are rarities in southern Britain. A closer examination reveals that these breed in much more overgrown conditions under a warm climate. Adonis Blue, Large Blue and Silver-spotted Skipper butterflies, for example, are found in ankle-deep limestone grassland in central France, and 'woodland' species thrive not only in shadier woods but also in hayfields and overgrown grasslands. Nor are these species confined to warm, south-facing slopes, as is often the case in England, but breed on all aspects and on flat ground. In other words, in regions where the climate is warmer, these heat-loving butterflies can breed in most of the places where their foodplants grow, and are not restricted to a few exceptionally warm spots.

It is reasonable to suppose that the Heath Fritillary had similarly less exacting requirements when it reinvaded Britain after the Ice Age, because the climate was then distinctly warmer. It may still have been fairly localised, for much of the country was under mature forest (the 'Wildwood'), but at least the butterfly would not have depended on fresh clearings for its survival.

Man-made habitats

About 7,000 years ago, humans started to fell the Wildwood. It is likely that Heath Fritillaries immediately colonised these Stone Age clearings. Then, in about 2500 BC, the climate cooled by 4°C or more. At this point, the artificial habitats created by man probably became vital refuges for our heat-loving butterflies. Thus Adonis Blues, Large Blues and Silver-spotted Skippers may have become restricted to heavily grazed patches on the south-facing southern downs, where the grass was so short that the soil baked in the sun. Silver-studded Blues were probably confined to hot, freshly burned areas of heath, and in woodlands several of our fritillaries may have found warm refuges in places that were regularly cleared through coppicing.

At this stage, it seems likely that the Heath Fritillary not only became trapped in man-made coppices (and on heaths that were burned back with similar regularity), but that it also became more sedentary. For from now on, any individual with a tendency to migrate would be likely to die without finding a new breeding site, whereas its less adventurous siblings were almost guaranteed to find a fresh clearing in the adjacent panel of managed woodland. We know from studies of other butterflies that the adults can become less mobile within ten to 30 generations when circumstances change in a way that puts migrants at a disadvantage. Having evolved a sedentary form, the Heath Fritillary was unlikely to become more dispersive during the last century, when the supply of new breeding sites diminished greatly across the country.

Permanent protection

It is hard to say just how widespread the Heath Fritillary was during the heyday of coppicing, for the practice was already in abeyance when the first comprehensive records of butterflies were compiled. It may always have been restricted to warm, light soils. Yet although clearly a local species in the 18th century, it was nonetheless familiar to the first collectors. It was, indeed, described as early as 1699, and called the 'May Fritillary' to distinguish it from the 'April Fritillary' (the Pearl-bordered). Both names became less appropriate after the calendar changed in 1751, and for much of the 18th and 19th centuries it was known instead as the 'Pearl Border Likeness'.

In retrospect, it appears that Warren's twin discoveries of the Heath Fritillary's feeble dispersal and short-lived specialised breeding habitat were made just in time to save its last populations from extinction. By the early 1980s, all sites were on borrowed time, and within a decade nearly every colony that was not receiving special management to encourage new breeding patches had, as predicted, become extinct. Not that the conservation effort was easy or continuously successful. With such a short-lived habitat, it was inevitable that in some localities there would be occasional breaks in the chain when creating new supplies, and not all advisors and wardens are equally adept at translating paper recommendations into appropriate site management. Like the conservation story of the Large Blue, it was frequently 'two steps forward and one step back', and occasionally *vice versa*.

On Exmoor, there was an unexpected bonus in the 1980s when a number of large colonies was discovered, many on National Trust land. These populations bred in sheltered heathland combes, and waxed and waned for 20 years, reaching a low point around 1999 when it was feared they might be lost. Fortunately, they came back from the brink following regular, small-scale burnings of the Bilberries or Gorse within inhabited and neighbouring valleys, whereby about a fifth of each site is now burned back every second or third year, and burnt patches are left to recover for ten to 15 years before the next firing. Equally crucial is the control of Bracken in the year or two after a fire, either by rolling or, in extreme cases, spraying. Together, these measures are creating flushes of fresh Common Cow-wheat among the Bilberries of Exmoor, and for the past 15 years, between ten and 15 Heath Fritillary colonies have been maintained and monitored by Butterfly Conservation across clusters of sites south and west of Minehead.

Elsewhere in the West Country, colonies were more isolated and became extinct at one stage or another. Today, the fritillary is confined to three sites in the Tamar and Lydford valleys of Cornwall and Devon, of which two are the result of successful reintroductions. One of these was on the Duchy of Cornwall land at Luckett, to which the butterfly was reintroduced in 2002 on rough, disturbed grassland and clearings where, on the Prince of Wales' instructions, conifers had been grubbed up to restore the butterfly's habitat; 'good numbers' were reported in 2013. The other is an abandoned railway track at Lydford, a site I first knew in the 1970s when it supported all five violet-feeding fritillaries, as well as Heath and Marsh Fritillaries. The spot was also known to an insatiable butterfly collector, whose bags included many Heath Fritillaries with striking aberrations to their wing markings. Although an old foe (whom I ejected several times from the last colony of Large Blues), we cannot attribute the extinction of the Lydford colony to over-collecting; it is more likely that the ground became overgrown. In 1993, after the site became a Butterfly Conservation reserve and was better managed, the butterfly was reintroduced and, in 2013, was 'doing very well'.

There have been losses and successes, too, in Kent. Much of the Blean woodland used to be in private hands, and the butterfly predictably disappeared from these areas. However, large areas have now been purchased and are managed as nature reserves by the RSPB, Natural England and the Kent Wildlife Trust. Following 25 years of annual coppice clearings, together with ride-broadening to encourage greater movement, there has been a welcome increase in the butterfly's numbers, and it is now flourishing again. Although enhanced by fair weather, the emergence of 2009 was exceptionally large, not least in the RSPB's newest clearing where, according to Nigel Bourn, adults 'swarmed'. By 2012 the butterfly was breeding over a larger total area of habitat in the Blean complex than at any time since 1980, with no fewer than 22 colonies located, of which six were large.

Equally heartening are several sites in Essex to which Heath Fritillaries were re-introduced in 1984 and the early 1990s after coppicing and wood clearances were restored. At least five of these populations were thriving in 2013, with 'hundreds' reported in 2013 in Essex Wildlife Trust and local Council reserves east of Basildon. Butterfly Conservation is therefore correct to conclude that 'The continued survival of this rare species in Britain is a notable conservation achievement.' Although survival is not yet assured, I would go further, and am optimistic for its medium-term future on two counts: first, management advice for nearly every landscape inhabited by the Heath Fritillary is now provided by the expert ecologists at Butterfly Conservation, whose understanding of this butterfly's needs is second to none; and second, as our climate warms, this delightful fritillary may extend its breeding into cooler, shadier or more overgrown habitats, including sheltered grassland, that require less frequent or less fine-tuned management than at present.

Speckled Wood

Pararge aegeria

The Speckled Wood has coped more successfully than any other butterfly with the fundamental changes that have occurred in British woodlands over the last century. A lover of dappled shade, this attractive brown has spread continuously since the 1930s through the abandoned coppices, mature deciduous woods and conifer plantations of Wales, southern England, Ireland and central Scotland. This represents a welcome return to most, but not quite all, districts from which the Speckled Wood had mysteriously disappeared in the second half of the 19th century. This species has even spread to some previously unrecorded regions.

Regional variation

Adult Speckled Woods can be seen at any time from late February until early November. Their chocolate-and-cream markings are unmistakable, although these vary somewhat with the season and in different parts of their range. Females have slightly larger pale patches than the males, and both sexes are darker later on in the year, when second and third broods of adults emerge. We illustrate the seasonal variation of the main subspecies, *tircis*, found in the British Isles, along with a subspecies called *insula* that is confined to the Isles of Scilly. In the latter, the pale patches are tinged with orange. The burgeoning Scottish populations belong to a third group, called *oblita*, which has larger wings and more contrasting markings than the other subspecies. In addition, its undersurfaces often have a hint of purple around the outer edges.

Few butterflies have been better studied than the Speckled Wood in recent years. Of particular interest is the way in which variation in adult wing markings and body shape at different times of year within a colony, or at different latitudes, makes individual adults better able to fly, disperse, reproduce and compete with rival Speckled Woods under prevailing conditions of weather and climate. For the most part, this account draws on the comprehensive studies of Tim Shreeves, Nick Davies and Jane Hill in Britain, and, inevitably, of those giants of European butterfly biology, Christer Wiklund and Hans van Dyck, who in turn had many co-workers.

Distribution
Expanding in range. Common in woods, scrub and tall hedgerows throughout Ireland, lowland Wales and much of England.

Overlapping generations

Speckled Woods are on the wing in the British Isles for eight or nine months of the year, although individual adults seldom live longer than a week. The species has the unique ability among our butterflies to overwinter either as a caterpillar or as a chrysalis. Hibernating chrysalises produce the first adults of spring, but these overlap with others that spent the winter as caterpillars. This results in two early peaks of emergence, the first in mid-May and the second in early June.

The offspring of both sets of adults then develop at variable rates, producing a second brood of butterflies that is strung out over a very long period indeed. The earliest-born and quickest developers produce adults that overlap with the last stragglers from the first brood, but several weeks pass before the second brood reaches its peak, usually in late August or early September. Speckled Woods are at their most abundant in these months, even though some springtime caterpillars do not produce adults that year, but instead enter a period of quiescence, or aestivation, before resuming growth in early autumn and forming chrysalises that hatch the following spring.

There is usually a final emergence of fresh adults in autumn. These are the offspring of the second brood, with which they also overlap. This third brood is seldom very numerous, because by late summer most caterpillars develop slowly and overwinter or form hibernating chrysalises. The main trigger that induces certain caterpillars to develop straight through to adulthood and others to hibernate is a combination of the length of daylight and the air temperature. As a rule, caterpillars that experience 12 or more hours of light a day go on to form adults in the same year, whereas those that receive less light either hibernate or form overwintering chrysalises. There is, however, considerable variation in the response of individuals to light and temperature, and the different races and subspecies also vary in their response. For example, the Speckled Woods of Scotland, Wales and southern England develop at markedly different rates when reared under the same temperatures and day-lengths, and by and large this makes them better suited to the climatic conditions that each experiences in the wild.

Perching and patrolling

Speckled Woods live in self-contained colonies, although the females wander further than most colonial butterflies. A large woodland colony contains several hundred adults in late summer, with the two sexes in roughly equal numbers. The males, however, are much more apparent because females hide on treetops or in bushes for much of the day. Both sexes roost at night and during cool weather, clinging to the undersurfaces of leaves, high up on tall shrubs or in the tree canopy. They crawl around to catch the morning sun, basking with their wings held wide open and pressed flat against the top of a leaf, before closing them to avoid the heat of the day. By this means, they regulate their body temperature to around 32-35°C. They also feed on treetops, drinking the sweet honeydew which coats many leaves in summer, descending to flowers in autumn and spring when honeydew is scarce.

When warm and well fed, the males float down to the ground, where they either perch in sunlit spots for the rest of the day or patrol back and forth through their breeding area. They are looking for females in both cases, and many studies have been made of how different individuals achieve this. In general, male mating behaviour varies with the weather, the time of day and the season, although some individual males are intrinsically more likely to be 'perchers' or 'patrollers', and have wing markings and body shapes better suited for one or other activity.

In the coolest weather, and especially in spring, the males perch in full sunshine in open rides, where they sit about 1m above the ground on a stem of Bramble or on a frond of Bracken. In slightly warmer conditions the rides become too hot, and males instead occupy sunlit spots where a gap in the canopy allows the sun to cast a pool of light onto the woodland floor. Small sunspots are inhabited by a single male, who fiercely protects his territory against intruders. Large patches may have two or more males in residence. Every passing insect is investigated. When this proves to be another male, he is ferociously attacked in an aerial skirmish in which both insects spiral around each other, bumping and clashing their wings as they ascend towards the canopy. After a minute or two, the intruder flies away, allowing the original occupant to descend to resume his vigil.

Virgin females are also attracted to these patches of sunlight. But they, when challenged, generally drop to the ground, where a courtship dance follows that is similar to that of the Grayling (see pages 255-6). The female then leads the excited male to a treetop, where mating occurs.

In the warmest weather most males become patrollers, and make slow, sustained flights through the canopy, searching sunlit gaps for mates on the wing before indulging in the same intricate courtship. Patrolling males have a higher strike rate of winning females than when they are perching, but since they lose heat gliding through the cooler air, many are reduced to perching in sunspots under all except the warmest weather conditions. The need to control their body temperatures at 32-35°C whilst competing for mates explains much of the physical variation seen in Speckled Woods. Males with dark wings heat up quickly, and are better able to patrol from one sunlit patch to another

▲ Male, first brood
Main subspecies, *P. a. tircis*; first brood has larger yellow spots than those that follow.

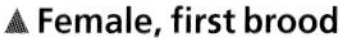

▲ Female, first brood
Main subspecies, *P. a. tircis*, has larger yellow patches than the male.

▲ Male, second brood
Subspecies *P. a. insula*, found on the Isles of Scilly, has more orange markings.

▼ Basking adult
Males with three spots on each hindwing often behave differently to those with four spots.

◄ Egg [× 15]
Laid singly on the underside of a grass blade.

▼ Caterpillar [× 2¼]
Caterpillars of all sizes can be found from spring until late autumn.

Single segment

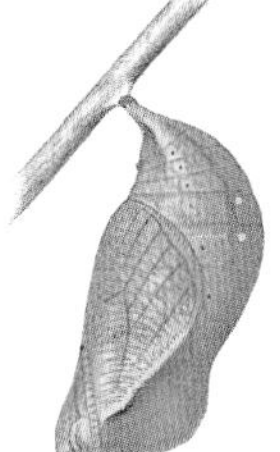

◄ Chrysalis [× 2¼]
Can vary in colour from pale green to dark olive. Usually formed in dense vegetation.

▲ Caterpillar on grass blade
Camouflaged caterpillars feed on a wide variety of grasses.

	Jan	Feb	Mar	Apr	May	Jun	Jul	Aug	Sep	Oct	Nov	Dec
Egg												
Caterpillar												
Chrysalis												
Adult												

in a shady wood; on the other hand, in hot weather these dusky males are apt to overheat if they perch for too long in a sunspot. Therefore in spring, when the weather mitigates against patrolling, most Speckled Woods have paler wings. They also have shorter, stumpier bodies, for brief though their investigative flights may be, they need great powers of acceleration to intercept a female from a standing position, or to succeed in the dog-fights and collisions with competing males. All this requires a broad thorax that is packed with wing muscles. So, by and large, the Speckled Woods that emerge in spring have paler wings and stocky bodies, whereas those of the summer generations are darker and more slender.

There is also some variation between individual males at any time of year that makes some more likely to be perchers than patrollers, and *vice versa*, and again their markings and body shape reflect their inclinations. In addition, males that possess four spots on the upperside of each hindwing are more inclined to perch, whereas those with three spots (like those in our illustration) have a tendency to patrol.

I have described at some length some of the variation in the shape, colour and behaviour of the Speckled Wood, because it adds greatly to the enjoyment of watching this commonplace butterfly if one understands a little of what it is doing. It is intriguing to determine whether an individual is a patroller or a percher, and whether his markings and the stockiness of his body are consistent with his behaviour.

Eggs in sun and shade

The female Speckled Wood generally mates once, usually within a few hours of emergence. She then rests while her eggs ripen, before venturing forth on slow, fluttering flights. Eggs are laid singly on a wide variety of grasses, including False Brome, Cock's-foot and Yorkshire-fog. Despite this lack of specificity, females are selective about the kind of plant that they choose, at least at the start of egg-laying. Small, isolated tufts receive the most eggs, particularly those growing in sheltered situations where the air temperature reaches 24-30°C. This restricts egg-laying to the sunny edges of woods, rides and hedges in spring and autumn, whereas during the heat of the summer most eggs are laid in fairly shady places, within the body of a wood or scrub. Older females with shorter life expectancies are less discriminatory, and produce more, but smaller, eggs that are quickly slipped out.

The egg hatches after seven to ten days. The little caterpillar stays on the same grass blade, resting on the undersurface and taking small bites, working from the edge inwards, on and off during day and night. The attractive, stumpy chrysalis is formed suspended beneath a grass blade or on nearby vegetation, usually within 20cm of the ground. Those that are not hibernating hatch after about ten days.

Despite its need for a warm environment in spring and autumn, the Speckled Wood can tolerate shadier woods than any other British butterfly. In summer, it breeds under canopies that exclude 40-90% of the light, and finds ideal conditions in a great many modern woods. In Scotland and much of England and Wales, colonies are more or less restricted to woodland, but in the south and across Ireland they also occur in many wooded lanes, hedgerows and patches of scrub, including along most of the southern coastline. The butterfly also breeds at low densities in country gardens.

There is little doubt that the gradual spread of the Speckled Wood has been greatly assisted by the decline in coppicing and the increased shadiness of modern woods. Yet this by no means explains all the changes: it has also benefited much from the milder autumns, winters and springs of recent decades.

A decline reversed

The earliest accounts of the 'Enfield Eye' or 'Wood Argus', as it was once called, indicate that although the Speckled Wood was seldom considered to be common, it was nevertheless widely distributed throughout the British Isles, with the main subspecies, *tircis*, extending into northern England and southern Scotland. A decline began from the 1860s onwards. This was very gradual, with local extinctions being continuously reported over the next 60 years. By the 1920s, the butterfly had disappeared from most parts of its former range, and was more or less restricted to lowland Wales, south-west England, Wiltshire, and the heavier, wetter soils of Dorset and West Sussex. The old Irish records are too sparse to say whether it declined there as well.

The recovery of the Speckled Wood was also very gradual. It was first noticed in counties such as Dorset, which contained residual populations, as early as the 1920s. Since then, the butterfly has spread to reoccupy most of its old range, and is now a common species, found in almost every wood in most regions shown on the map. This includes the whole of East Anglia, to which it returned as recently as the 1960s, and the whole of lowland Wales. Ireland is now fully occupied, but elsewhere the expansion continues: northern English populations of *tircis* are spreading towards Scotland at an ever increasing rate and, after an absence of about 150 years, have begun to recolonise the borderlands.

Meanwhile, the two former centres for *oblita*, in Argyll and Inverness in west Scotland, and from Moray to Aberdeenshire in the east, have also expanded greatly and now meet. The butterfly has also moved up the coast to the extreme north-west of the Scottish mainland, and has colonised several isles, even the wooded areas of Lewis.

It is clear from the work of Jane Hill that much of the recent expansion of the Speckled Wood has occurred in response to our warming climate. As a relatively mobile butterfly, it has been quicker than most to exploit new regions that have become warm enough for its races to inhabit, although it has spread at nothing like the pace of the free-flying Comma. An interesting feature of its spread is that there has been selection in favour of strong-flying individuals among the pioneering butterflies in frontier regions, as they spread from wood to wood founding new colonies. Thus the *tircis* Speckled Wood populations inhabiting newly colonised parts of northern England and the *oblita* populations that have spread into Banffshire both contain powerful flyers that possess broader, more robust thoraxes than are found among the Speckled Woods of the original populations from which their ancestors departed, many generations ago. The downside for these pioneers is that they also have smaller abdomens, for a greater proportion of their body matter is deposited in the thorax, and they consequently lay fewer eggs. In time, however, we can expect their offspring to settle down and revert to the fat, fecund forms of the south.

Wall

Lasiommata megera

The Wall has often been quite common and widely distributed in wild grassland throughout the lowlands of England, Wales and Ireland, but from time to time numbers plummet, leaving it restricted – sometimes for decades – to scattered centres where its favourite habitat abounds. These are extensive areas of dry, unfertilised grassland, with a rugged, broken terrain and an abundance of bare patches where this heat-seeking butterfly can toast in the sun. It is especially frequent along the coast.

The Wall is an easy species to identify. Flying adults are sometimes mistaken for Commas or fritillaries, but at rest the bright eye-spots indicate membership of the brown subfamily of butterflies. It is, nevertheless, quite ornate: no other British brown has such gleaming golden upperwings, or such an intricate pattern of zigzag stripes on the undersides. Petiver, who in 1699 coined its first English name, called the female 'The golden marbled Butterfly with black Eyes', and the male 'The London Eye'.

There are two full generations of Wall butterflies a year, the first emerging in late April and May to early June, with their offspring flying throughout August and often into September. In warm years the second emergence occurs earlier in the south, in July, and on these occasions there may be a third brood in autumn. Numbers are usually about three times higher in the second brood than in the first, but even then most colonies are small compared with those of other browns. The largest contain no more than a few hundred butterflies in their best years, and it seems likely that most colonies are reduced to a few dozen adults in springtime.

Flight and temperature

The Wall generally lives in self-contained colonies, but some individuals wander, enabling the species to spread quite quickly to reoccupy ground lost after one of its periodic retractions. There are at least two records of adults reaching the Outer Dowsing light vessel, 50km off the Norfolk coast, and it is not unusual to find Walls visiting garden flowers. Males typically live for just three days.

The behaviour of this striking butterfly has been well studied by Per-Olof Wickman in Sweden, and by Roger Dennis in Britain. The adult spends much of the daytime basking in warm patches of bare ground, with its wings held

Distribution
Much reduced in recent years, especially in central England and east Wales. Still widely distributed in dry grassland around coasts and in the west, and has recently been spreading northwards; also more coastal in Ireland and at the north of its range.

▲ Male
The oblique line of the sex-brand on the forewing is typical of the male.

▲ Female
Females are larger and brighter than males, and lack the sex-brand.

▲ Male
Markings in the rare aberrant form *bradanfelda* are creamy yellow, rather than orange.

▲ Female
Rare aberrant form *anticrassipuncta*, with larger and more distinct eye-spots.

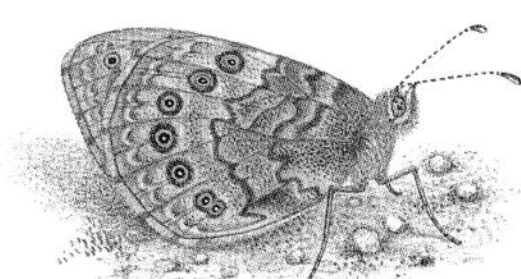

▲ Resting adult
Dull-coloured hindwings camouflage the butterfly when at rest on the ground.

◀ Egg [× 15]
Laid singly or in clusters on roots or leaves of grasses.

▼ Basking adult
Walls are often seen basking on bare, sunbaked ground, with their wings angled towards the sun.

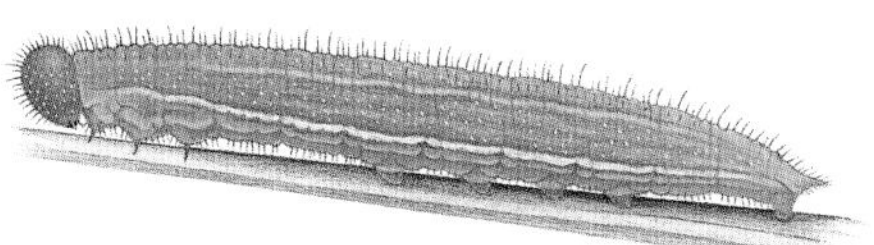

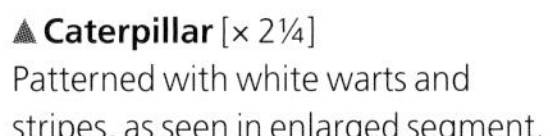

▲ Caterpillar [× 2¼]
Patterned with white warts and stripes, as seen in enlarged segment.

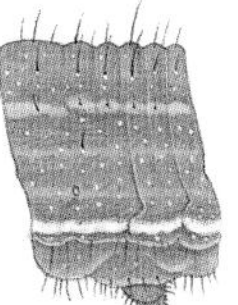

Single segment

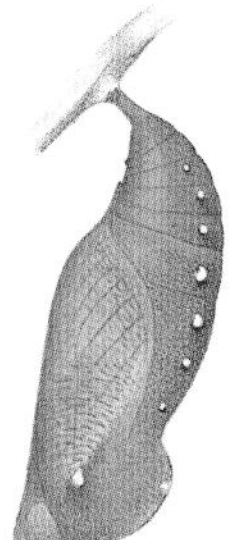

◀ Chrysalis [× 2¼]
Colour varies from bright green to almost black; suspended from a grass stem.

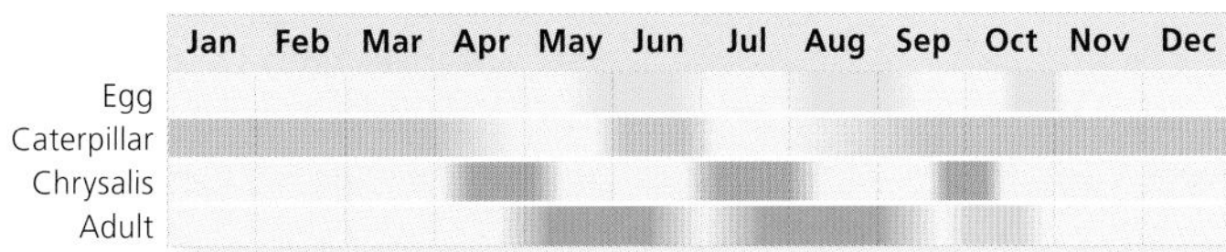

two-thirds open and angled towards the sun. In this way, it can raise its temperature to about 8-10°C higher than its surroundings. When the body reaches 25-30°C, it is warm enough to fly. Many of the local movements made by the butterfly on different days can be explained by the need to control its temperature. On warm, still days, males will be seen flying for long periods, particularly around hilltops. However, they cool down quite quickly during flight in windy weather, so exposed places are then avoided, and they frequently have to alight to warm up on a patch of earth. On the other hand, the temperature can get dangerously high on hot summer days, and then the Wall retreats into the shade.

A boisterous courtship

As with most butterflies, male Walls tend to emerge a few days before the females, although there is a considerable overlap. They gather at dusk in the sunniest parts of a site before settling upside-down to roost beneath fences, hedgerow leaves or under the lower boughs of trees. Most of the male's life is spent searching for females, punctuated by short bouts of feeding on whatever flowers are available. Females are encountered in one of two ways: by flying back and forth along sunny edges, such as paths, roadsides, hedges, banks and even fence lines, or by perching at regular intervals on the ground along these same edges. Many people will have disturbed a Wall when walking along a path, and have noticed that it always settles on the bare ground a few metres ahead, only to be flushed up again and again. This is the perching male waiting, while sunning himself, for a virgin female to fly down the track.

Patrolling males glide fast and low, generally 30cm above ground level, travelling 30m or so before backtracking along the linear feature. On other occasions they zigzag in drunken circles, continually scanning the earth and stopping to investigate likely objects. They skirmish with any flying insect that is encountered. If this proves to be a rival male, the two butterflies soar high into the sky in a spiralling dog-fight before diving towards the ground and separating about ten seconds later. Females are pursued with much greater persistence, and courting may begin in flight.

The courtship of the Wall is a brief, rumbustious affair. When pursued, the female soon alights on bare ground and starts fluttering her wings. The male crash-lands behind her and works his way to the front, beating his wings so violently that the female may be wafted off the ground in a cloud of dust and scales. Then, facing his hen, he hammers her head with his two antennae before buffeting and bombarding her with flapping wings in a series of head-butts. This is all very different from the elegant courtship of the Grayling (see pages 255-6), but the effect is the same: the male envelops his female with flying scent scales, and she in turn is mesmerised by their heavy chocolate-like odour. They quickly pair and disappear from sight into the surrounding vegetation.

Roughly one female in ten will mate for a second or even a third time in later life, but most fertilise all the eggs from the one pairing. Much of the female Wall's day is taken up by resting and basking in the sun, but when sufficiently warm she embarks on a short, fluttering flight in search of egg-sites.

Eggs and caterpillars

Walls lay eggs in very dry spots, and exclusively select grass that has an exposed vertical edge. In practice, this occurs in three distinct situations. The main one is where the turf has broken away to form a miniature cliff; hoof-prints, rabbit holes and spots where the soil has slipped down banks, sheepwalks or path edges are typical examples. The spherical white eggs are often placed on the exposed roots of grasses dangling down into these crumbling recesses. On downs that have smooth and uniform swards, they are laid solely along the upper edges of pathways, making them remarkably easy to find. The other two places sometimes chosen for egg-laying are the sides of large, isolated tussocks of wild grasses, such as Cock's-foot and the beautiful Wavy Hair-grass, or where bents such as Common Bent and Black Bent, or the downy blades of Yorkshire-fog form a wall of tall grass stems beneath fences and under the edges of shrubs.

As she grows older, the female Wall lays much smaller eggs than in her first days of life, but surprisingly the offspring survive equally well and grow just as large as those from early eggs. Whatever its size, the egg hatches after about ten days, the caterpillar nibbling a slice across the top of the shell until it lifts off like a hinge. It then eats the whole shell before embarking on a tender blade of grass. Tor-grass and False Brome are additional favourites to those listed above. The caterpillars from the summer brood of adults overwinter when half-grown, awaking on and off to feed on warm days. In contrast, those that hatch in spring take little more than a month to become full-grown. Although plain, the caterpillar is quite attractive, having a slightly furry blue-green body, with minute white warts and faint white stripes along the sides. The chrysalis, too, is pretty, but beautifully camouflaged and hard to spot as it hangs suspended for about two weeks on a grass stem. Its colour varies between individuals. Most chrysalises are bright grass-green, but some are pale and a few almost black; we illustrate a typical example.

Periodic decline and its causes

Thirty-five years ago, small colonies of Wall could be found on almost any patch of open, unfertilised grassland that lay within the butterfly's range. Due to its fastidiousness when egg-laying, this brown was – and still is – most frequent on dry, sparse soils with low fertility, such as grassy heathland or chalk and limestone downs, and is common only where the terrain is broken and contains sheltered hummocks and hollows, or has been perturbed by cattle, ponies or man. Abandoned railway lines, quarries, derelict land, dunes, cliffs and undercliffs support some of the finest colonies. Indeed, the coastline provides a refuge during periods of national decline, and coastal populations have been stable both in the number and size of colonies in recent decades. In contrast, a great many smaller inland colonies went extinct during this period, and those that survived oscillated on dramatic seven- to eight-year cycles of severe decline and partial recovery, maintaining a steep, long-term downwards trend. Chief among the casualties during this period were the many colonies that bred in small numbers on the dry edges of tussocks on heavier soils, including clays.

The status of the Wall appears to have fluctuated regularly over the centuries. For example, it disappeared from vast areas during a series of exceptionally cold, wet summers in the early 1860s, but recovered much lost ground during the next half century. Since the Second World War, the butterfly must have declined enormously on a local scale owing to the widespread application of synthetic fertilisers and herbicides to most lowland grasslands, coupled with the abandonment of unfertilised pastures and, in the 1950s to 1980s, the loss of rabbit colonies, which not only grazed its fine swards but also scraped and disturbed the soil. Nevertheless, the Wall remained widespread and sometimes common over large areas of the country wherever suitable habitat survived, and recovered to some extent in the 1970s and early 1980s. There was then a dramatic nationwide decline in the mid-1980s, coinciding with four wet summers, followed by a considerable recovery on the best sites in southern England around the end of the decade. Since then, it has been a story of progressive decline, at least on inland sites, and the butterfly has virtually disappeared from very large areas of central southern England – regions where its presence was once taken for granted. Against this, the Wall has persisted well along all occupied coastlines within its range, and has spread north during recent warmer seasons to occupy new sites in Yorkshire, Northumberland and south-east Scotland. These northern specimens are particularly beautiful and dusky, and were once considered to be a separate subspecies. They were much more widely distributed in early Victorian times, reaching as far north as Aberdeen and Glasgow.

Today, the Wall is still a widespread species in lowland England, Wales, Ireland and the Isle of Man, and just extends into the south-west and south-east corners of Scotland. It remains frequent along all coasts within its range and, with some exceptions, on inland heaths, downs and impoverished hillsides. The exceptions include large areas of the Cotswolds, Chilterns and North Downs. It is virtually absent from inland sites on heavier soils. No-one, on present knowledge, has satisfactorily explained the cause of the Wall's recent decline, and we can only hope that, as in the past, it will recover its former status.

Mountain Ringlet

Erebia epiphron

This small and dusky butterfly is confined nowadays to a few bleak mountain-tops in the English Lake District, and to the Grampians and other mountainsides in Scotland. It is our only true montane species, the sole relic of the first wave of cold-hardy butterflies that repopulated our lands when the last Great Ice Age receded, roughly 12,000 years ago. Although there was still another little Ice Age to come, the so-called Loch Lomond Re-advance of 11,000-10,200 years ago, the ice flows and polar desert were then restricted to northern uplands, leaving a barren landscape of permafrost, tundra and steppe across the south. Most butterflies would have been eliminated during this period, but there is every reason to believe that the Mountain Ringlet flourished throughout the chilly wastelands of latter-day Surrey, Sussex and Hampshire, and more or less continuously across the continental land-link to central Europe.

The Loch Lomond Re-advance gave way to a climate that was substantially warmer than our present one. From that time onwards, the European populations of Mountain Ringlets became separated into isolated groups, each restricted to zones of moist grassland at high altitude in the Pyrenees, the Vosges, the mountains of Harz, and to others in the Alps, Balkans, Apennines and Dolomites.

Distribution
Confined to high mountain grassland in the Lake District and Scotland; abundant where it does occur.

19th-century discovery

Not surprisingly, our own Mountain Ringlets, having inbred for 10,000 generations, have evolved into a subspecies distinct from those now living in mainland Europe. Ours is called *mnemon,* and is smaller and duller than most of its continental counterparts. The distinction, however, is slight, for this is a highly variable butterfly and adults within every colony differ in size, brightness and in the number of spots. Nevertheless, minor differences exist even between the Scottish and English races, suggesting that they too have been isolated from one another for several thousand years. Scottish specimens tend to have longer wings than those in England, with larger black spots.

Despite being our oldest surviving butterfly, the Mountain Ringlet was one of the last British species to be discovered, the first specimen being caught on 25th June 1809, on the slopes of Red Screes, above Ambleside, in the Lake District.

▲ **Male, Scottish form**
The Scottish form is generally larger and brighter than specimens found in England.

▲ **Female, Scottish form**
Orange spots are brighter in the female, and the wing ground-colour lighter.

▲ **Male, Lake District form**
Males fly close to the ground, searching for females.

▲ **Female, Lake District form**
Females of both forms rarely fly, and so are seen less often.

▼ **Feeding adult**
On grassy hillsides, Mountain Ringlets often use hawkweeds as a source of nectar.

▲ **Resting adult**
Adults rest inconspicuously with the forewings concealed by the hindwings.

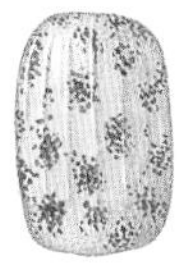

▲ **Egg** [× 15]
Laid singly on the leaves of Mat-grass. The rusty brown blotches develop after a few days.

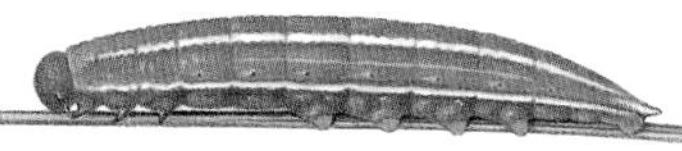

▲ **Caterpillar** [× 2¼]
Sluggish caterpillar hibernates when half-grown.

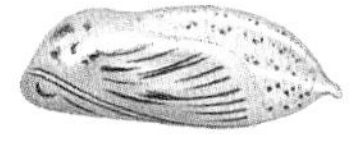

▲ **Chrysalis** [× 2¼]
Formed in a loose cocoon, deep within a grass tussock.

	Jan	Feb	Mar	Apr	May	Jun	Jul	Aug	Sep	Oct	Nov	Dec
Egg												
Caterpillar												
Chrysalis												
Adult												

It took another 35 years before the first Scottish colony was found, at Rannoch in Tayside. But this is an easy species to overlook. Typical colonies are restricted to small parts of their mountains, and the flight period seldom lasts more than three weeks; one vast colony flew for just 12 days in the hot, dry summer of 1976. The exact date of emergence varies with the season and altitude, and may be two weeks earlier at low levels in the north of Scotland. However, in most colonies the first males will be seen in late June after a warm spring, and from mid-July onwards in a cool year.

Life at high altitude

Little is known about the natural history of this elusive butterfly, apart from a study made by Keith Porter at Seathwaite Fell, in the Lake District. The main colony that he investigated bred on a barren, undulating plateau, 600m above sea level and surrounded by crags, rocky outcrops and large expanses of scree. The butterflies were restricted to little more than 0.5ha of east-facing grassland, containing drier slopes of Mat-grass surrounding springs, boggy hollows of *Sphagnum* moss, and stagnant peaty pools.

In just 12 days, 650 Mountain Ringlets were marked on this small plateau, and the results confirmed that the butterflies live in strict colonies with little or no mixing between neighbouring hillsides. The numbers on the peak day – about 2,870 adults – corresponded to a total emergence of between 8,500 and 9,000 adults that year. This is typical of Mountain Ringlets: their sites may be few and far between, but where they do occur the butterfly is often found in great abundance. One population in central Europe was recently estimated to contain 100,000 adults.

Adult Mountain Ringlets are greatly dependent on sunshine. They rest deep among the Mat-grass on cloudy days, with their forewings tucked down so that only the well-camouflaged hindwings are exposed. They are extraordinarily easy to overlook until the sun shines. Then, hundreds suddenly emerge from the tussocks, all with their dark velvety wings spread wide to absorb the maximum warmth. Females seldom fly, being weighed down by clusters of large eggs that distend their hairy black bodies almost to breaking point. On Seathwaite, at least 17 males were seen flying for every female, although the two sexes were apparently present in even numbers.

On sunny days, male Mountain Ringlets patrol back and forth across their restricted breeding grounds, fluttering or gliding slowly around each tussock, and seldom rising more than 30cm above the ground. They settle to investigate any brown object, including other males, dead moss, and the droppings and brown wool of the Herdwick sheep that half-heartedly graze these nutrient-poor Cumbrian uplands. Virgin females are courted and quickly mated, then left to mature on a warm clump of Mat-grass. They emerge only to lay eggs or feed. Tormentil, Heath Bedstraw and Bilberry are often the only flowers available, but if mineral flushes exist there may be a carpet of Wild Thyme where the butterflies cluster to drink nectar.

The average lifespan of the adults has only once been measured, in the hot summer of 1976. It was then a mere day and a half, but is likely to be two to four times longer under normal conditions. Skylarks, Whinchats, Pied Wagtails and other birds are believed to be the main predators, and the Mountain Ringlet's distinctive behaviour on settling has evolved to minimise their impact. On alighting, it sits for a few seconds with the wings wide open, so that any bird that had been attracted by its flight is likely to peck the wings' eye-spots rather than the vulnerable body. It then closes its wings, keeping the forewings raised, again so that any following bird will strike a harmless area. Finally, the forewings are withdrawn between the hindwings, so that no further predators are likely to notice them. Pecked wings are visible on many adults.

Surviving the winter

Female Mountain Ringlets lay up to 70 eggs a day, given warm, sunny conditions. These are placed singly on the wiry grey-green blades of Mat-grass, and are quite easy to find on good sites if you search the bases of tufts in late July and early August. They are surprisingly large for a butterfly of this size, and are creamy yellow at first, soon developing the rusty brown blotches found on most browns' eggs.

Each egg hatches after about three weeks. The caterpillar eats the entire shell, then starts feeding on the tender tips of Mat-grass. By September, the half-grown caterpillar settles down to hibernate deep in the dense tussock, where it remains for six or seven months, protected from the snowdrifts that invariably blanket the breeding sites. It can spend two years as a caterpillar if spring is late and the summer short, but most complete their growth in April and May, sluggishly crawling to the nutritious tips of Mat-grass between periods spent basking on the sides of tussocks.

Typical caterpillars pupate in late May or June, in a loose cocoon of silk and grass blades. The chrysalis is particularly attractive and well worth seeking in the base of tussocks. Look early in the season, though, for it is thought that the majority are killed by the Field Voles that teem among the loose scree alongside most Mountain Ringlet sites.

A curious distribution

There has been much controversy over whether Mountain Ringlets exist anywhere other than in Scotland and Cumbria. Mat-grass is the dominant plant at high altitudes on most British mountains, yet it now seems certain that the butterfly is absent from the Pennines, Cheviots and, most curiously of all, from the massif of Snowdonia. Ireland poses a greater problem: three 19th-century records exist from Croagh Patrick, the eastern shores of Lough Gill, in County Leitrim, and from Nephin Beg, in County Mayo, although considerable doubt surrounds the last two records.

No such doubts surround the Cumbrian colonies, which can be abundant despite some local losses as a result of recent climate-warming. Huge populations still exist in suitable patches alongside streams and damp, swampy grassland down to 180m above sea level, although most are between 460m and 760m altitude; famous sites include Red Screes, Langdale Pikes, Stye Head Tarn and Helvellyn.

The great majority of colonies are in Scotland, breeding on damp mountain pastures and, especially, on the sides of steep gullies. They are commonest between 450m and 730m above sea level, but also occur down to 275m and as high as 915m. The main concentrations are in the Grampians, on many mountains south of the Great Glen, from Glen Clova in the east to Ben Vane and Ben Nevis in the west. Further north the Mountain Ringlet still breeds south of Newtonmore in Inverness-shire, and there is a single colony on Ben Lomond. There are also unconfirmed reports from Sutherland and, more dubiously, from the mountains of Galloway in the south-west. The latter have been quite well explored in recent years and seem unlikely still to support a colony, if they ever did. Although still under-recorded in the Scottish Highlands, there are ominous reports of losses in response to recent climate-warming, a portent of things to come. For example, a survey of selected sites by Alidna Franco in 2004-5 found that a third of known colonies had disappeared, almost certainly as a result of climate change, with a disproportionately high level of extinctions at lower elevations. At present, it is unknown whether these have been compensated by the colonisation of new sites above the Mountain Ringlet's historical altitudinal limit of around 900m above sea level.

Scotch Argus

Erebia aethiops

This handsome, dusky satyrine was first discovered in the 18th century on the Isle of Bute, and has been recorded from many hundreds of localities since that time. As the name implies, virtually all are in Scotland, where it replaces the Meadow Brown as the dominant butterfly in many upland grasslands, and can be seen flying in thousands over sheltered, moist grasslands in the August sunshine. It is also common in mountainous areas on the Continent, but is curiously absent from the rest of Britain except for the Lake District, where it is now confined to just two colonies.

Like most species of *Erebia*, the dusky upperwings are adorned with intense white spots, hence its common name of argus, after the giant in Greek mythology that possessed a hundred eyes; in Gaelic this translates as *argus albannach*.

The adults live in well-defined colonies, and seldom fly far from their traditional breeding sites. They are usually the commonest butterflies where they do occur; Roger Dennis estimated that one Lakeland colony contained thousands, if not tens of thousands, of adults, and it is clear that many Scottish populations are equally large. However, the single flight period is short, even on good sites. The first males emerge in the last days of July, and only a few faded females last into September.

A love of sunshine

For a butterfly that inhabits some of the wettest regions of the British Isles, this species is extraordinarily dependent on sunshine. Adults occasionally fly in warm, overcast weather, but float to the ground the moment a cloud appears, vanishing deep into the tussocks. They are then surprisingly hard to spot because they so closely resemble dead leaves. But they emerge in numbers the moment the sun reappears, stretching their black wings to absorb as much heat as possible before launching themselves on jerky, rolling flights.

Males fly more often than females, and are especially active on bright, windless days, when they weave tirelessly around the tussocks, searching for mates. They will investigate any brown object, including withered leaves caught among the grass. This, however, is not their only method of finding mates; at other times males simply perch on grass clumps, resting just below the flowerheads, ready to chase any female that flies past.

Distribution
Reduced to two large colonies in the Lake District, but widespread and locally abundant in sheltered grassland in Scotland.

Egg-laying

Once mated, the female Scotch Argus feeds frequently on hawkweeds, Bramble and heathers, but spends most of her life resting or basking in open areas, slightly apart from the males. Only in the hottest weather does she embark on ponderous egg-laying flights, stumbling from one clump of Purple Moor-grass to another. Until recently, this was the only known foodplant in Scotland, but it is now recognised

▲ **Male**
Upperwings are almost black when the butterfly emerges. Ground-colour of underwings grey and deep brown.

▲ **Female**
Upperwings are more brown than those of the male. Underwings mid-brown; hindwing barred with ochre.

▲ **Colour variant**
Aberrant *croesus* form, with larger eye-spots.

▲ **Resting adult**
Adults fly only when the sun shines, otherwise they hide in vegetation.

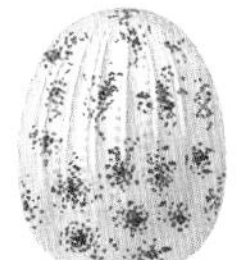

▲ **Egg** [× 15]
Speckled pattern develops after five days; caterpillar hatches after two weeks.

▲ **Basking female**
Females spend much of the time basking, interrupted by bouts of egg-laying.

▲ **Caterpillar** [× 2¼]
Young caterpillar hibernates at the base of grass tussocks.

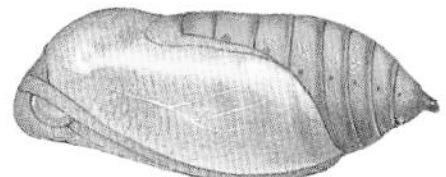

▲ **Chrysalis** [× 2¼]
Formed at the base of grass tussocks, in moss or leaf litter.

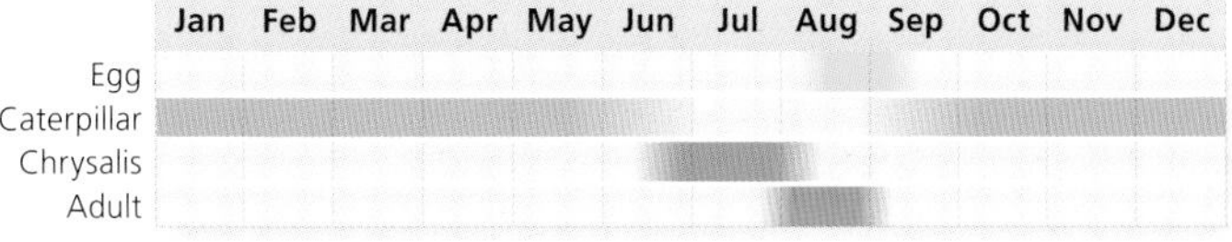

that caterpillars will also eat Tufted Hair-grass, Wavy Hair-grass, Sheep's-fescue, Common Bent and Sweet Vernal-grass. In Cumbria, the butterfly invariably chooses Blue Moor-grass.

Whatever the grass, the attractively speckled eggs are laid singly, deep among tussocks, on plants growing in open, though sheltered areas. As with the eggs of all species of *Erebia*, they are large and barrel-shaped, but unlikely to be found in the wild because of the dense vegetation in which they occur. They are more easily obtained in captivity: like many moths, a female Scotch Argus will lay even in the dark of a cardboard box.

This ability to lay eggs in the dark suggests that the female Scotch Argus might not be quite so dependent on sunshine as is popularly supposed, and may be able to crawl among vegetation to lay her eggs on overcast days. As with most butterflies, there are substantial gaps in our knowledge of this species' behaviour, and there is scope for amateur entomologists to make novel observations.

Caterpillars and chrysalises

Scotch Argus eggs hatch after about a fortnight. Like the Marbled White, the tiny caterpillar eats a neat groove around the shell, until it can push open a lid and squeeze out. More of the shell is then eaten before it switches to feeding on grass blades. This continues for about four weeks, then the caterpillar settles down to hibernate on a dead piece of vegetation deep in the tussock. It is still very small at this stage, having usually completed a single skin moult. Feeding resumes in spring, and as it grows older the caterpillar becomes increasingly nocturnal. Although I have not searched myself, I am told that the fully grown caterpillar can be found in late June, hiding deep in a tussock by day or out on the grass-tips at dusk.

Finding the caterpillars may require a degree of exertion. As long ago as 1895, a Mr Haggart of Galashiels warned that,

> 'No artificial light can be used as the larvae immediately drop down among the grass... the method is by no means enviable, even to the most ardent entomologist, as in the uncertain light it necessitates crawling on one's hands and knees amongst the grass, and there is always the risk of grasping those little brown slugs in mistake, which resemble the larvae very much in shape and colour.'

The caterpillar is, however, distinct from that of any other Scottish butterfly, except perhaps the Ringlet, which has brown rather than green stripes.

The chrysalis is formed in moss or under leaf litter, in a thin web of silk spun at the base of the grass clump. It lasts for two to three weeks, and can be found with difficulty in the wild by searching at the best Scotch Argus sites.

After the Ice Age

An ability to live in cool climates means that the Scotch Argus was probably one of the first butterflies to recolonise Britain after the last Ice Age, 11,000-12,000 years ago. The amelioration of the climate was followed by a temporary return to Arctic conditions a few hundred years later, which would again have eliminated all but the most cold-hardy of our current species. In the extreme south, the Scotch Argus may have been among the few to survive, and should therefore have been among the front-runners to regain ground when the present temperate climate returned. It is surprising, therefore, that it has never been recorded from Ireland, and that it is absent from a number of apparently suitable islands, such as Islay and Jura. The lack of any Welsh colony is also a puzzle, as is the cause of its early to mid-20th century decline in England. At present, two large colonies are known in the Lake District, breeding on ungrazed tussocky grassland that is dominated by Blue Moor-grass and sheltered by scattered shrubs. There were once several colonies in northern England, for example at Castle Eden Dene near Durham, at Fawdon in Northumberland, and at Grassington in Yorkshire.

The large Scotch Argus colony at Grassington was famous because it supported a highly distinctive race. The males had almost no orange on their upperwings, while that on the females was little more than on a normal male. Numerous collecting trips were made to this site – one Victorian excursion by the Yorkshire Naturalists' Union tells of hundreds being secured by the members, and of hundreds more that could easily have been taken. It is doubtful, however, that collecting caused the loss of this unique colony, which became extinct in 1955.

North of the border

Scottish colonies have been very much more stable, although there is a worrying trend of local extinctions occurring at low altitudes during warmer recent decades in southern Scotland. Nevertheless, the Scotch Argus is still common in numerous places where its habitat occurs, from sea level to altitudes of about 500m. It can be seen, particularly, in warm sheltered valleys, among open scrub, and in rough ground beside woodland edges. Smaller colonies can be found in broad woodland rides and in young plantations, but they soon disappear from shady woods. It is less common on open moorland, but is sometimes found where there is a little shelter. There are also colonies among the rough grass and rushes on raised beaches along the west coast.

Marbled White

Melanargia galathea

This is one of the loveliest insects to be seen in high summer on southern English downs and, increasingly as the climate warms, on sheltered grassland further north. Despite its name and appearance, it belongs to the browns – the subfamily Satyrinae – and the old English names of 'Marmoress' or 'Marmoris' (both ancient terms for marbling), 'Our-Half Mourner' or 'Marbled Argus' are all more appropriate than Marbled White, and more attractive.

Curious patterning

No other British butterfly has such a striking black-and-white pattern. The reasons for its singular appearance are only partly understood. Angela Wilson discovered that the cream markings, which can vary from white to yellow, depend for their tint on the caterpillar's unusual ability to break down two flavanoid chemicals. These are found in the grass that they eat and are composed of seven derivatives which they store in the body and deposit, after pupation, in the pale areas of their wings – the higher the concentration, the yellower the wing. Curiously, the flavanoids from which these chemicals derive have been found only in one of the wild grasses eaten by caterpillars, the Red Fescue, perhaps providing a clue to an essential ingredient of its diet.

Distribution
Fairly common on dry, unfertilised grassland across much of southern England, but scarce and local towards the north and east of its range.

During a 40-year friendship, Miriam Rothschild regularly told me that adult Marbled Whites must be toxic and that their markings were aposematic – that is, deliberately designed to be conspicuous so as to advertise their unpalatability to enemies, like the yellow-and-black banding of wasps. In fact, flavanoid chemicals are, at most, only mildly toxic, and their main use in colouring the wings may be to increase the camouflage of the settled butterfly when it rests among grass, while making them more conspicuous when flying in direct light, including to other Marbled Whites, by the reflection of ultra-violet light from these areas, which amplifies the contrast with the black patches.

Rothschild's hunch was confirmed, nevertheless, by her demonstration that the chrysalis and adult Marbled White are indeed obnoxious to birds, although this results from a different chemical, ioline, in their bodies and wings. Ioline is a pyrrolizidine alkaloid produced by a fungus called *Acremonium* that often infests the Red Fescues on which the caterpillars feed. This too is presumably stored in the body of the growing insect to ward off enemies.

◀ **Red mites**
Like most browns and skippers that inhabit grassland, Marbled Whites often carry the parasitic larval stages of red mites.

Whatever the full story behind the markings of the

Marbled White, the effect is beautiful. The butterfly is seen at its best in weak sunlight, first thing in the morning or in late afternoon, when the wings are held open to absorb maximum warmth from the sun. Large groups bask like this on good sites, perched in clusters on tall Field Scabious and knapweeds, generally in sheltered pockets towards the base of a hill. Too lethargic to move, basking adults can be approached very closely, and this is undoubtedly the time to watch, paint or photograph them.

The flying season is short compared with other common browns. The first adults appear in late June, reach a peak in mid-July and disappear before mid-August. They live in sharply delineated colonies ranging in size from a handful of adults to several thousand on the finest sites. As Roger Dennis and Tim Shreeve have pointed out, although essentially colonial, the adult butterfly behaves more like a white than a brown, with both sexes flying considerably more often than related species, such as the Wall, Gatekeeper or Meadow Brown. Moreover, in comparison to these browns the Marbled White has a gliding flight, and the males continuously patrol their grasslands seeking mates on the wing, whereas most other browns perch in wait for passing females, or flutter weakly around the grass-clumps. Could this white-like behaviour also be symptomatic of a toxic butterfly seeking to advertise its unpalatability? Dennis and Shreeve argue convincingly that it might.

The female Marbled White does share one trait with other browns: she is not at all fussy over egg-laying. In my experience, she merely sits on a tall plant, pulsates her abdomen until an egg appears at the tip, then, with a little wriggle, flies off. The egg simply falls to the ground. There is no evidence that she selects spots where Red Fescue is growing, but it is possible that she detects its scent if this grass is, indeed, found to be essential to the caterpillar's diet.

Caterpillar foodplants

Marbled White eggs are quite large, unpatterned and white, similar to those of the Speckled Wood and Wall, but less shiny. They hatch after about three weeks, with each caterpillar neatly nibbling a slit around the top until the lid is pushed open. It then squeezes out, eats the shell, and crawls into a small piece of dead vegetation where it hibernates. This may be a crucial period during the life cycle, because after a warm August the emergence of Marbled Whites in the following year is much larger than usual, and *vice versa*.

The first proper feed occurs in early spring. To begin with, the caterpillar sits along a grass blade, nibbling at the leaf during the daytime. This changes as it grows larger, and after the third and final moult the caterpillar becomes nocturnal, hiding head-downwards in the body of a grass clump by day and ascending to feed at night. Several species of grass may be eaten by older caterpillars, including Sheep's-fescue, Tor-grass, Cock's-foot and Timothy. The possibility of a dependency on Red Fescue at an earlier stage in the caterpillar's life has already been raised.

A puzzling distribution

The Marbled White has a most curious distribution in Britain that has yet to be satisfactorily explained. It is a common butterfly over large areas of the south-west, especially along the coast. Here, it occurs on all soils except the most acid, and on almost every unfertilised chalk or limestone hillside west of Surrey and from the Cotswolds southwards. On warm, south-facing slopes, where the sward has been left to grow quite tall, it becomes particularly abundant, but the butterfly also breeds in very small areas and in a wide range of situations, including woodland glades, sunny rides, and many of the wider road verges. In Dorset, I have found high densities flying and breeding along the 2-5m-wide strip of the central reservation of a busy dual carriageway, with the adults apparently oblivious to the noise and fumes of the traffic streaming past on either side.

The Marbled White is much scarcer elsewhere in southern Britain. Although locally common along several stretches of the South Downs, it is absent from much of the North Downs apart from the eastern end, in Kent. It is also a rarity in Wales, with colonies mainly centred on Gower. Until recently, colonies were few and far between in central England, although large ones bred along some woodland rides in Oxfordshire and Buckinghamshire.

This is a butterfly, however, that has benefited from the warmer climates of recent years, and numbers on monitored sites have fluctuated in close synchrony with the weather. David Roy and I have shown that historical records of the Marbled White's dramatic swings between rarity and abundance were also closely correlated with fluctuations in the weather during the last 250 years. In the past two decades there has been a strong upward trend in numbers, which has coincided with a substantial spread to new sites in the Midlands, in the south-east and northwards. Similar increases have occurred at its former edge-of-range in the North Yorkshire Wolds, where the once isolated colonies have spread well beyond the traditional south-facing slopes in dry river valleys and roadside verges that served as breeding sites. This welcome expansion appears set to continue, for successful introductions made in Durham, tens of kilometres north of its range, indicate that this slow-moving butterfly has yet to colonise many vacant grasslands that have recently become suitable for it as a result of the warmer weather.

▲ Male
Both sexes have similar upperwings; the male's hind underwings have only a tinge of yellow.

▲ Male
Form *nigricans*, a rare aberration with much heavier black markings.

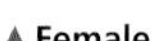

▲ Female
The hind underwings have a marked yellow tinge, as does the leading edge of the forewing.

▼ Basking adult
Adults sit on prominent flowers and grass heads early and late in the day.

◄ Egg [× 12]
Eggs are dropped rather than laid, near to the foodplants.

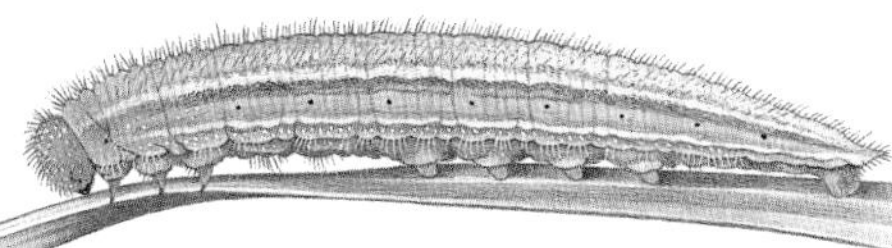

▲ Caterpillar [× 2¼]
Covered in short hairs; colour varies from yellow-brown to light green.

▲ Resting adult
The wings are kept tightly closed during the heat of the day, and when roosting.

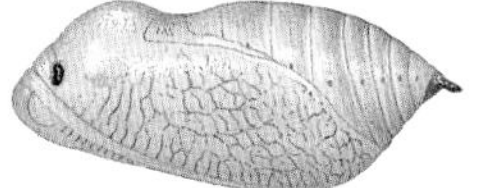

▲ Chrysalis [× 2¼]
Formed at the surface of the ground, under soil or moss.

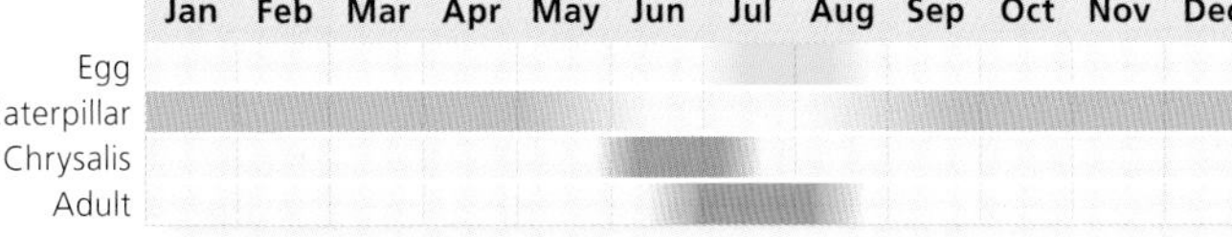

Grayling

Hipparchia semele

The Grayling, more than any of our butterflies, is confined to dry, dusty places where the soil is poor and thin, the vegetation sparse, and the terrain so rutted or broken that the sun's rays bake the ground. Arid soils of all types may support a colony, yet it is scarce in most counties, being common only on southern heaths and on cliffs and dunes around the coast.

Like all browns, the Grayling lives all year on clearly defined breeding sites, although wandering adults are occasionally seen. Colonies vary enormously in size. In 1976, I spent one hot summer's day on Dartmoor marking the adults to see how many lived in a typical colony. There were 55 alive on that day, equivalent to a total emergence of around 150 adults. That was on 3ha of acid grassland, but I have often found Graylings on much smaller sites, such as abandoned quarries, which must contain no more than 20 or 30 adults. On the other hand, some heath and dune populations are vast, and my Dartmoor colonies each increased to more than 2,000 adults after spring and autumn grazing was introduced to encourage the Large Blue. It is clear that other sites support tens, if not hundreds, of thousands of individuals.

This is one of the later butterflies to emerge. There is one generation a year, typically starting in early July and reaching a peak towards the end of the month, with a few stragglers into September. The timing is a little later in north-east Scotland, while at Great Ormes Head, north Wales, there is a curious dwarf race that regularly emerges at the beginning of June.

Reluctant feeders

The female Grayling is secretive, except when laying eggs, so sightings tend to be of males. It is often said that the adult never feeds. This is untrue, but feeding is seldom a major activity, occuring mainly first thing in the morning or in late afternoon. Graylings visit a wide range of flowers, depending on the habitat. Bell Heather is a favourite on heaths, and if there is any sap oozing from a tree-trunk, a salty puddle or a wooden stake sticky with resin, they find these irresistible. Once fed, males disperse within their breeding sites to perch singly, usually on lichens in the early morning, then on a patch of bare sand, rock or earth once it gets warmer. Most of their sites are treeless, but where they can find elevation they do, resting a metre or so up on the sunny side of a tree-trunk or boulder. I have often had them settle on my leg, especially when wearing jeans. They even probe the material for sweat if the day is hot and sticky. Males are remarkably tame when drinking, and can be gently stroked without taking fright.

Distribution
Locally common on dry coastal grasslands and dunes, and southern lowland heaths. Rare and declining elsewhere in its range.

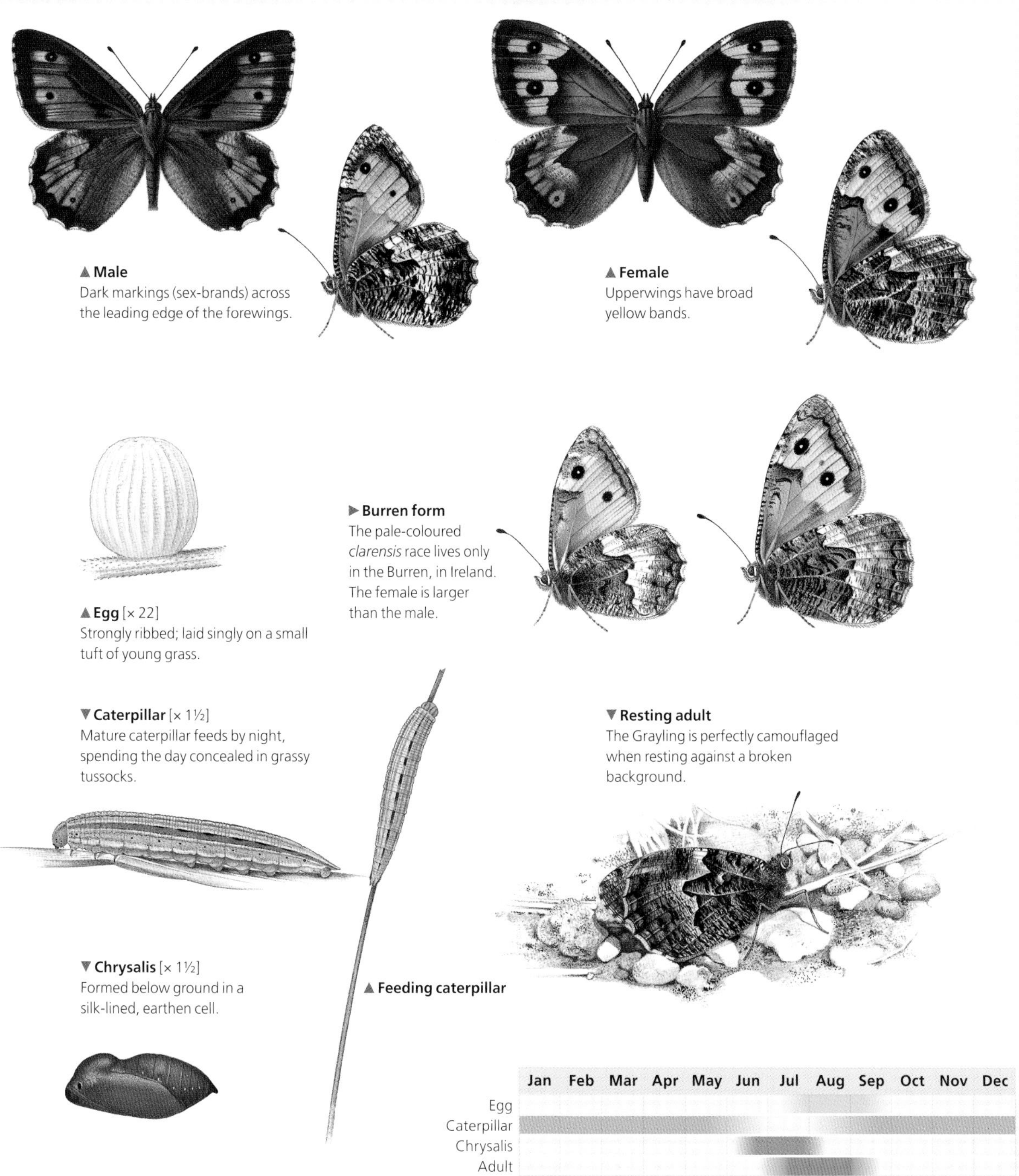

▲ **Male**
Dark markings (sex-brands) across the leading edge of the forewings.

▲ **Female**
Upperwings have broad yellow bands.

▲ **Egg** [× 22]
Strongly ribbed; laid singly on a small tuft of young grass.

► **Burren form**
The pale-coloured *clarensis* race lives only in the Burren, in Ireland. The female is larger than the male.

▼ **Caterpillar** [× 1½]
Mature caterpillar feeds by night, spending the day concealed in grassy tussocks.

▼ **Resting adult**
The Grayling is perfectly camouflaged when resting against a broken background.

▼ **Chrysalis** [× 1½]
Formed below ground in a silk-lined, earthen cell.

▲ **Feeding caterpillar**

Courtship and mating

Each male remains at his station for much of the day, awaiting the freshly emerged females that occasionally fly past. He then soars up and gives chase. In fact, the male Grayling will intercept any moving object, be it a butterfly, bird, person or falling leaf. If it proves to be a female, she soon alights and the male lands behind her, before walking around so that the two are face to face. An unreceptive female then flaps her wings with great vigour to drive the male away. But if, instead, she stays still, he takes this as an invitation to begin the charming courtship that was first described by Niko Tinbergen in one of the classic studies of animal behaviour.

Courtship
Towards the end of his elaborate courtship, the male (right) bows to the female with his wings apart. He catches her antennae and draws them over his scent glands.

First of all, while still facing the female, the male jerks his wings upwards and forwards in quick succession, so that the beautiful orange patch on the underside, with its two white-pupilled eye-spots, is clearly visible. Next, with the forewings and hindwings still held separate, he repeatedly flicks the forewings open and shut for perhaps a minute. This culminates with the quivering forewings being held wide open as he executes a deep bow to the facing female. Gently he folds his wings together, catching the female's two antennae and drawing them through his forewings as he straightens up. The tips of a butterfly's antennae are the main organs of smell, and these are dragged across the male's scent glands (sex-brand), which are conspicuous as a dark ridge across each forewing. Their scent is just detectable to our senses, and has been variously described, for the Grayling, as having the faint aroma of sandalwood or an old cigar box. This is a powerful aphrodisiac to the female, who responds immediately. Indeed, to all intents and purposes the male has now seduced her, and he quickly withdraws his wings and walks around behind her to pair. Mating takes place for 30-45 minutes before they go their separate ways.

Temperature regulation

Naturalists often wonder what a butterfly is doing when they see one perched or flying around. The male Grayling is preoccupied with his vigil for mates, while the female forays for egg-sites with equal persistence. But both sexes also spend much of their lives regulating their body temperature. The bare terrain that typifies Grayling sites can bake in the sun of high summer, reaching temperatures of 40-45°C. Yet these barren sites become very cool in overcast weather, and to be active a Grayling needs to keep its body at around 32°C. It achieves this by exposing different areas of its wings and body to the sun. This is, in fact, a species whose physiology is adapted to inhabiting warmer micro-habitats than almost any other British butterfly. In Sweden, Christer Wiklund and Bengt Karlson found that female Graylings lived longest, and laid the maximum number of eggs, when kept in an environment of 30°C. While this is also true of the Small Heath, butterflies from less arid habitats, such as the Ringlet and Speckled Wood, show peak longevity and fecundity in sites that are 5°C cooler.

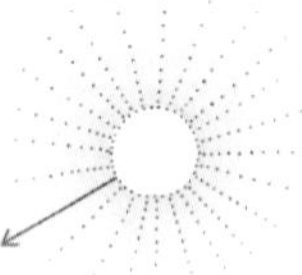

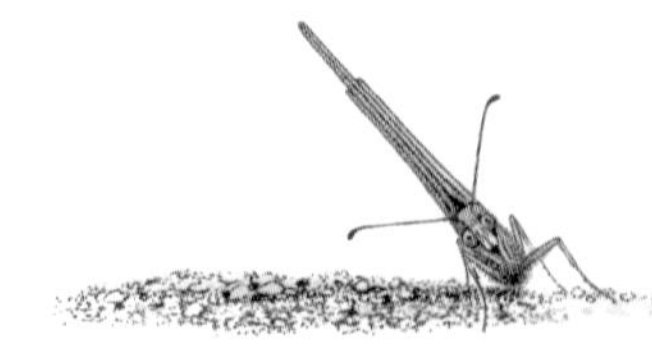

Regulating temperature
When it is too cold, the Grayling leans sideways on to the sun (above) to expose the maximum wing and body area to its heat. In hot weather (below), it stands head-on to the sun, on tiptoe.

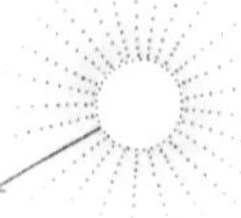

Egg-laying

A female Grayling lays eggs in warm weather throughout the day. Each is placed singly on the caterpillar's foodplant or nearby, generally on a small tuft of young grass growing in a sheltered, sunny pocket of bare ground. Several of our native grasses are used, depending on the site. On chalk, lime and shales, Sheep's-fescue is a favourite, while many coastal and dune populations breed largely or exclusively on Marram. My own searches have mainly been on acid shales and heaths, where I have found eggs almost entirely on the wiry tufts of Bristle Bent. They are quite easy to find if you search small grasses growing in the hottest sun-baked spots; all the examples recorded on Dartmoor occurred in niches that reached temperatures of 40°C or more during sunny days in August, whereas none was found on the abundant semi-shaded or sheltered tufts that formed the majority of Bristle Bents available on the site. Despite choosing warm dry spots in which to lay, the resulting caterpillars evidently dislike grass that has become really desiccated. Ideal conditions are for a wet July to be followed by a warm sunny August, a combination which results in abnormally high numbers of Grayling the following year.

The egg is very beautiful, almost spherical and white, with a faint glow as if made of porcelain. It can be distinguished from those of most browns by its clear colour that lacks rust-like patches, and from those of the Wall and Marbled White by the distinct ribs that run from top to bottom. It hatches after ten to 20 days. The small, cream-coloured caterpillar feeds on the tender tips of the grass, and sheds its skin twice before settling down to hibernate deep in its little tussock. It resumes feeding next spring, moulting twice more before attaining full growth in mid-June. By now it is feeding exclusively at night, but is quite easy to pick out by torchlight if you scan the beam over the tips of grass blades. This must be done gently, however, for the caterpillar releases its grip and curls up into a ball at the merest rustling of the clump. It is then almost impossible to find, as it is by day, concealed in the depths of a tussock.

The caterpillar is beautifully camouflaged to match the grey, parched grasses among which it lives. Ringlet and Scotch Argus caterpillars may at first appear similar, but neither has such distinctive brown, yellow and white stripes, nor so smooth a skin. The chrysalis is formed in a silk-lined cell, hidden below the soil surface, and like the caterpillar is well camouflaged against its background.

A perfect camouflage

It is the adult, though, that is the master of disguise. With its wings closed and the forewing tucked down so that both eye-spots are hidden, it is extraordinary how difficult this butterfly is to spot. Whether resting on sand or chalk rubble, and especially on lichen or bark, the grey-brown and black marbling of the underwing blends imperceptibly into its barren background. But when startled, the forewing is immediately raised, exposing the bright eye-spot to frighten or at least distract any enemy. Indeed, when watching this I often think how much more apt, and attractive, were the early English names for this butterfly before it became the 'Tonbridge Grayling', and then simply the Grayling, like that unappealing fish. Both the 'Rock-Eyed Underwing' and 'Black-eyed Marble', as it was once known, better express the character of this fine satyrine.

For this is our largest brown. It scarcely appears so when huddled, wings closed, on the ground, but it seems considerably larger on the wing, enhanced by a peculiar looping, gliding flight, more like a vanessid than a brown. This enables it to soar and swoop at high speed up and down rocks and cliff-faces, while its smaller relatives flap weakly beneath. It also appears very much paler when flying, for the straw-coloured bands of the upperwings are then exposed and offset the darker tones of the underside.

Distribution and status

Graylings vary slightly in appearance within any colony, and certain forms predominate in different parts of the British Isles. The main type, *anglorum,* is illustrated. This is found throughout England, Wales (except on Great Ormes Head), and southern Scotland. The finest colonies are on the southern heaths of Dorset, the New Forest and Surrey, where it is the commonest brown and is to be found everywhere that dry heath survives. It is also common in the more open, grassy areas, and in sunny rides among forestry plantations on these soils. It does particularly well where fine grasses are beginning to invade the sandy strips that are ploughed as fire-breaks.

Colonies were also common on steep, unfertilised chalk downs in the days when rabbits kept the turf short and scraped bare patches of soil. Almost all of these populations disappeared following the near extinction of rabbits by myxomatosis, from the mid-1950s to the late 1980s. Sadly, despite the recovery of rabbits, few if any of the butterflies have returned. Nevertheless, occasional colonies survive on inland chalk or lime, and it is always worth searching for this butterfly in abandoned quarries.

The inland colonies on other soils also virtually disappeared during the last century, except in Devon, Cornwall, and parts of Wales, where unfertilised rough pasture on mica-schists and shales, such as those surrounding Dartmoor, support a few relics. But the main British populations, other than those on heathland, are to be found along the coast. Graylings may be present on any eroding cliff or undercliff, and are especially abundant among sand dunes; the adults fly throughout these areas, but it is only worth looking for the young stages among Marram grass on the driest summits.

Regional variation

A curious race of Grayling lives solely on the south-west edge of the Great Ormes Head peninsula of north Wales. Its markings are much the same as other Graylings, but it is no larger than a Ringlet or a Speckled Wood. Another feature is that it consistently emerges a few weeks earlier than Graylings elsewhere. It is thought that this race, called *thyone,* originated from butterflies that colonised this peninsula about 10,000 years ago, but were then cut off from other populations for 3,000 years or so when, as a result of rising sea levels, these limestone stacks became islands.

Another very beautiful form of Grayling is found in the northern half of Scotland. The markings show greater contrast than the southern type, with wider pale bands on the uppersides, and undersurfaces richly marbled with thick jet-black lines on a paler white background. The contrast is greater still on the Hebridean islands, where the broad pale bands are a bright yellow. These so-called *scota* Graylings are virtually confined to the coast, where they may be found in abundance.

The other named races of this variable butterfly live in Ireland. There are two types. The commonest, *hibernica,* is in many ways similar to the Scottish form, but can be distinguished by its warmer brown ground-colour; it, too, is restricted to the coast. Finally, there is a further form that occurs only on the limestone pavement of the Burren. Even in England, the Graylings that live on chalk and lime tend to be paler and more grey, but here the adults are so pallid that they have their own name, *clarensis.*

Gatekeeper

Pyronia tithonus

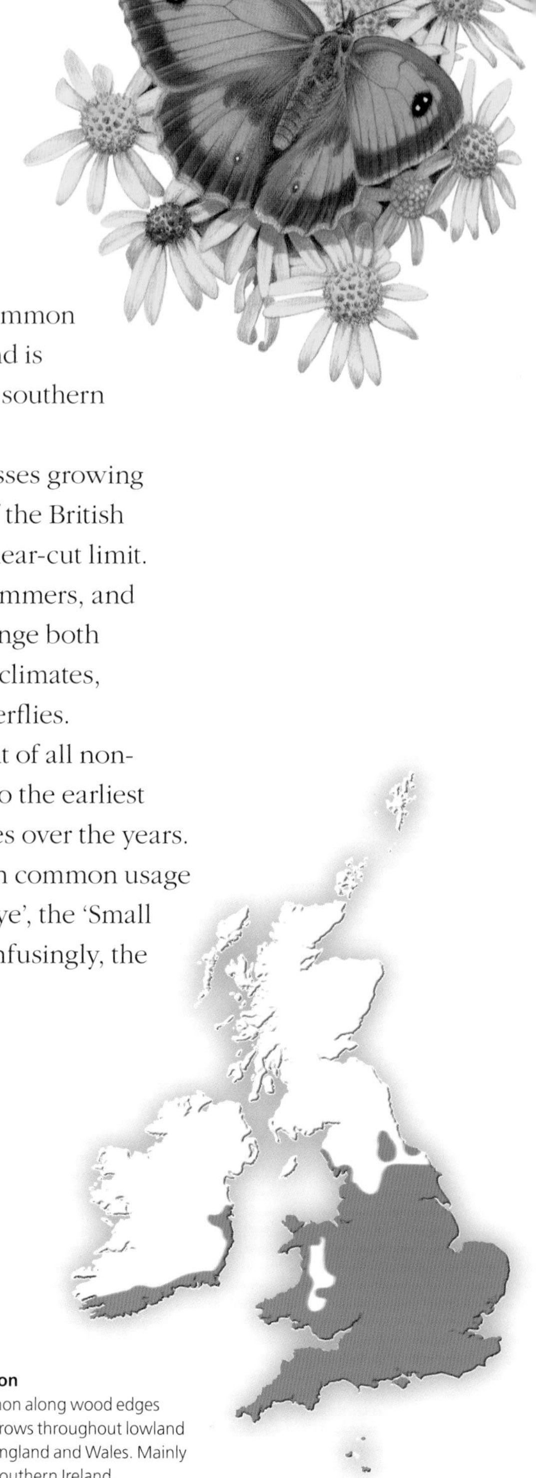

The Gatekeeper is one of those butterflies which, like the Small Skipper, is considered to be either commonplace or extremely rare, depending on where one lives. As a Dorset man, I find it impossible to imagine high summer without seeing scores of Gatekeepers jinking along hedgerows and woodland rides, or jostling for nectar on the flowers of Bramble, Common Ragwort and Common Fleabane. Yet it abruptly becomes scarce travelling north, and is absent from the northern half of Britain and from all but the southern coastal areas of Ireland.

The Gatekeeper's habitat – a combination of tall, wild grasses growing beneath sunny, sheltered shrubs – is abundant over much of the British countryside, but the butterfly can exploit it only south of a clear-cut limit. This reflects its need for comparatively warm springs and summers, and hence it is another species that has been able to spread in range both northwards and into higher altitudes under recent warming climates, although to nothing like the same extent as some other butterflies.

In the south, the Gatekeeper is perhaps the most abundant of all non-migratory butterflies in high summer, and was well known to the earliest British naturalists. Not surprisingly, it acquired various names over the years. The 'Hedge Brown' is more apt and is the only other name in common usage today, but the species was once also known as the 'Hedge Eye', the 'Small Meadow Brown', the 'Lesser Double-eyed Butterfly' and, confusingly, the 'Large Heath'.

Eye-spots and wing patterns

The Gatekeeper is not a difficult butterfly to identify. The two sexes are different, with the male distinctly smaller and brighter, and possessing a dark broad band of scent scales across the orange on each forewing. Beginners sometimes confuse it with the larger, but duller, Meadow Brown. One small but clear-cut distinction is that the tiny dots on the undersides of the hindwings are white on the Gatekeeper whereas they are black on the Meadow Brown. In addition, Meadow Browns generally have just one white pupil in the large eye-spot on each forewing, whereas the Gatekeeper's eye-spot always contains two. This last feature is not an absolute rule, for one occasionally finds a Meadow Brown with two white pupils, but these generally occur in

Distribution
Very common along wood edges and hedgerows throughout lowland southern England and Wales. Mainly coastal in southern Ireland.

▲ Male
Males have a broad sex-brand on each forewing, and a fuller orange colour.

▲ Female
Females are larger and paler than males, and lack the sex-brand on the forewings.

▲ Egg [× 22]
Laid singly; gradually becomes mottled and then brown.

◄ Eye-spot variant
Aberrant *multiocellata* form, one of a range of eye-spot variants.

▼ Caterpillar [× 2¼]
Hairy and of variable colour; green specimens are numerous on some sites.

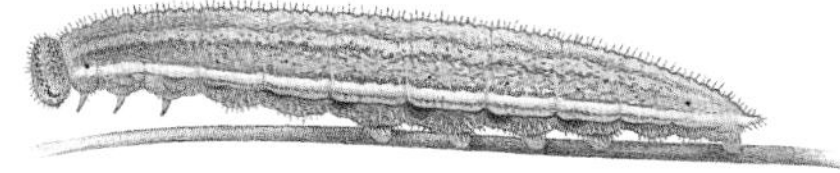

► Basking adult
Adult male basking on a leaf of bindweed, with its wings exposed to the sun.

▼ Resting adult
Male at rest on flower of Traveller's-joy.

▲ Chrysalis [× 2¼]
Slung beneath a leaf at the base of a shrub.

	Jan	Feb	Mar	Apr	May	Jun	Jul	Aug	Sep	Oct	Nov	Dec
Egg												
Caterpillar												
Chrysalis												
Adult												

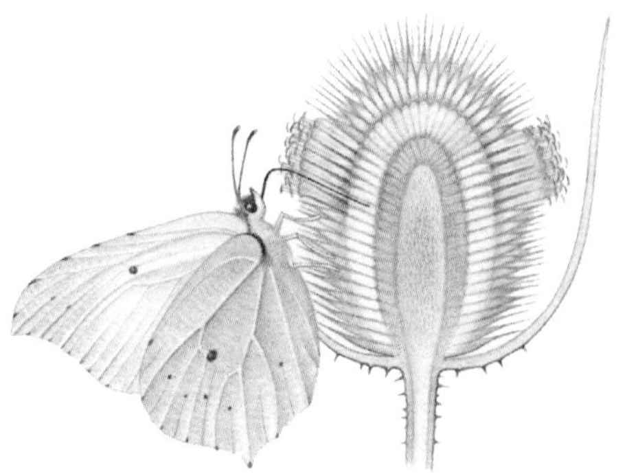

Long-tongued and short-tongued feeders

The Brimstone (left) has an unusually long proboscis, about 15mm in extent. It can reach the nectaries at the bottom of Wild Teasel flowers. In contrast, the Gatekeeper's proboscis (right) is just 6mm long. As a result, it can feed only on flat flowers.

Scotland, far north of the Gatekeeper's range.

Unlike several of our browns, the Gatekeeper shows little variation in wing pattern in different regions of the country, although it is not uncommon to find the occasional adult with extra spots in any colony. Some heavily spotted varieties are very beautiful. We illustrate the form *multiocellata*, an example that is by no means an extreme case, and which is quite common in certain colonies in some years.

Gatekeeper colonies

The Gatekeeper's flight period is sharply defined and occurs on the same dates across the country, in contrast to that of the Meadow Brown, which emerges a good month earlier and which may still be on the wing long after the last tattered Gatekeepers have vanished. The Gatekeeper first appears in early July, reaches a peak in the first week of August, and all but disappears by the end of the month, although some curious colonies on the southern chalk regularly last for two or three weeks longer.

Less is known about the natural history of this common little butterfly than of any other brown. Like its close relatives, it undoubtedly lives in clearly defined colonies, and I have found little or no migration between nearby sites. Thus when adults are seen in country gardens, this usually results from the presence of a colony breeding in an adjoining hedge-bank rather than from individuals flying any distance to the flowers. It is very rare for Gatekeepers to appear in city gardens, even in counties where the species is abundant.

Gatekeeper colonies vary enormously in size, from a few dozen adults up to several thousand. This largely reflects the size of the breeding area, the smallest populations being confined to narrow strips along the bottoms of hedgerows, whereas the largest occur in open but scrubby woodland, where the terrain is criss-crossed by broad, sunny rides. Numbers also fluctuate within a colony from one year to the next, generally in synchrony from site to site. They tend to be most numerous after August temperatures have been high the previous year.

Gatekeepers are almost always associated with shrubs. Their caterpillars feed in the sheltered grasses below, while the adults fly strictly around the bushes, except in woodland, where they periodically ascend to the tree-tops to drink honeydew. Most, however, feed on hedgerow and woodland flowers, and although they visit a wide variety of species, they are restricted to flat, open blossoms. This is because the Gatekeeper's proboscis is exceptionally short and unable to penetrate the deep corollas of Wild Teasels and other tubular flowers, a constraint illustrated above.

A gradual development

Gatekeeper eggs are laid singly at the bases of shrubs. Some are deposited directly on leaf blades but most are attached to bark or simply ejected into the air, in the same way as the eggs of the Marbled White, Ringlet and Meadow Brown. Unless an egg-laying female is followed, its pretty eggs are impossible to find, but they are easily obtained in captivity by caging a freshly emerged female over a half-shaded clump of grass. They hatch after about three weeks, during which time the colour changes from pale yellow to white with rust-coloured patches, and eventually to a uniform brown.

The tiny caterpillar first eats part or all of its eggshell, and then nibbles tender grass blades for the rest of its life. It is unknown whether it finds certain species more palatable than others, but there is reasonable evidence that a wide range of fine and medium-bladed species is eaten, including Common Couch, various bents, fescues and meadow-grasses. In every case, the egg-laying females select tallish plants growing in sheltered, sunny spots.

The caterpillar is sluggish and slow-growing, taking eight months between hatching and forming a chrysalis. It hibernates after making the first of four skin changes, choosing dried, curled leaves deep in the body of a grass clump. Growth resumes in March or April, with the caterpillar's daytime spent head-down and hidden within the clump before ascending to nibble the tender growing tips at dusk. This is not a particularly easy caterpillar to find, but it can be discovered on good sites by shining a torch on the taller grasses along the bottom of shrubs on warm evenings in May. Caterpillars vary somewhat in colour, but most are grey-brown with darker stripes; all are slightly hairy. They are similar to those of the Ringlet, except that there is less pinkness to the pale stripes along their sides.

In a typical year, the grey cylindrical caterpillar completes its growth in early June, and settles down to form a much prettier chrysalis, suspended beneath a leaf blade, again towards the base of a shrub. It hatches after three or four weeks, and is unlikely to be found in the wild.

Southern abundance

Gatekeeper colonies can be expected almost anywhere in the south in warm, sheltered places where shrubs exist with tall, wild grasses growing beneath them. A great many of the hedgerows of southern England and Wales support small numbers along their bases, and populations occur in almost every southern wood, breeding either along the outer edge or, where these exist, along shrubs bordering sunny, open rides. Enormous colonies breed in many shrublands, including on heathland, sand dunes and the scrubby undercliffs of southern coastlines. They are, however, almost completely absent from the surroundings of some towns: for example, few Gatekeepers occur in London's parklands, although the occasional colony exists in copses such as the London Wildlife Trust's Gutteridge Wood.

The Gatekeeper is also common in suitable habitats throughout the lowlands of Wales, but is absent from all mountains. In England, it is probably the commonest summer butterfly in south-western counties such as Devon and Dorset, but it becomes increasingly confined to woods in the Midlands. Colonies have spread well into the southern Pennines in recent decades, and extend further north on either side in sheltered coastal habitats, petering out in Cumbria to the west and in north Yorkshire and County Durham to the east.

Old records exist for southern Scotland, but their veracity has been questioned. There is little doubt that no genuine Scottish colony exists today, despite interesting reports of a few individuals that bred for a year at least following releases or escapes in south-west Scotland. Across the Irish Sea, colonies are almost entirely confined to the southern Irish coastline and to a few kilometres inland.

Meadow Brown

Maniola jurtina

This large but unspectacular brown has declined greatly over the past century as a result of the agricultural improvement of most lowland grasslands. No longer is it possible to write – as did C G Barrett in Victorian times – that there is 'hardly a grassy field in the United Kingdom from which it is wholly absent'. Yet this remains an abundant and widespread species. Colonies survive in almost every patch of wild grassland that is not very closely grazed, and the Meadow Brown still deserves its reputation of being the commonest butterfly in the British Isles.

Despite its abundance, this species is frequently confused with other browns when on the wing, especially the Gatekeeper and the Ringlet. The Meadow Brown also varies somewhat in appearance in different parts of its range. The largest and most beautiful forms occur in the west and north – the race *iernes* in south-west Ireland and the Atlantic Isles, *cassiteridium* in the Isles of Scilly, and *splendida* in the Hebrides, Orkney and mainland Scotland north of the Great Glen.

Variation and natural selection

One apparently trivial way in which individual adults and whole colonies of Meadow Browns differ is in the position and number of the tiny black dots on the undersurface of each hindwing. Some individuals – particularly females – have no dots at all, whereas others have up to five dots per wing. In the 20th century, this variation was used by E B Ford and W H Dowdeswell as the subject of some of the pioneering research on animal genetics. Working first in the Isles of Scilly, Ford and Dowdeswell found that Meadow Browns on the small, isolated islands were much more uniform in their pattern of dots than were those on the three large islands. They attributed this to the fact that the larger islands contained a greater variety of grassland habitats, and that certain forms of Meadow Brown survived better in some localities, while other patterns were favoured elsewhere. The end result, produced by natural selection, was a more diverse population on each large island, and one that was better able to exploit the full range of conditions. Hitherto, many scientists had believed that this sort of variation between isolated colonies was largely the result of chance,

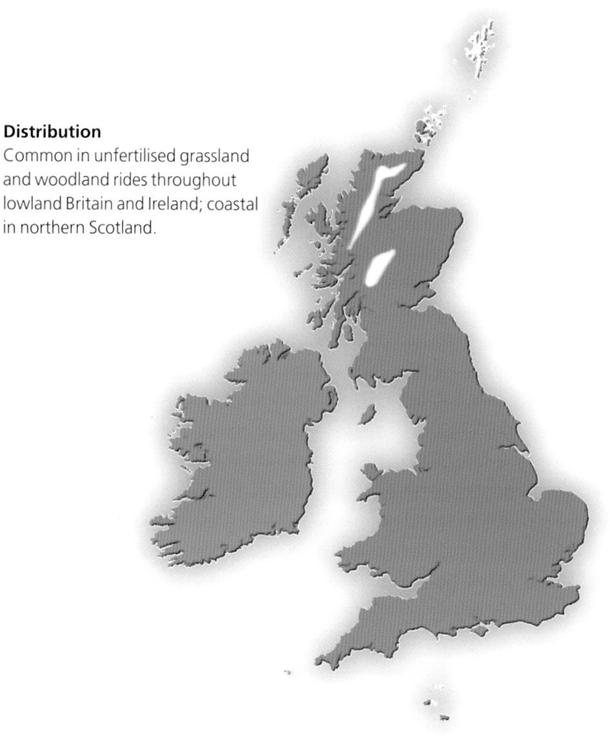

Distribution
Common in unfertilised grassland and woodland rides throughout lowland Britain and Ireland; coastal in northern Scotland.

▲ **Male**
Typical males have virtually no orange on the upperwings. Black dots on the hindwings vary in number.

▲ **Female**
In both sexes, there is much variation in eye-spots and other markings.

▲ **Irish female**
The Irish subspecies *M. j. iernes* is large and brightly marked.

◄ **Female in late summer**
Towards the end of the summer, adults become faded and tattered.

▲ **Feeding adult**
Creeping Thistle is a favourite source of nectar in grassy places.

◄ **Egg** [× 22]
Laid singly, usually on grass blades, but sometimes dropped from the air.

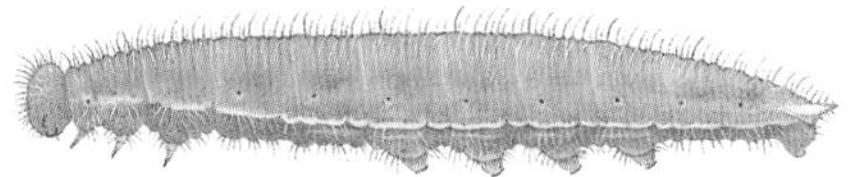

▲ **Caterpillar** [× 2¼]
Sluggish, hairy caterpillar feeds on a variety of grasses.

◄ **Chrysalis** [× 2¼]
Markings vary from highly striped to plain green.

	Jan	Feb	Mar	Apr	May	Jun	Jul	Aug	Sep	Oct	Nov	Dec
Egg												
Caterpillar												
Chrysalis												
Adult												

or was caused by the characteristics of the females that first colonised each site.

Ford and Dowdeswell also found differences in the presence and pattern of these same black dots on Meadow Browns in various parts of England. Colonies in the north tended to have fewer dots, whereas the Meadow Browns of west Devon and Cornwall were nearly as heavily spotted as those on the larger Isles of Scilly. By and large, this regional variation alters gradually from one part of the country to another, but there are many local pockets containing colonies with atypical adults.

Wing patterns and survival

Geneticists argued for many years over the significance of these findings. Few thought that the presence or absence of the tiny dots could make much difference to the chances of a Meadow Brown surviving in the wild, yet at the same time many believed that very powerful forces must be favouring one pattern or another to produce such clear-cut differences. It came as something of a relief, therefore, when Paul Brakefield found that this apparently trivial variation in dotting could be linked to real differences in the Meadow Brown's survival under different circumstances.

Brakefield discovered that the black dots were controlled by large groups of genes, called polygenes, and that some of these also controlled other characteristics of the butterfly. For example, certain genes have the dual effect of causing the caterpillar to reach maturity quickly and also to develop into a large adult possessing a large number of dots, whereas Meadow Browns without these genes emerge late in the season and tend to be small and spotless. If either the early- or late-developing caterpillars in a colony suffer unusually heavy losses during spring, the colony will tend to switch from a preponderance of one pattern of spotting to another.

In addition, it is now clear that not only are these dots large enough to distract enemies in their own right, but that those butterflies that possess several dots on their hindwings also tend to have larger eye-spots on their forewings. The latter clearly attract the attention of predators, and any circumstance that favours the survival of Meadow Browns with large or small eye-spots inevitably also affects the number of hindwing dots in the colony. From this rationale, Paul Brakefield made an ingenious suggestion that accounts for the differences in the markings of male and female Meadow Browns, and for much of the local variation in their dot patterns.

Distraction and camouflage

Male Meadow Browns are active creatures that spend much of the day searching for females, either by launching themselves from prominent perches on low vegetation or, more often, by weaving between grass clumps on erratic investigative flights. This makes them conspicuous to birds and mammals. It pays for the males to have a variety of dots near the outer edges of their wings, for these distract predators, deflecting their attacks towards this area and usually allowing the butterfly to escape with no more than a peck-mark.

Females, on the other hand, spend most of the day resting near the ground. They sit with their wings closed and overlapping, so that only the undersurfaces of the hindwings are visible. The emphasis now is on camouflage. Survival is highest among individuals that do not attract any predators in the first place, and their camouflage is enhanced if there are no or few dots on the hindwing. Females also have a second line of defence. If discovered, they quickly raise their forewings, exposing the gleaming eye-spot in a flash, in the hope of startling the predator into flight. To achieve this females tend to have much larger eye-spots on their forewings than males, even though the hindwings contain fewer dots.

Brakefield also noticed that the nature of the habitat could account for some of the variation in dotting between individual males and females. Butterflies that have few or no dots are much better camouflaged on sites that contain uniform grassland, such as those that predominate in the north, whereas Meadow Browns with several dots merge better into mixed, scrubby backgrounds. In addition, butterflies that breed at low densities over large areas of mediocre habitat are forced to roam much more than those that emerge at high densities on ideal breeding sites. Thus, the Meadow Browns living in the first situation are more likely to be noticed – and therefore need dots that deflect attack – than those that live at high densities.

High-density living

Meadow Browns may be crammed together in remarkable numbers on the best sites. The ideal habitat for this species consists of warm, open grassland, 0.5m or so in height, containing an abundance of summer flowers and medium- or fine-leaved grasses. This situation exists on many lightly grazed downs, on undercliffs, in recently abandoned grassland, and in unfertilised hay meadows that are cut late in the season, after the adult butterflies have emerged. Up to 2,000 Meadow Browns can emerge in a single hectare on these sites. These populations are very stable from year to year in the south, so long as the habitat remains unchanged. However, as one approaches the butterfly's northern climatic limit, at altitude in north Scotland, we found that its populations increasingly oscillate over a number of years, gradually building up to high numbers before falling back to very low densities, with each cycle lasting seven to ten years.

Escaping attack

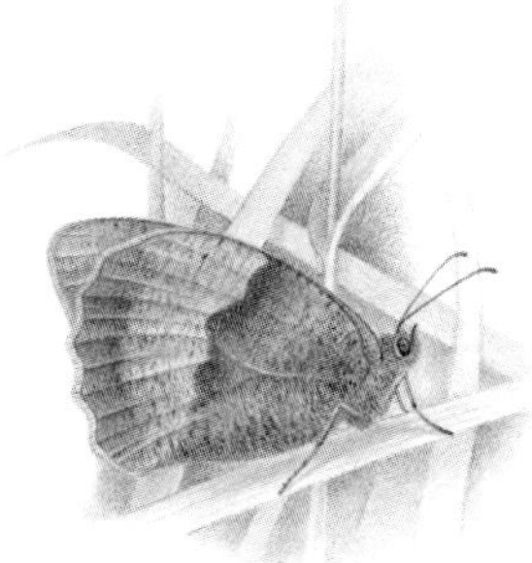

◀ **Concealment**
Females that breed in grassland spend most of their lives out of sight in the uniform vegetation. They hide their forewings between the hindwings, which generally have no dots to betray them to birds.

◀ **Survival in the air**
The dots on a Meadow Brown's hindwings may trick birds into pecking at the wings rather than at the vulnerable body. Meadow Browns that keep on the move – such as those living in poor habitats – have evolved heavily dotted hindwings.

◀ **Scare tactics**
Females without hindwing dots can also raise their forewings, exposing gleaming eye-spots to frighten a predator away. These eye-spots tend to be larger than on more mobile Meadow Browns.

◀ **Survival in the open**
Sedentary females that live in patchy, open habitats are also inclined to be conspicuous. They, too, tend to have dots on their windwings to deflect bird attacks.

High densities of Meadow Brown are also found along the verges of some busy roads, and in July the butterfly can be seen by the hundred in almost every sunny woodland ride in the British Isles. Very much smaller numbers occur in marginal habitats, such as tall, coarse grassland or turf that is grazed fairly short. A few dozen butterflies is also the normal complement on innumerable small sites, such as the thin strips of grasses that grow beneath almost every hedgerow.

Taking to the air

By and large, the Meadow Brown is a sedentary insect, which strays from its breeding sites only when these are disturbed or are of low quality. Adults will, however, readily fly distances of several hundred metres within their breeding areas, and periodically cross short stretches of unsuitable habitat, for they can recognise new breeding patches, presumably by scent, up to 150m away. Dual carriageways are not an obstacle; Miguel Munguira and I found that remarkably few are killed when crossing even the busiest roads, for the featherweight adults are merely swept into the air streaming over modern ergonomically-designed cars and deposited, safe but flustered, on the nearest grass verge.

On southern sites, the first male Meadow Browns emerge in mid-June in a normal year, but some fly as early as May after a warm spring. The date of peak numbers varies from one year to the next by up to four weeks, depending on June temperatures, but is usually reached in late July. There is then a gradual decline, with the butterfly disappearing on most sites by the end of August. However, the emergence is greatly prolonged on warm, southern chalk downs, where adults continue to emerge throughout September; it is by no means unusual to see fresh females in mid-October. Why this occurs is unknown. There is no evidence that some individuals aestivate – that is, enter a summer hibernation to sit out a hot dry season – such as occurs in southern Europe; they simply appear to emerge gradually over a very long period on English chalk.

Male and female Meadow Browns usually roost together in tall clumps of grass, then bask with their wings spread wide to absorb the early morning sunshine. When warmed, they are capable of flying in the dullest weather, and even in light rain. Both sexes feed avidly on a range of summer flowers.

Courtship and egg-laying

The males, as described above, spend much of the day hunting virgin females. Once found, there is a brief courtship during which he envelops his partner in an unpleasant scent that has been variously described as resembling an old cigar box, musty hay or dirty socks. Humans vary in their sensitivity to this; some find it exceedingly strong, while others can scarcely detect it. The female Meadow Brown, however, is entranced, and settles on a firm piece of vegetation while the male grips her with his claspers.

Female Meadow Browns generally mate on the first active day of their lives, and rarely need to pair a second time. Like the males, they live on average for five to 12 days, lasting longer in cool, humid seasons. The eggs ripen quite quickly, with most being laid between two and four days after the female has mated. As with many butterflies, the female first embarks on distinctive fluttering flights, flapping slowly just above the grass. She frequently alights in warm, sheltered patches of fine or short turf, and walks jerkily with wings closed in a sort of rocking movement, twitching her way excitedly through the grass tufts.

Moses Harris watched this behaviour over two centuries ago, and wrote that 'The hens, when impregnated, cast forth their eggs but I cannot be certain whether they fix them to the blades of grass, or scatter them loose on the ground.' In fact, they do both: some are deposited with the utmost care, whereas others are simply squirted out, some even in flight.

Early development

It is not known exactly what prompts the female Meadow Brown to lay in certain patches of turf. Perhaps she selects a favourable scent within the field or chooses particular grasses. For, although the caterpillar can feed on a wide range of species, it has distinct favourites: all are medium- or fine-leaved species, especially meadow-grasses, bents and rye-grasses. It is possible, too, that the egg-laying females can detect patches where the grass is infested with certain fungi, for Miriam Rothschild found that the Meadow Brown caterpillar sequesters chemicals that act as powerful bactericides, from the fungi that they unwittingly consume.

The young caterpillars feed on grasses by day in high summer and autumn, but are too small to be found with ease. They settle down for the winter in a grass clump, emerging during mild spells to nibble a few blades. They resume feeding in earnest in spring, and by March many have moulted for the second time and reached a size at which it is safer to feed by night.

Fully grown caterpillars are quite easy to find on the best sites, where there may be as many as ten per square metre. They are also quite easy to identify, being hairier than the caterpillars of most browns and possessing 'tails' that are distinctly white. They can be found by day deep in the base of favoured grasses, such as Smooth Meadow-grass, by looking for tussocks with blades that have been eaten back. More will be seen after dusk if you scan the upper grass blades with a torch, especially on warm, damp evenings in May. This must be done very gently, for the caterpillars are sensitive to the merest rustling and instantly drop to the ground, curled up in a ball.

Most caterpillars pupate by early June, and last two to four weeks before hatching. The chrysalis is extremely variable in appearance, ranging from pale green with few dark markings, to almost white but heavily striated with broad black bands.

A broad distribution

Although many former colonies have disappeared because of the spraying and ploughing of ancient grassland, the Meadow Brown remains an extremely common lowland butterfly in almost all suitable habitats in the British Isles. It is easier to list the regions where it is scarce or absent. These include northern Scotland, where it is distinctly local throughout; Orkney, where it is confined to warm, south-facing slopes; and Shetland, where it has never been reliably recorded. Colonies are seldom found at altitudes above 200m in the Highlands, Pennines and Welsh hills, or above 300m in the warmer climate of Dartmoor.

Ringlet

Aphantopus hyperantus

The Ringlet is an enchanting inhabitant of humid grassland over much of the British Isles. Flying mainly in July, it avoids the summer's heat by living in woods and cool moist places, where the air is damp and still. F W Frohawk aptly described it as a 'peaceful' butterfly, and there are few more refreshing sights than the dusky adults fluttering silently along rides, crossing glades or bobbing among dense tussocks on a sticky summer's day.

This is a comparatively common butterfly, frequently overlooked because of its resemblance in flight to the male Meadow Brown. The Ringlet has much darker wings that are bordered by a fine white fringe that shines as it catches the sun. There should be no confusion when the butterflies settle: Ringlet uppersides are soft, velvety and uniformly dark, with none of the orange of a Meadow Brown, while the undersides are adorned with distinctive gleaming eye-spots. It is these that gave this butterfly its name two centuries ago, when 'ringlet' was in common usage to describe any small circle, such as the fairy rings caused by a fungus on a leaf. Other early names included the 'Brown-eyed Butterfly' and the 'Brown Seven Eyes'. In France it is 'Le Tristan' (The Sorrowful), in Germany 'Brauner Waldvogel' (Brown Wood-bird).

Eye-spot variations

The eye-spots may vary in size, shape and number on different individuals. The most striking form, called *lanceolata*, has large and elongated spots, as can be seen in our illustration. It is a beautiful variety that is sometimes quite common in certain colonies in the south. It is also easily bred in captivity, as this characteristic is controlled by a single recessive gene that ensures that at least one in nine butterflies will turn out to be *lanceolata* if the offspring from one are paired with each other.

Other forms have few, small or even no spots. We illustrate an extreme example called *arete*, but every intermediate can be found between this and the normal spotting. Another form, *caeca*, merely possesses the inner white dots. This, like the *arete* Ringlet, becomes increasingly common as one travels north. The butterfly also tends to be smaller and greyer in Scotland, as it is in Scandinavia and at high altitudes further south.

Distribution
Common and often overlooked in woodland rides and damp grasslands in Ireland, lowland Wales and southern lowland England.

All-weather flight

Ringlets are invariably found in self-contained colonies. Standardised counts indicate that these range enormously in size from a few dozen individuals up to many thousands per colony on the best sites. Numbers also fluctuate considerably from one year to the next, typically in close synchrony on different sites. In general, the butterfly benefits from a rainy summer but declines in the year after a dry one. There was, for example, a severe reduction almost everywhere after the exceptional drought of 1976, and a rapid recovery during the cool wet years that followed. This species is one of the most consistent of all butterflies in its time of emergence. The first males are seen on the last days of June but the main flight occurs in July, peaking in the third week. There is then a rapid decline, with a few faded stragglers surviving to mid-August. They do, however, make the most of their short lives, and remain surprisingly active in gloomy weather when every other butterfly is grounded, even flying in light rain.

Male Ringlets spend much of their lives fluttering around grass heads and jinking between tussocks in a relentless search for mates. Both sexes also spend long periods at flowers, where they jostle for nectar among Meadow Browns and Gatekeepers on Bramble, thistles or any other blooms that are available. They visit orchids on downland sites, which can cause problems because the sticky sacs of pollen adhere to their proboscises, making it impossible to curl them away after feeding.

Aerial egg-laying

Female Ringlets appear to be indiscriminate when it comes to egg-laying. They either sit high up on grass heads squirting eggs into the air or, less often, lay them while hovering in a stuttering, stumbling flight. I suspect that there is more to this than meets the eye, and that they are responding to the scents of specific – and possibly fungus-infected – coarse grasses that waft upwards on the cool, humid air.

I have watched Ringlets laying on many occasions but have yet to find their eggs on the ground. They are, however, easy to obtain in captivity if a female is caged over a pot of moist grasses. Each egg is triangular in outline, with a wet, glossy sheen unlike that of any other brown. It hatches after a fortnight and, if necessary, the little caterpillar can survive without food for a day or two while it searches for a suitable plant. The full range of grasses eaten in Britain is unknown, but all who know the butterfly agree that the list is short, often just one species being used at a particular site. Cock's-foot and False Brome are great favourites, but only where these coarse grasses grow as lush, uncropped tussocks.

The caterpillar lives about ten months, hibernating while quite small after the second of its four skin moults. It is impossible to find in the autumn, but can be seen quite easily by searching tall grass clumps by torchlight on warm evenings in May. It is rather like the grey forms of the Marbled White and Gatekeeper caterpillars, but is hairier than the latter and lacks the former's distinctive pink tails. It pupates in mid-June, forming an attractively streaked chrysalis that rests at the bottom of its grass clump in a small cocoon of silk. Miriam Rothschild maintained, and to a certain extent showed, that the caterpillars of this butterfly sequester poisons from the fungi that often infect their foodplants, and pass them on to the adult stage; she believed – and on these matters was invariably right – that the rings on the adult's wings function primarily as aposematic markings that warn birds of their unpalatable bodies, as explained for the Marbled White (see page 251).

Woodland and hedgerow habitats

The best places to find this dusky brown are the rides, edges and glades of woods, where wild grasses have been left to grow into a dense strip or as isolated tussocks. In the southern half of Britain, most large or medium-sized woods contain a colony, so long as there are open or semi-shaded spaces. Although it avoids a closed canopy, the Ringlet is more tolerant of shade than any other butterfly that breeds on the woodland floor, apart from the Speckled Wood. But the species is by no means confined to woods, and can be encountered on any patch of rank, unfertilised grassland. Colonies are commonest on heavy soils, and on our southern clays it is even a hedgerow species, being found along many ditches, hedge bottoms and road verges, especially those that are not mown in summer. It is much less common on light soils, and although a few downland colonies exist, these tend to be on damp, north-facing slopes or in combes where the soil is deep and where some scrub has encroached. These sub-optimal open habitats generally support small colonies, as do the more shaded woodland sites, and both are apt to disappear after a summer drought.

The status of the Ringlet has improved dramatically in the British Isles during the past 40 years. This may partly be due to warmer climates in the north, but is probably in response to increased growths of coarse grasses in many woods. Ernie Pollard analysed a hundred-fold rise in Ringlet numbers during a 20-year period in Monks Wood, and concluded that the most likely cause was the greater deposition of nitrogen in rain, which encouraged coarse grasses at the expense of finer-leaved species, together with some increase in the shadiness of rides. Another theory is that the butterfly is sensitive to air pollution, and that this explained its curious absence, until recently, from the Midlands and northern England, as well as around London, Glasgow and Edinburgh, regions that all experienced local extinctions of Ringlets during the Industrial Revolution. It is tempting to speculate that the fungi infesting its foodplants declined under these sooty atmospheres, but this is sheer conjecture.

Whatever the cause of the late 19th-century declines, the Ringlet has regained most of its post-industrial range during the past 40 years. The striking gaps that were a feature of its distribution in the 1970s – notably in the central Midlands and north-east Wales extending to north-east and north-west England, and a second area from southern Scotland to Perthshire and Fife – have now largely filled, although the butterfly is still absent from a sizeable area of the north-west

▲ **Male**
Upperwings of fresh males are almost black; their colour fades slightly with age.

▲ **Female**
The ground-colour of the female is slightly paler than that of the male. Wings of both sexes have distinctive white fringes.

▲ **Eye-spot variants**
Size and shape of the underside eye-spots varies greatly.

▼ **Feeding adult**
Wild Privet is a frequent nectar source on downland sites.

◀ **Egg** [× 22]
Dropped, rather than laid, among grasses.

▼ **Caterpillar** [× 2¼]
Overwintering caterpillar feeds occasionally on mild nights.

▲ **Orchid pollen**
Pollen sacs from orchids may prevent the proboscis being coiled up.

◀ **Chrysalis** [× 2¼]
Formed at the base of grass tussocks.

Midlands, extending through Cheshire, Greater Manchester and Lancashire into Cumbria. Although it continues to spread northwards in Scotland, it is still largely absent from the north-west, having colonised the Great Glen and several locations along the River Spey in the last five years. Apart from Islay and Jura, the Ringlet is absent from most islands, including the Isle of Man. It is, however, a common butterfly throughout Ireland.

Small Heath

Coenonympha pamphilus

The Small Heath is one of the smallest and most successful of the 117 species of browns found in Europe. Although seldom as numerous, for example, as the Meadow Brown or Scotch Argus, it nevertheless occurs over a wider geographical range and in a greater variety of habitats than any other member of its family. It also has an extremely long flight period in the south, where a complicated sequence of broods ensures that the tawny adults can be seen bobbing around wild grasses almost continuously from May to October. Sadly, although still as widely distributed as ever, many Small Heath populations have declined in size over the past 20 years, and many sites have lost their colony altogether. It tends to be more numerous when there has been high rainfall in the July of the previous year, but an increase in summer droughts only partly explains its long-term decline.

Drunken flight

Size and colour distinguish this common species from other brown butterflies. Among our small species, it is both lighter and more golden than the Dingy Skipper or the female blues, and its pastel shades have nothing of the gaudy brightness of the four golden skippers. A drunken pattern of flight is also diagnostic, as it flops and bobs erratically just a few centimetres above the ground. Blues have a more purposeful flight, whereas skippers characteristically dart back and forth at high speed.

Small Heaths vary in appearance, with both unusually bright and dull adults often flying in the same colony. The size of spots on the forewings and hindwings also differs. There is a tendency for adults to be duller and paler as you travel north and west in Scotland, and the butterfly has been described as a separate subspecies or form, called *rhoumensis*, at the extreme of this range. This is also illustrated, although its status as a true subspecies is doubtful.

Distribution
A declining species found in low numbers in most patches of dry grassland throughout its British range. Absent from Orkney and Shetland.

Protective eye-spots

Whatever the ground-colour, both sexes appear much brighter in flight, because the settled butterfly always rests with its wings closed. In dull weather, or when the butterfly has been sitting for a while, the forewings are tucked well down between the hindwings, so that only the grey hind undersurface is visible. However, like the Grayling, the Small Heath usually rests for a short while after landing with

▲ Male
Generally smaller and brighter than the female; sexes otherwise similar.

▲ Female
Ground-colour and size of spots is highly variable in both sexes.

▲ Male
Form *rhoumensis* from Scotland; duller than the forms found further south.

▲ Male
Colour variants with brighter wings may occur alongside duller forms in the same colony.

◀ Egg [× 22]
Large, in proportion to butterfly; laid singly on a grass blade.

▼ Caterpillar [× 2¼]
Camouflaged caterpillar lives low down in grass, feeding by night.

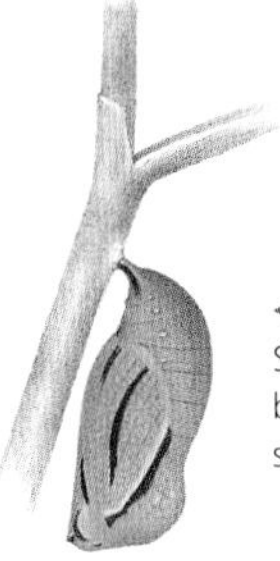

◀ Chrysalis [× 2¼]
Suspended beneath a grass stem; black streaking varies, and is sometimes absent.

◀ Roosting adult
At night, adults rest inconspicuously on vegetation.

▼ Mating
Mating takes place within territories established and defended by the males.

	Jan	Feb	Mar	Apr	May	Jun	Jul	Aug	Sep	Oct	Nov	Dec
Egg												
Caterpillar												
Chrysalis												
Adult												

In Scotland, typically single brooded (flying from June to August).

the forewings held aloft, which exposes a gleaming eye-spot set in a tawny triangle; this no doubt was responsible for Petiver's early English name, 'The Golden Heath Eye'. As with other browns, the eye-spot is a ruse to trick any insectivorous bird that may have noticed the butterfly land into pecking at the wrong place. You will often find tattered Small Heaths with V-shaped cuts, made by a bird's beak, around the eye-spot. By deflecting the attack from head to wing, they at least live to continue breeding.

Colonies and territories

The Small Heath lives in self-contained colonies of limited size, with most adults flying and breeding in the same patch of grassland for their entire lives. Yet the occasional individual of both sexes wanders, rather more than is the case with most sedentary species. Hence isolated patches of fresh breeding habitat tend to be quickly colonised. Sometimes they make longer journeys, and it is not unusual for the butterfly to be seen in gardens; a few have even been caught on lightships.

Little was known about the breeding behaviour of this butterfly until Per-Olaf Wickman made a classic series of studies in Sweden: his findings tally so well with most casual observations in Britain that I have no doubt that our Small Heaths act in the same way. Males live, on average, for about seven days and spend most of their lives searching for mates. They gather in leks beside landmarks in the open grassland, such as bushes and trees. The dimensions of the landmark are important. Males (and virgin females) avoid pyramid-shaped bushes that taper towards the top, but are strongly attracted to those with narrow bases and which widen two or more metres up, growing more in the shape of a cornet.

Each male tries to establish his own territory by perching on the ground around a suitable landmark, competing with rivals to sit in the lee. Unfortunately, there are seldom enough perching points to go round, so males engage in prolonged aerial battles for possession. The larger-winged individuals generally win, condemning smaller males to fly further afield, where they will have little chance of mating. Wickman found that nearly 90% of the pairings on his Swedish site were accomplished by the 60% of males that managed to establish territories. This pattern broke down only on exceptionally hot days, when many territorial males deserted their perches in favour of flights.

Male territories seldom overlap with breeding areas, which are found in more open grassland. Virgin females emerging from these patches fly directly towards male leks, and then make themselves conspicuous by flying back and forth, about a metre above the ground, soliciting the attention of any perching mates. These, in fact, need little invitation, and will launch themselves after any insect that passes by. Having attracted her mate, the female lands on the ground and the male advances with a combination of head butts and fluttering wings, almost certainly showering her with an aphrodisiac, as described for the Wall (see page 242).

Females that have mated avoid male territories, and their rare flights are short, fluttery affairs, just above the ground. They search for open, wild grassland in which to lay eggs, landing where the sward is short or sparse. Each then wanders for a while, testing the vegetation before laying a single egg, usually on a blade of fine grass. In England, a range of wild species, including meadow grasses, may be used, although Sarah Meredith, working in Co Durham and the south coast, found that Sheep's-fescue was greatly preferred. Wickman similarly found that four-fifths of eggs were laid on Sheep's-fescue on his Swedish study sites.

Hidden in the grass

The egg of the Small Heath is large for the size of the butterfly, and is highly attractive. Young females lay especially large eggs, and eject them more rapidly than do the older females; they also lay eggs with green shells, whereas after 100 or so have been laid, the shells become yellow. No-one knows the significance of this change, and both forms are well camouflaged against the grass blades on which they are laid.

Eggs last for about a fortnight, during which time the shell turns pale and freckled, becoming almost transparent just before hatching. The caterpillar lives low down in the sward, mainly in a small tuft of young grass. It emerges at night to nibble the nutritious tips, although I have also found final-stage caterpillars eating by day on dull days. They are not easy to spot, for the green-and-white striped body blends beautifully with the narrow leaf blades.

Overlapping broods

Small Heaths have a complicated growth pattern in the south of the British Isles that is not fully understood. Most caterpillars are quite large by the time they hibernate, although a few may be considerably smaller. They resume feeding in spring, and by late April the first of the pretty striped chrysalises have been formed, dangling beneath grass stems. These produce a first batch of adults in mid-May. Numbers build up to reach a peak a month later, then gradually fall during July, with the last stragglers overlapping with a second brood of adults in early August. The second peak is reached towards the end of that month. In warm years, there may be a third emergence in autumn. But not every egg that is laid develops into an adult the same year: some offspring of the first brood hibernate instead, as do many or all from the second brood, and after a warm summer three generations of caterpillars may be hibernating together. Hibernation probably explains why fewer Small Heaths emerge in the second brood than in the first in most years, with fewer still in the third brood, when this exists. This is especially the case after cool, wet summers; hot seasons tend to result in large emergences in August, followed by small ones the following spring.

Northern Small Heaths belong to a different physiological race, and are less likely to produce second and third emergences, even when they are reared under artificially warm conditions. Thus, there is generally a single emergence each year in Scotland. This varies in timing according to the warmth of the season and the locality, but typically begins in the first week of June and lasts for three months, until the end of summer.

A ubiquitous but declining species

Colonies of Small Heath are found throughout the British Isles, apart from Orkney and Shetland, or on mountainsides higher than 750m; it is also rare in the more intensively farmed regions of Ireland, especially in the east, the south and the midlands. Numbers fluctuate considerably from one year to the next, and are consistently large only on open, well-drained grasslands where wild, fine-leaved grasses dominate the sward. These habitats include dunes, heaths, and chalk or limestone downs that are grazed fairly short but not nibbled down to the ground. Smaller numbers may be found almost everywhere else where wild grasses grow in open, sunny conditions, for example in unfertilised meadows, verges, wasteland and woods. Nor is this butterfly restricted to dry soils. It is often seen in ones and twos on heavy clays and in marshes, although in these places breeding is probably restricted to local dry spots. For example, I have found many caterpillars in the glades of a swampy wood in the Surrey Weald, but only on the sides of ant hills protruding as islands above the waterlogged ground.

Unfortunately, the Small Heath has declined in numbers throughout its range during the past 20 years, although there was a temporary recovery in 2002-2004. Monitored populations have fallen to about half the average size of the 1970s and 1980s, and there have been many local extinctions among the smaller colonies, especially in woodland. Unlike the Wall Brown, which shares similar habitats, this has yet to be reflected in any serious contraction in range, but it is a worrying trend, nevertheless.

Large Heath

Coenonympha tullia

The Large Heath occurs in a string of isolated colonies that stretch from central Wales to Orkney. It is one of our few true wetland species, and in high summer flutters over countless small bogs throughout the north. It is also one of our most variable butterflies, so much so that it was once considered to be three distinct species, known as the 'Scarce Heath', the 'Marsh Ringlet' and the 'Small Ringlet'. These roughly correspond to the three forms of Large Heath recognised today – *scotica*, *polydama* and *davus* – which are illustrated opposite. Other early names were the 'Manchester Argus', commemorating the first known colony of this butterfly, long since destroyed by the city's sprawl, the 'July Ringlet' and the 'Silver-Bordered Ringlet' – all more attractive names than 'Large Heath' and less confusing, for it is not so long ago that another satyrine butterfly, the Gatekeeper, was known by this name.

A range of forms

The variation in the markings of this butterfly, both in different parts of its range and – except in northern colonies – within individual sites, was a great attraction for Victorian collectors. The largest and most uniform form is *scotica*. Its wings are pale, grey and virtually spotless, giving it the appearance of a very large Small Heath. It is the form found in all the isles and throughout northern Scotland, down to an abrupt boundary shown on the map. The *scotica* form is now known to constitute a separate subspecies.

Travelling south, the *polydama* form predominates. This is slightly smaller and darker than *scotica*, and possesses appreciable eye-spots. It is found in Cumbria, Northumberland and southern Scotland, as well as on one site in the Pennines, on the Lincolnshire and Yorkshire Moors, and on several sites in mid- and north Wales. It is also the predominant type in Ireland.

The third form, *davus*, is the most beautiful of all. It has richer, redder wings and an array of large eye-spots, many with gleaming white pupils. It is the predominant type in lowland England but, sadly, most of its colonies have been lost. Nevertheless, it is still to be found in good numbers along the coastal plain south of the Lake District, on the lowland mosses of north-west England, in a small area of

Distribution
Very local species in peat bogs and wetlands, but often abundant where it does occur. Commonest in north Scotland.

▲ Male
Subspecies *scotica*; the underside of this form has few spots. Found in Scotland.

▲ Female
Subspecies *scotica*; the palest of all the variants.

▲ Male
Form *polydama*; an intermediate form found in northern England, Wales and Ireland.

▲ Female
Form *polydama*.

▲ Male
Dark form *davus*, found in north-west England.

▲ Female
Form *davus*; female's coloration is slightly lighter than the male's.

◀ Egg [× 22]
Initially yellow, later developing dark blotches.

▼ Caterpillar [× 2¼]
Can withstand extreme cold; may hibernate for two winters.

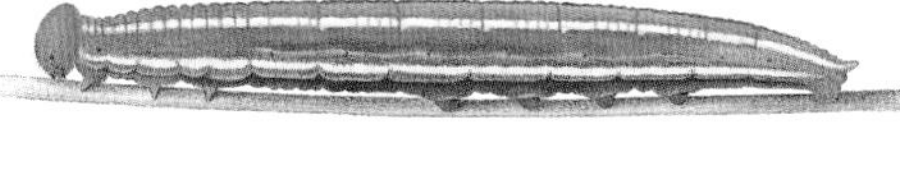

◀ Chrysalis [× 2¼]
Suspended from a stem; the boldness of the black stripes varies.

▲ Feeding adult
Adult clinging to the flowers of Cross-leaved Heath, the main nectar source.

	Jan	Feb	Mar	Apr	May	Jun	Jul	Aug	Sep	Oct	Nov	Dec
Egg												
Caterpillar												
Chrysalis												
Adult												

Shropshire, and at one site in lowland Lancashire. It is fair to say, however, that these last two groupings are largely for convenience. There is much variation within both forms and, where they meet, one very much resembles the other. The *scotica* Large Heaths, on the other hand, form a distinct and invariable group, and have probably been isolated from the southern populations for 10,000 generations or more.

Quite why the Large Heaths in various regions should have evolved such strikingly different markings was explained by Tim Melling, on whose work much of this account is based. He examined many colonies in the wild, and concluded that the reason was similar, but more clear-cut, to that suggested for the Meadow Brown. It centres on which is the safest method of protection against bird attack – to be exceptionally well camouflaged while sitting with closed wings on the ground, or to develop wing eye-spots which make the butterfly more conspicuous, but which deflect any attack away from the vulnerable body. Melling found a close correlation between the number and size of spots on different Large Heaths and the hours of sunshine that their various sites typically received in June and July, the months when adults are on the wing. On warm, sunny, southern lowland bogs, and also on sheltered scrubby ones, the adults tended to be much more active and flew more frequently than on cool sites. Their movements drew them to the attention of Meadow Pipits, which pecked particularly at the outer edge of the lower wings, if there were gleaming eye-spots there. This, Melling showed, gave the butterfly a good chance of escaping.

Over the years, the less heavily spotted individuals tend to be caught by birds, while the spotted ones survive to do more of the breeding. This eventually results in a colony made up of the *davus* form of Large Heath. At the other extreme, on cool northern sites it is seldom warm enough for the butterflies to take to the air. Instead, they spend long periods perched on the ground. For these butterflies it is a better strategy to escape the notice of birds in the first place. Under these conditions, unspotted *scotica* individuals are more likely to survive. On intermediate sites, there is probably a balance between the advantage of being camouflaged and that of surviving bird attacks. This leads to the *polydama* form, with its intermediate wing markings.

This is, perhaps, an oversimplification of the system, but even so it is typical of many of the subtle forces that affect the appearance and behaviour of insects. It certainly adds greatly to the fascination of watching this little butterfly, trying to weigh up whether each is a 'hider' or a 'deflector' in its battle for survival.

Bogland breeding sites

Large Heath colonies can be found from sea level up to altitudes of about 760m. Typical habitats are lowland raised bogs, waterlogged peat mosses, upland blanket bogs and damp acid moorland – all flat, wet areas where the caterpillar's main foodplant, Hare's-tail Cottongrass, grows in immense shaggy tussocks. Most sites also contain an abundance of Cross-leaved Heath, the favourite nectar source of the adults.

Adult Large Heaths emerge in mid-June on most lowland sites, but not until a month later at high altitudes. They usually reach a peak in mid-July, lingering well into August, although individual adults live, on average, just three to four days. The main cause of death is thought to be attack by Meadow Pipits, which hop between the tussocks in search of prey. In one study, about a third of all the Large Heaths examined had beak marks across their wings, but had escaped being eaten. Of more than 100 Meadow Pipit droppings teased apart, practically all contained the remains of what may be presumed to be Large Heaths.

Meta-populations and habitat quality

The Large Heath is a fairly sedentary species, with little interchange between the butterflies of nearby bogs. Nevertheless, the species shows clear evidence of a meta-population structure in certain landscapes, indicating low-level dispersal and eventual recolonisations of vacant sites after a local extinction. In a pioneering study in Northumberland, Roger Dennis and Harry Eales found that three-quarters of the 154 suitable sites supported a colony at any one time, and that the unoccupied sites tended to be more isolated and smaller than those that contained the species. However, irrespective of a site's isolation and size, the quality of the butterfly's breeding habitat was equally important: the Large Heath was much more likely to be present in a bog where the vegetation was in optimum condition for the caterpillars, compared with sites where the habitat was only marginally suitable.

In the best habitat numbers can grow very large, with up to 15,000 adults recorded on one bog. Roughly equal numbers of both sexes emerge, but at first sight males seem to predominate as the females have a habit of hiding within the tussocks. The males sit with wings tightly closed, leaning sideways to the sun to warm up, for like the Grayling, this is a species that never basks with its wings held flat.

The flight of both sexes is a slow, bobbing affair that occurs in dull as well as sunny weather. Cold weather grounds the butterflies, as does high wind, but even on a windy day it is usually possible to flush up a few by shuffling through the undergrowth. Unlike most browns, they dart away through the air rather than drop into deep vegetation when a potential enemy disturbs them.

The egg is laid on dead leaves at the base of immense tussocks of Hare's-tail Cottongrass or, on a few sites, of White

Beak-sedge. It is shiny yellow at first, but develops rust-coloured blotches after a week. After a further week it darkens, and the little caterpillar emerges. This starts feeding on the tender tips of its foodplants, and in late September, after two moults, it hibernates deep in the tussock. It is quite resilient to flooding and even to being frozen, but prolonged immersion for three to four months is generally harmful.

Loss of the Large Heath's habitat

Although often classified as a 'northern' butterfly, the Large Heath is common in warm bogs in central southern Europe, and its British distribution probably reflects the predominant distribution of its habitat rather than any intrinsic need for a cooler climate. It is, indeed, our only 'northern' butterfly whose recent declines cannot be attributed to climate warming. The butterfly is, nevertheless, well adapted to cool climates: on northern sites the caterpillars may spend a second winter in hibernation, taking two years to develop from egg to adult. But more often they are fully grown in June, and go on to form attractive, stumpy green chrysalises.

Today, colonies of Large Heath can be found locally throughout the bogs of Ireland, where it is still under-recorded, and in considerable numbers across the northern half of Scotland. They are present, too, on many islands, including the Outer Hebrides and Orkney, but almost certainly not in Shetland. Although the species is much more localised in the southern half of Scotland, it is plentiful enough in parts of Northumberland, where a recent survey identified 107 colonies, representing roughly three-quarters of those known in England and Wales. Other colonies are found in Cumbria and the more mountainous areas of north and central Wales.

There was a time when the Large Heath was common and widespread on the mosses around Manchester and Liverpool, but the vast majority of these were drained and converted to farmland. There has been some confusion about the rate of decline of this butterfly in recent decades. Two separate surveys of former colonies in different regions showed that roughly a quarter and a half of the listed sites, respectively, no longer support the butterfly. The latter is serious and suggests a substantial loss of habitat; the former is about what one would expect to find at any one snapshot in time of a meta-population of this species. No-one doubts, however, that the widespread loss of peatlands in Ireland is causing numerous local extinctions in the species' stronghold, and is a very great cause for concern.

Monarch

Danaus plexippus

The Monarch, or Milkweed, is the largest and most spectacular of the world's great migratory butterflies. It dwarfs every other species seen in Britain and, like most really large butterflies, looks very different from them in flight. Instead of launching into the air using 'clap-and-fling' wing-beats (see pages 176-7), like the darting or fluttery butterflies to which we are accustomed, it flaps its black-and-tawny wings slowly and deliberately, like a lumbering bird of prey. Once aloft, however, it soars and glides as gracefully as any species. An airborne Monarch is a truly wonderful sight, and one that is commonly seen in North America, from whence most, if not all, our butterflies originate.

After the many declines chronicled in this book, it is pleasant to report that Monarchs are being seen with increasing frequency in the British Isles, with roughly four-fifths of all sightings having been made during the past 50 years. Most were in the five great immigration years of 1968 (c. 65), 1981 (c. 100), 1995 (150-200), 1999 (c. 300) and 2001 (40-50). Although each coincided with freak weather patterns in North America, the underlying trend has also been upwards. It is little more than a century since the first British sighting was made, at Neath, in south Wales, on 6th September 1876. Three more were seen that year – two in Sussex and one in Dorset – and by the end of the 19th century a total of 39 Monarchs had been reported. There was then a lull for about 30 years when just the occasional individual was spotted, followed by a gradual, if erratic, increase ever since.

To date, roughly 900 specimens have been seen in the British Isles, of which about 800 were sighted in September or October, mostly around the turn of the two months. There were even periods in the autumns of 1981, 1995 and 1999 when you could more or less guarantee to see a Monarch on certain stretches of coast on the Isles of Scilly, Cornwall or Dorset.

Distribution
A rare immigrant from America, mainly seen in the West Country and Isles of Scilly. Seen more frequently in recent years.

The Monarch in America

To appreciate this butterfly properly, you must visit America. Each continent has its own subspecies, but those seen in Britain are invariably the North American form. There, the Monarch is one of the commonest and most spectacular of butterflies, breeding in weedy meadows, damp pasture and wasteland almost everywhere as far north as the Hudson

▲ **Male**
Male has a small sex-brand near the centre of each hindwing.

▲ **Male underside**
Markings similar to those of upperside but hindwing paler.

◀ **Egg** [× 15]
Laid singly on the leaves of milkweeds.

▼ **Caterpillar**
Poisons obtained from the foodplant protect the caterpillar from predators.

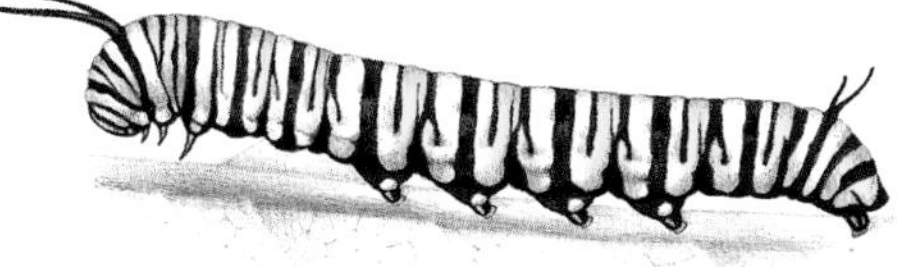

▲ **Female**
Similar to male; veins often more heavily marked with black.

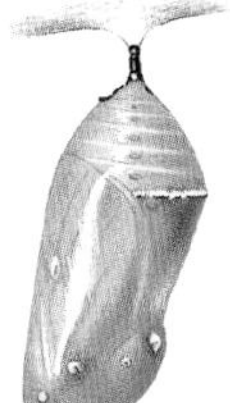

◀ **Chrysalis**
Wing patterns are visible just before emergence.

	Jan	Feb	Mar	Apr	May	Jun	Jul	Aug	Sep	Oct	Nov	Dec
Adult												

Bay. The beautiful striped caterpillars are also conspicuous as they chew the poisonous leaves of milkweeds. They strip and devour whole plants, impervious to the milky latex that deters other herbivores.

This immunity is responsible for the striking colours of the butterfly. The cocktail of chemicals, including heart-stopping agents and emetics, that is found in milkweed leaves is absorbed by the caterpillar and passed on to the adult, making both unpalatable to potential enemies. A few species of Mexican mice can apparently eat Monarchs, and some birds have learned to strip off the more poisonous parts, but by and large the butterfly is obnoxious, causing immediate vomiting and distress to anything that eats it. The body is also extremely tough, and for every Monarch killed, many more are spat out by potential predators, which soon learn to recognise and avoid its warning colours.

Migrations and roosts

The Monarch is a feature of the American countryside throughout spring and summer, yet this is essentially a sub-tropical butterfly that is unable to withstand intense winter cold. In September, adults start to fly south, first in groups of twos and threes, gradually gathering numbers all the way, and finally culminating in gigantic swarms. One awed observer in the 1880s wrote that these were 'almost beyond belief... millions is feebly expressive... miles of them is no exaggeration'.

Migrating Monarchs fly mainly by day, resting at night by the thousand on trees. Two major streams soon develop, divided by the Rocky Mountains. Western Monarchs settle in California, where they hibernate in many different places, mainly near the sea. Tens of thousands roost together in groves or on individual trees, dangling below the branches with their wings closed, as if the tree itself were a gigantic butterfly, and they the scales. The greatest concentrations occur at Muir Beach, and Santa Cruz and Santa Barbara, where roughly 180,000 adults overwinter. These wonderful roosts have long been a feature of Californian culture, celebrated in statues, children's parades and Beach Boys' songs. Needless to say, not everyone is appreciative, and there are heavy fines for anyone who interferes with a roost.

The mass roost of middle Mexico

Californian Monarch roosts are one of the natural wonders of the world, but they are surpassed by the almost unbelievable wintering site of the east and central American populations. Their destination had always been a mystery, and it was not until 1975 that Fred Urquhart, following a lifetime's research on this species, traced the migrations to the Sierra Madre, the mountain forests of middle Mexico. Here, in contrast to California, the entire east-continental swarm, consisting of 23 million butterflies in a poor year and 175 million in the best, hibernates in just 5-10ha of forest patches. Monarchs smother everything that can be perched upon. My good friend, Bob Pyle, who has tracked the migration through its entire cycle, eloquently describes the extraordinary spectacle:

> 'Imagine yourself basking... on a Mexican mountainside... it is snowing, snowing butterflies. Monarch butterflies fill the sky and eclipse the sun; Monarchs falling, drifting, sailing, and gliding; Monarchs every shade of orange, as the sun backlights or falls full upon them: pumpkin, salmon, or flame. They alight on every surface, and every purple *Senecio* supports a dozen drinkers, every bough a hundred baskers. Monarchs alight on your boots, your belly, your face, give you a physical and mental massage of softly beating wings. So many butterflies flutter that a soft, rushing whir fills the mountain air – yes, you can actually hear the Monarchs.
>
> 'The rustle keeps up, unabated for a second, until the sun begins to fall below the Firs. Only then do the millions of migrants start to settle back onto the Oyamels for the night, clinging to every needle so thickly that they form a pelage like reddest fox fur. Then it becomes a still world of orange butterflies – walls of Monarchs, curtains, solid tree trunks, boughs, and the whole forest groves of Monarchs, with scarcely a green needle showing through. The sun goes down and a chill rises. Having caught the last beams... the masses close their wings and become ashen scales against the dark foliage of Firs.'

Monarchs remain in their roosts until spring, before embarking on a return flight of up to 3,600km north to breed. Few, in fact, retrace their steps completely: the first wave usually lays eggs in the central southern states, and these develop within three or four weeks to form a second generation that in turn progresses north. This sequence continues until southern Canada is colonised in June, where Monarchs breed for three months before embarking on the great trek southwards.

Orientation by antennal clocks

It is still unknown how all the Monarchs from a continent manage to locate, with unerring accuracy, the exact few hectares of Californian or Mexican overwintering sites that their ancestors had vacated two to five generations earlier. A first step towards elucidation has been provided by Steven Reppert of the University of Massachusetts. It had long been thought that the butterflies were using the sun to guide them to their targets, and that, because the sun changes position throughout the day, they must also possess a 24-hour clock or GPS system to interpret the information coming from the sun-compass in their brains. This corrective device has finally been located, not in the butterfly's head but in its antennae. In a series of elegant experiments in which Reppert painted the antennae of some Monarchs black and others with translucent paint, he found that the latter navigated in the correct direction to reach Mexico, whereas those with blacked-out (or no) antennae flew in different directions, misguided by uncorrected signals from a molecular clock in their brains. This does not explain how Monarchs pinpoint the same trees used by their ancestors, but is perhaps the beginning of us understanding the phenomenon.

Transatlantic visitors

There is convincing evidence that our British immigrants stem from the southward autumnal migrations of the Monarch, and represent individuals that are blown out over the Atlantic in years of strong westerly winds. As well as arriving mainly in September or October, peak Monarch years have coincided with unusually high numbers of rare immigrant birds from America. Bill Shreeves has made a meticulous study of the great immigration of 1995. The story began on the north-east coast of America, where there was an abnormally large build-up of adults as they turned to migrate south. Reports of 'lift-offs', especially at Cape May in New Jersey, coincided with a flow of warm air across the Atlantic, and these exceptional winds blew many Monarchs off course. With wind speeds of 30-35 knots, individuals took perhaps four days to reach the British Isles. In America, the Monarch build-up peaked around 28th September, and the westerly airflow lasted until 2nd October. The first Monarch records here were on 3rd October, at Curlston in Cornwall and Durdle Door in Dorset; on 5th October, the first Devon Monarch was seen at Slapton, and the first from Hampshire was reported two days later. By 8th October they were seen in the Isles of Scilly, Sussex and the Isle of Wight, and also invaded Dorset on a broad front that stretched from Chideock to Bournemouth, arriving in groups at five main beachheads. Over the next six days, Monarchs were being reported from most southern counties and as far north as Cumbria.

This pattern of sightings is similar to those in other years. Although an occasional individual may be found almost anywhere in the British Isles – there is even a record from Shetland – the large majority are seen in the south-west, with most butterflies remaining near the southern coastlines where they made landfall. In Ireland, where it is poorly recorded, there are records from the entire southern coast.

Escapes and new colonies

There is some evidence that Monarchs also arrive as stowaways on ships, but these are almost certainly a small minority, as are the recent escapes from butterfly houses. A very few might come from the Azores, Canary Islands or Spain. The first two archipelagos were colonised by Monarchs in the 1860s during a worldwide expansion that took the species to several Pacific islands, Australia, New Zealand and the East Indies. The butterfly has colonised Spain in the last decade, with breeding recorded on introduced milkweeds in many localities.

If current trends in weather and wind patterns continue, we may look forward to the arrival of growing numbers of this magnificent butterfly in the British Isles in future years. That is, so long as its Mexican wintering sites are spared. These are a matter of acute concern for conservationists, for much of the Sierra Madre is being cleared by illegal logging, despite being declared an international nature reserve. The problem, to date, has been not so much the loss of trees on which to roost, but that the depleted forests are less able to shelter the butterflies from icy winds: literally millions of adult monarchs are now freezing to death during the cruellest winters.

In the immediate future, we can expect and enjoy an increasing number of Monarchs on our shores, although they will never establish colonies here. In the first place, the autumn arrivals cannot survive our winters, and any that did would be unable to breed, for milkweeds are not indigenous and are found only in a few sheltered gardens. It is interesting to note, however, that one record of a sort for British breeding exists for this species. This occurred in 1981, when a female escaped from a south London butterfly farm and laid on the milkweeds of Kew Gardens. The eggs were reared indoors, and successfully produced the only known example of home-bred Monarchs.

Further reading

Readers wishing to learn more about the biology and natural history of butterflies, or of the history of collecting and studying them, will find much of interest in the wealth of literature available for the British and Irish species. We recommend the following:

History

Many classic accounts have been published since Thomas Moffet (father of 'Little Miss Muffet') compiled the first British book of insects, *Insectorum sive minimorum Animalium Theatrum*, which was published in 1634 and contains details of 18 native species. In *The Butterflies of Britain & Ireland* we have drawn heavily from this literature, particularly the works of James Petiver, Benjamin Wilkes, Moses Harris, AH Haworth, HN Humphries & JO Westwood, FO Morris, HT Stainton, Edward Newman, CB Barrett, JW Tutt, Richard South, FW Frohawk, Edmund Sandars, PBM Allen, CB Williams and EB Ford.

Ford's wonderfully erudite *Butterflies* (1945) was the first, and many claim the greatest, in Collins' The New Naturalist series: Miriam Rothschild considered it to be the best natural history book ever written. It was certainly an inspiration to me. Ford's book is still well worth reading today; although inevitably out of date in some areas, the opening chapter 'The history of British Butterfly Collecting' is a masterpiece.

More recently, the history of British butterflies and their collectors has been described in a book that is both an enchanting read and a work of scholarship. It also contains Peter Marren's research into old English names of species. For its wealth of information and historical anecdote, I strongly recommend: Salmon, M A 2000 *The Aurelian Legacy* Harley Books, Colchester.

Modern field guides

These are legion. At the risk of immodesty, we recommend:

Lewington, R 2003 *Pocket Guide to Butterflies of Great Britain and Ireland.* British Wildlife Publishing, Oxford

Lewington, R, & Tolman, T 2008 *Collins Butterfly Guide.* Collins, London

Thomas, JA, 2014 (3rd edition) *Philip's Guide to Butterflies of Britain and Ireland.* Philip's, London

Tolman, T, & Lewington, R 1997 *Collins Field Guide Butterflies of Britain and Europe.* HarperCollins, London

Distribution and status

Asher, J, Warren, M, Fox, R, Harding, P, Jeffcoate, G, & Jeffcoate, S 2001 *The Millennium Atlas of Butterflies in Britain and Ireland.* OUP, Oxford

Fox, R, Asher, J, Brereton, T, Roy, D, & Warren, M 2006 *The State of Butterflies in Britain and Ireland.* Pisces Publications, Oxford

Settele, J, *et al.* 2008 *Climatic Risk Atlas of European Butterflies.* Pensoft, Sofia, Bulgaria

Ecology and natural history

Settele, J, Shreeve, T, Konvicka, M, Van Dyck, H 2009 *Ecology of Butterflies in Europe.* Cambridge University Press, Cambridge

Dennis, RLH 2010 *A resource-based habitat view for conservation – butterflies in the British landscape.* Wiley-Blackwell, Chichester

Barkham, P 2010 *The Butterfly Isles.* Granta Books

For more expert lepidopterists, the journal *Atropos* is a mine of information, especially for information about migrant butterflies.

Local publications

Some of these publications are out of print, but we have included them because they may be available secondhand or in libraries.

Avon

Barnett, R, Higgins, R, Moulin, T, & Wiltshire, C 2003 *The Butterflies of the Bristol Region.* Bristol/Avon Regional Environmental Records Centre

Bedfordshire

Arnold, VW, Baker, CRB, Manning, DV, & Woiwod, IP 1997 *The Butterflies and Moths of Bedfordshire.* Bedfordshire Natural History Society

Nau, BS, Boon, CA, & Knowles, JP 1987 *Bedfordshire Wildlife.* Castlemead Publications

Berkshire, Buckinghamshire and Oxfordshire

Asher, J 1994 *The Butterflies of Berkshire, Buckinghamshire and Oxfordshire.* Pisces Publications

Asher, J, Bowles, N, Redhead, D, & Wilkins, M 2005 *The State of Butterflies in Berkshire, Buckinghamshire and Oxfordshire.* Pisces Publications

Baker, BR 1994 *The Butterflies and Moths of Berkshire.* Hedera Press

Cambridgeshire

Field, R, Perrin, V, Bacon, L, & Greatorex-Davis, N 2006 *Butterflies of Cambridgeshire.* Butterfly Conservation

Cheshire
Shaw, BT 1999 *The Butterflies of Cheshire*. National Museums Liverpool
Cleveland
Smith, JK, & Smith, H 1984 *Butterflies in Cleveland*. Cleveland Nature Conservation Trust
Cornwall and The Isles of Scilly
Jones, S 2010 *Insects of Cornwall and the Isles of Scilly (Pocket Cornwall)*. Alison Hodge
Penhallurick, RD 1996 *The Butterflies of Cornwall and the Isles of Scilly*. Dyllansow Pengwella
Smith, FHN 1997 *The Moths and Butterflies of Cornwall and the Isles of Scilly*. Gem Publishing Co
Smith, FHN 2002 *A Supplement to the Moths and Butterflies of Cornwall and the Isles of Scilly*. Gem Publishing Co
Spalding, A 1992 *Cornwall's Butterfly and Moth Heritage*. Twelveheads Press
Wacher, J, Worth, J, & Spalding, A 2003 *A Cornwall Butterfly Atlas*. Pisces Publications
Derbyshire
Harrison, F, & Sterling, MJ 1985 *Butterflies and Moths of Derbyshire, Part 1*. Derbyshire Entomological Society
Harrison, F, & Sterling, MJ 1986 *Butterflies and Moths of Derbyshire, Part 2*. Derbyshire Entomological Society
Harrison, F, & Sterling, MJ 1987 *Butterflies and Moths of Derbyshire, Part 3*. Derbyshire Entomological Society
Devon
Bristow, CR, Mitchell, SH, & Bolton, DE 1993 *Devon Butterflies*. Devon Books, Tiverton
Dorset
Thomas, J, Surry, R, Shreeves, B, & Steele, C 1998 *New Atlas of Dorset Butterflies*. Dorset County Museum
Thomas, JA, & Webb, N 1984 *Butterflies of Dorset*. Dorset County Museum
Essex
Benton, T, & Firmin, J 2005 *The Butterflies of Colchester and North East Essex*. Colchester Natural History Society
Corke, D 1997 *The Butterflies of Essex*. Lopinga Books
Emmett, AM, Pyman, GA, & Corke, D 1985 *The Larger Moths and Butterflies of Essex*. Essex Field Club
Payne, RG, & Skinner, JF 1982 *Butterflies of Essex – Provisional Maps*. Essex Biological Records Centre
Gloucestershire
Mabbett, R, & Williams, M (eds) 1992 *The Butterflies and Moths of the West Midlands and Gloucestershire: A Five Year Review*. Butterfly Conservation
Walsey, R 2008 *40 Butterfly Walks in Gloucestershire*. Butterfly Conservation
Hampshire
Oates, MR, Taverner, JH, & Green, DG 2000 *The Butterflies of Hampshire*. Pisces Publications
Norriss, T, & Barker, L 2009 *Hampshire and Isle of Wight Butterfly and Moth Report 2008*. Butterfly Conservation
Hertfordshire
Sawford, B 1987 *The Butterflies of Hertfordshire*. Castlemead Publications, Ware
Kent
Philp, E G 1993 *The Butterflies of Kent: An Atlas of their Distribution*. Kent Field Club, Sittingbourne
Lincolnshire
Johnson, R, & Smith, C (eds) 2006 *The Butterflies and Moths of Lincolnshire*. Lincolnshire Naturalists' Union
Leicestershire
Morris, R 1990 *The Butterflies of the Hinkley District*. Occasional Publications Series, No. 6. Leicestershire Entomological Society
Robertson, T 1994 *Some Mid-Century Leicestershire Butterflies*. Leicestershire Entomological Society
London
Herbert, C 1993 *Butterflies of the London Borough of Barnet: A Provisional Atlas*. ARM Conservation Ltd
Plant, CW 1987 *The Butterflies of the London area*. London Natural History Society, London
Manchester
Hardy, PB 1998 *Butterflies of Greater Manchester*. PGL Enterprises
Norfolk
Watts, BR, & McIlwrath, BJ 2002 *Millennium Atlas of Norfolk Butterflies*. Butterfly Conservation
Northamptonshire
Goddard, D, & Wyldes, A 2012 *Butterflies of Northamptonshire*. Butterfly Conservation (Beds & Northants Branch)
Northumberland
Cook, NJ 1990 *An Atlas of Butterflies of Northumberland and Durham*. Biological Records Centre, Hancock Museum
Dunn, TC, & Parrack, JD 1986 *The Moths and Butterflies of Northumberland and Durham Part 1: Macrolepidoptera*. Northern Naturalists' Union
Shropshire
Riley, AM 1991 *A Natural History of the Butterflies and Moths of Shropshire*. Swan Hill Press, Shrewsbury
Staffordshire
Warren, RG 1984 *Atlas of the Lepidoptera of Staffordshire – Part 1: Butterflies*. Staffordshire Biological Recording Scheme
Suffolk
Mendal, H, & Piotrowski, S 1986 *The Butterflies of Suffolk*. Suffolk Naturalists' Society
Stewart, R 2001 *The Millennium Atlas of Suffolk Butterflies*. Suffolk Naturalists' Society

Surrey

Willmott, K, Bridge, M E, Clarke, H E, & Kelly, F 2013 *Butterflies of Surrey revisited.* Surrey Wildlife Trust

Sussex

Gay, J, & Gay, P 1996 *Atlas of Sussex Butterflies: with a commentary of their changing conservation status.* Butterfly Conservation

Pratt, C 2014 *A Complete History of the Butterflies and Moths of Sussex, 3rd Supplement*

Warwickshire

Smith, R, & Brown, D 1987 *The Lepidoptera of Warwickshire, Parts 1 and 2.* Warwickshire Museum

Warmington, K, & Vickery, M 2003 *Warwickshire's Butterflies.* Butterfly Conservation (Warwickshire Branch)

Wiltshire

Fuller, M 1995 *The Butterflies of Wiltshire.* Pisces Publications

Worcestershire

Green, J 1982 *A Practical Guide to the Butterflies of Worcestershire.* Worcestershire Nature Conservation Trust

Harper, M, & Simpson, T 2001 *The Larger Moths and Butterflies of Herefordshire and Worcestershire : An Atlas.* Butterfly Conservation (West Midlands Branch)

Yorkshire

Frost, H M (ed) 2005 *The Butterflies of Yorkshire.* Butterfly Conservation (Yorkshire)

Frost, H M (ed) 2008 *Yorkshire Butterflies and Moths: An assessment of the status and distribution of Yorkshire's Butterflies and Moths in 2008.* Butterfly Conservation (Yorkshire)

Scotland

Barbour, D, Moran, S, Mainwood, T, & Slater, B 2008 *Atlas of Butterflies in Highland and Moray.* Butterfly Conservation (Highland Branch)

Futter, K, *et al.* 2006 *Butterflies of South West Scotland: An Atlas of their Distribution.* Butterfly Conservation

Mercer, J, Buckland, R, Kirkland, P, & Waddell, J 2009 *Butterfly Atlas of the Scottish Borders.* Atropos Publishing

Northern Ireland

Nelson, B, Thompson, R 2006 *The Butterflies and Moths of Northern Ireland.* Ulster Museum

Societies to join

Butterfly Conservation
www.butterfly-conservation.org

Butterfly Conservation is the foremost charity dedicated to conserving butterflies and moths. Membership is essential for anyone interested in the natural history of Britain's butterflies or in contributing to their conservation. Butterfly Conservation has a network of local branches and runs internationally renowned survey and monitoring programmes, as well as more than 70 landscape-scale conservation projects throughout the UK.

The Wildlife Trusts **www.wildlifetrusts.org**

Forty-seven Wildlife Trusts cover the whole of the UK, the Isle of Man, Alderney and the Isles of Scilly. With more than 800,000 members and 2,300 nature reserves, covering about 100,000ha, they represent the largest UK voluntary organisation dedicated exclusively to conserving all UK habitats and species.

National Trust **www.nationaltrust.org.uk** *and*
National Trust for Scotland **www.nts.org.uk**

It is hard to exaggerate the importance of the National Trust and the National Trust for Scotland in preserving and conserving the landscapes and semi-natural ecosystems of the United Kingdom. With 3.7 million and 310,000 members, respectively, and owning 330,000ha of countryside, including 950 miles of coastline, the two National Trusts have the power and holdings to make a real difference in the conservation of British and European wildlife. This includes the preservation and enlightened management of key sites for almost every British butterfly species.

Buglife **www.buglife.org**

The Invertebrate Conservation Trust is the only organisation in Europe devoted to the conservation of all invertebrates. It leads in saving Britain's rarest bugs, snails, bees, wasps, ants, spiders, beetles and other invertebrates.

Royal Entomological Society **www.royensoc.co.uk**

For the more serious entomologist, both amateur and professional, the Royal Entomological Society plays a major national and international role in disseminating information about insects. Activities range from local meetings and events, including initiatives for schoolchildren, to an internationally renowned scientific symposium held every second year.

Websites, DVDs and apps

www.ukbutterflies.co.uk is the essential and most comprehensive website on British butterflies. Its inspiring pages provide beautiful images, news and information about butterflies for everyone who is interested in them.

British Butterflies – an interactive guide. A comprehensive multimedia DVD and app from Birdguides that covers 61 British species, with high quality video-clips of species and interviews with experts.

Patrick Barkham's Guide to British Butterflies 2013 An enchanting DVD from the acclaimed author of *The Butterfly Isles.*

Names of plants

These are the nectar and foodplants mentioned in the text. Nomenclature follows Stace, C A 1997 *New Flora of the British Isles.*

Agrimony *Agrimonia eupatoria*
Alder Buckthorn *Frangula alnus*
Ash *Fraxinus excelsior*
Aspen *Populus tremula*
Autumn Gentian *Gentianella amarella*
Barren Strawberry *Potentilla sterilis*
Bell Heather *Erica cinerea*
bents *Agrostis*
Bilberry *Vaccinium myrtillus*
birches *Betula*
bird's-foot-trefoils *Lotus*
Bitter-vetch *Lathyrus linifolius*
Black Bent *Agrostis gigantea*
Black Medick *Medicago lupulina*
Blackthorn *Prunus spinosa*
Bladder-senna *Colutea arborescens*
Blue Moor-grass *Sesleria caerulea*
Bluebell *Hyacinthoides non-scripta*
Bog-myrtle *Myrica gale*
Bracken *Pteridium aquilinum*
Bramble *Rubus fruticosus* agg.
Bristle Bent *Agrostis curtisii*
Broom *Cytisus scoparius*
Buck's-horn Plantain *Plantago coronopus*
Buckthorn *Rhamnus cathartica*
Buddleia *Buddleja davidii*
Bugle *Ajuga reptans*
Bullace *Prunus domestica insititia*
buttercups *Ranunculus*
Charlock *Sinapis arvensis*
clovers *Trifolium*
Cock's-foot *Dactylis glomerata*
Common Bent *Agrostis capillaris*
Common Bird's-foot-trefoil *Lotus corniculatus*
Common Couch *Elytrigia repens*
Common Cow-wheat *Melampyrum pratense*
Common Dog-violet *Viola riviniana*
Common Fleabane *Pulicaria dysenterica*
Common Nettle *Urtica dioica*
Common Ragwort *Senecio jacobaea*
Common Rock-rose *Helianthemum nummularium*
Common Sorrel *Rumex acetosa*
Common Stork's-bill *Erodium cicutarium*
Cow Parsley *Anthriscus sylvestris*
Cowslip *Primula veris*
Creeping Cinquefoil *Potentilla reptans*
Creeping Soft-grass *Holcus mollis*
Creeping Thistle *Cirsium arvense*
Cross-leaved Heath *Erica tetralix*
Cuckooflower (or Lady's Smock) *Cardamine pratensis*
currants *Ribes*
Devil's-bit Scabious *Succisa pratensis*
Dog-rose *Rosa canina*
Dogwood *Cornus sanguinea*
Dove's-foot Crane's-bill *Geranium molle*
Dwarf Thistle *Cirsium acaule*
Dyer's Greenweed *Genista tinctoria*
elms *Ulmus*
English Elm *Ulmus procera*
False Brome *Brachypodium sylvaticum*
Fennel *Foeniculum vulgare*
fescues *Festuca*
Field Maple *Acer campestre*
Field Pansy *Viola arvensis*
Field Scabious *Knautia arvensis*
Foxglove *Digitalis purpurea*
Garlic Mustard *Alliaria petiolata*
Germander Speedwell *Veronica chamaedrys*
Globe Artichoke *Cynara cardunculus*
Goat Willow *Salix caprea*
Gorse *Ulex europaeus*
Great Burnet *Sanguisorba officinalis*
Great Fen-sedge *Cladium mariscus*
Greater Bird's-foot-trefoil *Lotus pedunculatus*
Green-winged Orchid *Orchis morio*
Grey Willow *Salix cinerea*
Hairy Violet *Viola hirta*
Hare's-tail Cottongrass *Eriophorum vaginatum*
hawkweeds *Hieracium*
Hawthorn *Crataegus monogyna*
Heath Bedstraw *Galium saxatile*
heathers *Erica* and *Calluna vulgaris*
hebes *Hebe*
Hedge Mustard *Sisymbrium officinale*
Holly *Ilex aquifolium*
Honesty *Lunaria annua*
Honeysuckle *Lonicera periclymenum*
Hop *Humulus lupulus*
Horseshoe Vetch *Hippocrepis comosa*
Ice Plant *Sedum maximum*
Ivy *Hedera helix*
Kidney Vetch *Anthyllis vulneraria*
knapweeds *Centaurea*
Lesser Trefoil *Trifolium dubium*
Ling *Calluna vulgaris*
Lucerne *Medicago sativa*
lungworts *Pulmonaria*
mallows *Malva*
maples *Acer*
Marram *Ammophila arenaria*
Marsh Thistles *Cirsium palustre*
Marsh Violet *Viola palustris*
Mat-grass *Nardus stricta*
Meadow Vetchling *Lathyrus pratensis*
meadow-grasses *Poa*
melilots *Melilotus*
Michaelmas-daisies *Aster*
Milk-parsley *Peucedanum palustre*
milkweeds *Asclepias*
Musk Thistle *Carduus nutans*
Nasturtium *Tropaeolum majus*
nettles *Urtica*
oaks *Quercus*
Pale Dog-violet *Viola lactea*
Pear *Pyrus communis*
Pellitory-of-the-wall *Parietaria judaica*
plantains *Plantago*
poplars *Populus*
Primrose *Primula vulgaris*
Purple Moor-grass *Molinia caerulea*
Purple-loosestrife *Lythrum salicaria*
Ragged-Robin *Lychnis flos-cuculi*
Red Campion *Silene dioica*
Red Clover *Trifolium pratense*
Red Fescue *Festuca rubra*
Reed *Phragmites australis*
restharrows *Ononis*
Ribwort Plantain *Plantago lanceolata*
rye-grasses *Lolium*
Salad Burnet *Sanguisorba minor*
sallows *Salix*
scabiouses *Knautia, Scabiosa, Succisa*
Sea Radish *Raphanus maritimus*
Sheep's-fescue *Festuca ovina*
Sheep's Sorrel *Rumex acetosella*
Small-leaved Elm *Ulmus minor*
Small Nettle *Urtica urens*
Smooth Meadow-grass *Poa pratensis*
snowberries *Symphoricarpos*
Spear Thistle *Cirsium vulgare*
Spindle *Euonymus europaeus*
sundews *Drosera*
sweet rockets *Sisymbrium*
Sweet Vernal-grass *Anthoxanthum odoratum*
teasels *Dipsacus*
Timothy *Phleum pratense*
Tor-grass *Brachypodium pinnatum*
Tormentil *Potentilla erecta*
trefoils *Lotus*
Tufted Hair-grass *Deschampsia cespitosa*
Tufted Vetch *Vicia cracca*
valerians *Centranthus*
vetches *Vicia* and *Anthyllis*
Water Dock *Rumex hydrolapathum*
Water-cress *Rorippa nasturtium-aquaticum*
Wavy Hair-grass *Deschampsia flexuosa*
Western Gorse *Ulex gallii*
White Beak-sedge *Rhynchospora alba*
Wild Cabbage *Brassica oleracea*
Wild Carrot *Daucus carota*
Wild Cherry *Prunus avium*
Wild Marjoram *Origanum vulgare*
Wild Mignonette *Reseda lutea*
Wild Pansy *Viola tricolor*
Wild Plum *Prunus domestica*
Wild Privet *Ligustrum vulgare*
Wild Strawberry *Fragaria vesca*
Wild Teasel *Dipsacus fullonum*
Wild Thyme *Thymus polytrichus*
Wood Avens *Geum urbanum*
Wood Small-reed *Calamagrostis epigejos*
Woolly Thistle *Cirsium eriophorum*
Wych Elm *Ulmus glabra*
Yorkshire-fog *Holcus lanatus*

Index

Acknowledgements

Much of the information given in this book is based on the research or first-hand observations of the author and artist. We have also drawn heavily on the published and unpublished accounts of many other scientists and naturalists, past and present. Although, for reasons of space and readability, we have not been able to mention all sources throughout the texts, we wish to acknowledge our very great debt and gratitude to those who have knowingly or indirectly contributed to this book.

In addition to authors of the historical textbooks on butterflies listed on page 282, we should like to record our particular thanks to the following, whose work has been of great assistance in the preparation of this book:

Chequered Skipper NOM Ravenscroft, RV Collier, DA Joyce, AS Pullin, MA Salmon; **Small Skipper** SA Corbet, D Lowry, S Warrington, TH Brackenbury, HA Ellis, RE Freber, RF Pywell, JW Dover; **Essex Skipper** D Louy, GE Bucher, AK Brodsky; **Lulworth Skipper** DJ Simcox, R Smith, CD Thomas, NAD Bourn, M Pfaff, MA Salmon, R Jones, K Pradel, S Woodley; **Silver-spotted Skipper** DJ Simcox, R Smith, CD Thomas, MS Warren, RJ Wilson, ZG Davies, JK Hill, OT Lewis, TM Jones, EJ Bodsworth; **Large Skipper** RLH Dennis, WR Williams, H Dreisig; **Dingy Skipper** CD Thomas, D Gutiérrez, JL León-Cortés, NAD Bourn; **Grizzled Skipper** TM Brereton; **Swallowtail** JP Dempster, ML Hall, C Wiklund, L Svärd, T Järvi, B Sillén-Tullberg, K Honda, MA Salmon; **Wood White & Cryptic Wood White** C Wiklund, MS Warren, M Friberg, P Réal, B Nelson, M Hughes, S Jeffcoate, N Vongvanich, AK Borg-Karlson, M Olofsson, D Berger, B Karlsson, M Bergman, J Kullberg, N Wahlberg, DJ Kemp, S Merilaita, K Fiedler, A Freese, L Svärd, H Descimon, E Pollard; **Clouded Yellow** CB Williams, MA Salmon, J Asher, MJ Skelton, TH Sparks, RLH Dennis; **Brimstone** TJ Bibby, E Pollard, J Forsberg, C Wiklund, JP Dempster, KH Lakhani, V Lindfors, S Andersson, HV McKay; **Black-veined White** CR Pratt, M Baguette, S Petit, F Queva; **Large White** OW Richards, JSE Feltwell, M Rothschild, RR Baker, NE Fatouros, ME Huigens, JA van Loon, M Dicke, M Hilker, ND Mitchell, RT Aplin, R d'Arcy Ward, MW Chun; **Small White** OW Richards, JP Dempster, M Rothschild, RR Baker, JE Moss, JH Myers, TW Gossard, RE Jones, CJ Bissoondath, C Wiklund, N Wedell, PA Cook, RT Aplin, R d'Arcy Ward, M Watanabe, S Ando, A Hern, G Edwards-Jones, I Kandori, N Ohsaki; **Green-veined White** J Forsberg, C Wiklund, CJ Bissoondath, A Kaitala, V Lindfors, J Abenius, F Stjernholm, B Karlsson, D Goulson, JS Cory, N Wedell, PA Cook, TG Shreeves, S Yata, HA Leeds, FS Chew, WB Watt, J Mevi-Schütz, A Erhardt, E Pollard, BC Wood, AS Pullin; **Bath White** HBD Kettlewell, J Forsberg; **Orange-tip** SP Courtney, C Wiklund, J Forsberg, AE Duggan, JP Dempster, RLH Dennis, B Karlsson, O Leimar, L Svärd, B Wijnen, HL Leertouwer, DG Stavenga; **Green Hairstreak** W Vanreusel, H Van Dyck, RB Morris; Brown Hairstreak KJ Willmott, David Redhead; **Purple Hairstreak** D E Newland; **White-letter Hairstreak** M Davies, KJ Willmott, M Brookes, SA Bailey, RH Haines-Young, C Watkins; **Black Hairstreak** AE Collier, M Goddard, KJ Willmott, A Croft, D Elias, NAD Bourn, D Redhead, G Roberts, R Clarke, R Parslow, MA Salmon, C Steele, B Woodell, D Wilton, M Wilkins, S Hodges; **Large Copper** E Duffey, F Bink, R Collier, H Short, AS Pullin, MR Webb, CN Nicholls, J Joy, W Scisoft, MA Salmon; **Small Copper** JP Dempster, KH Lakhani, Y Suzuki, K Endo, Y Kamata, JL Leon-Cortes, MJR Cowley, CD Thomas, MF Wallis De Vries; **Long-tailed Blue** B Wurzell, D Bevan, C Pratt, N Hughes, P Eeles; **Geranium Bronze** C Holloway, C Pratt; **Small Blue** AC Morton, M Baylis, RL Kitching, J Rawles, J Krauss, I Steffan-Dewenter, T Tscharntke, B Binzenhöfer, R Biedermann, J Settele, B Schröder, M Baguette, S Petit, F Queva; **Silver-studded Blue** CD Thomas, D Jordano, J Rodriguez, K Schönrogge, K Murray, K Horstmann, M Shaw, K Fiedler, MJ Read, NOM Ravenscroft, OT Lewis, JK Hill, MI Brookes, H Mendel, KJ Willmott, YA Graneau, P King, OC Rose, SWT Glen, RJ Rose; **Northern Brown Argus** FVL Jarvis, O Hoegh-Guldberg, S Ellis, K Aagaard, BJ Selman, CD Thomas, J Hill, R Fox, MG Telfer, SG Willis, J Asher, B Huntley, RJ Wilson, RLH Dennis, TG Shreeve, AS Pullin; **Brown Argus** NAD Bourn, FVL Jarvis, O Hoegh-Guldberg, K Aagaard, K Fiedler, R Menendez, V Hummel, K Hindar, AS Pullin, C D Thomas, E J Bodsworth, R J Wilson, A D Simmons, Z G Davies, M Musche, L Conradt, RJ Wilson, S Ellis, A Gonzalez-Megias, OT Lewis; **Common Blue** RLH Dennis, K Fiedler, J Rawles, F Burghardt, H Knüttel, M Becker, U Schittko, V Wray, N Janz, A Bergstrom, A Sjogren, O Leimar, M Rothschild, P Proksch, M Goverde, A Bazin, JA Shykoff, A Erhardt, R Menendez, OT Lewis, AH Axén, D Gutiérrez, JL León-Cortés, E Pollard, B Hölldobler, JFD Frazer; **Chalkhill Blue** K Fiedler, U Maschwitz, A Wilson, DJ Simcox, M Pfaff, T Schmitt, A Seitz, GM Hewitt, JC Habel, J Besold, T Becker, J Krauss, I Steffan, TM Brereton, MS Warren, DB Roy, K Stewart, R O'Connor, JFD Frazer; **Adonis Blue** DJ Simcox, DB Roy, M Pfaff, AD Bourn, KEJ Whitfield, GS Pearman, R Smith, CD Thomas, GL Harper, N Maclean, D Goulson, K Fiedler, R O'Connor; **Holly Blue** RC Revels, KJ Willmott, E Pollard, TJ Yates; **Large Blue** DJ Simcox, S Meredith, TA Chapman, EB Purefoy, F W Frohawk, GM Spooner, A Spalding, ME McCracken, J Settele, F Barbero, K Schönrogge, S Bonelli, E Balletto; **Duke of Burgundy** M Oates, KJ Porter, KJ Willmott, TH Sparks, JN Greatorex-Davies, ML Hall, RH Marrs, NAD Bourn, MS Warren, T Fartmann, N Anthes, G Hermann, EC Turner; **White Admiral** E Pollard, AS Cooke; **Purple Emperor** KJ Willmott, IRP Heslop, GE Hyde, RE Stockley, M Oates, D Dell, TH Sparks, RLH Dennis, L Goodyear, A Middleton; **Camberwell Beauty** LH Newman, C van Swaay, R Fox, J Bouwman; **Painted Lady** E Pollard, MS Warren, CB Williams, RL Nesbit, JW Chapman, JK Hill, IP Woiwod, D Sivell, KJ Bensusan, RF Bretherton, JM Chalmers-hunt, J Asher, P Marren, MA Salmon, CAM van Swaay, C Stefanescu, KE Lundsten, D Maes, JN Greatorex-Davies, MJP Blackford, L Dinan, R Fox, C Stephanescu, J Chapman; **Small Tortoiseshell** RR Baker, RLH Dennis, AS Pullin, C Wiklund, A Vallin, M Friberg, J Lind, OT Lewis, N Hamer, S Gripenberg, P Marren, S Jakobsson, S Andersson, MJP Blackford, M Niehaus, M Gewecke, S Vandewoestijne, G Neve, M Baguette, S Bryant, C Thomas, J Bale; **Large Tortoiseshell** J Asher, MS Warren, R Fox, MA Salmon, RLH Dennis, TH Sparks, P Eeles; **Red Admiral** RB Srygley, ALR Thomas, E Pollard, JN Greatorex-Davies, C Stefanescu, O Bratttström, N Kjellén, T Alerstam, S Akesson, CB Williams, S Benvenuti, P Dall'Antonia, P Loale, RJ Bitzer, KC Shaw, R Nesbit, J Chapman; **Peacock** RR Baker, C Wiklund, A Vallin, S Jakobsson, J Lind, AS Pullin, AK Brodsky, A Erhardt, HP Rusterholz, JJ Windig, B Mohl, LA Miller, S Bryant, C Thomas, J Bale; **Comma** S Nylin, N Janz, C Wiklund, GH Nygren, JJ Windig, A Bergström, N Wedell, G Gamberale-Stille, BS Tullberg, C Stefanescu, PO Wickman, A Vallin, S Jakobsson, J Lind, C Pratt, P Holland, MA Salmon, JK Hill, MS Warren, J Asher; **Violet-feeding fritillaries** DJ Simcox, RG Snazell, I Moy, Z Randle, RT Curtis, MS Warren, R Fox, J Asher, SA Clarke, PA Robertson, RE Feber, TM Brereton, M Oates, O Tudor, RLH Dennis, JN Greatorex-Davies, NAD Bourn, J Plackett; **Queen of Spain Fritillary** PBM Allen, MA Salmon, N Hulme; **Silver-washed Fritillary** DBE Magnus, W Vielmetter, P Ockenfels, M Boppré, OW Fischer, S Schulz, RR Askew; **Marsh Fritillary** KJ Porter, DJ Simcox, MS Warren, CR Bulman, OT Lewis, MJ Klapwijk, RJ Wilson, N Wahlberg, T Klemetti, I Hanski, C Hurford, AP Fowles, RG Smith, N Anthes, T Fartmann, G Hermann, G Kaule, DA Joyce, AS Pullin, ML Munguira, J Martín, NAD Bourn, J Plackett; **Glanville Fritillary** DJ Simcox, RT Curtis, KJ Willmott, MR Shaw, I Hanski, M Kuussaari, M Nieminen, I Saccheri, MC Singer, K Enfjall, O Leimar, O Ovaskainen, S Haikola, I Pen, GC Lei, V Vikberg, T Pakkala, M Kankare, S van Nouhuys, O Gaggiotti, MA Salmon, CR Pope; **Heath Fritillary** MS Warren, CR Bulman, CD Thomas, DJ Simcox, B Schwarzwälder, M Lörtscher, A Erhardt, GJ Holloway, NAD Bourn, J Plackett, C Kelly, R Jones; **Speckled Wood** TG Shreeve, NB Davies, H Van Dyck C Wiklund, S Nylin, PO Wickman, K Gotthard, A Persson, B Karlsson, JJ Windig, A Persson, K Berwaerts, S Van Dongen, E Matthysen, A Dhondt, RM Sibly, L Winokur, RH Smith, P Aerts, L Svärd, E Vints, CJ Breuker, M Gibbs, MJ Goddard, DJ Kemp, JK Hill, CD Thomas, DS Blakeley HA Leeds, JA Downes; **Wall** RLH Dennis, PO Wickman, C Wiklund, B Karlsson, J Mevi-Schütz, J Forsberg, A Erhardt, H Elligsen, B Beinlich, H Plachter; **Mountain Ringlet** KJ Porter, AM Franco, JK Hill, CD Thomas, T Schmitt, GM Hewitt, P Muller; **Scotch Argus** RLH Dennis, JEH Blackie, PK Kinnear, P Kirkland, AM Franco, JK Hill, CD Thomas; **Marbled White** A Wilson, M Rothschild, RJ Nash, EA Bell, RLH Dennis, TG Shreeve, M Baguette, S Petit, F Queva, E Pollard, AC Morton, ML Munguira; **Grayling** N Tinbergen, DJ Simcox, R Findlay, MR Young, JA Findlay, RLH Dennis, TG Shreeve, TH Sparks, H Dreisig, B Karlsson, C Wiklund, D Maes, A Ghesquiere, M Logie, D Bonte, J Joy; **Gatekeeper** PM Brakefield, JW Dover, RE Feber, H Smith, DW Macdonald, E Pollard, L Conradt, TJ Roper; **Meadow Brown** PM Brakefield, WH Dowdeswell, EB Ford, ML Munguira, E Pollard, M Rothschild, RE Feber, H Smith, DW Macdonald, JW Dover, TH Sparks, JN Greatorex-Davies; **Small Heath** PO Wickman, E Garcia-Barros, C Rappe- George, P Jansson, L Svärd, C Wiklund, B Karlsson, J Forsberg, E Pollard, JN Greatorex-Davies, JW Dover, HA Leeds, S Meredith; **Large Heath** T Melling, JRG Turner, RLH Dennis, HT Eales, J Joy, AS Pullin; Ringlet M Rothschild, E Pollard, OL Sutcliffe, CD Thomas, D Peggie, A Valtonen, K Saarinen; **Monarch** RM Pyle, FA Urquhart, LP Brower, PM Tuskes, P Ackery, RI Vane-Wright, M Rothschild, WG Shreeves, C Merlin, RJ Gegear, SM Reppert; **Rare species and migrants** MA Salmon, J Asher, R Fox, EB Ford, J Heath, RF Bretherton, CB Williams, RR Baker; **Vernacular names** P Marren.

We are also extremely grateful to colleagues for commenting on parts or all of the text, or for general discussions or for providing additional information. These include, in alphabetical order: J Asher, NAD Bourn, PM Brakefield, RT Curtis, M Davies, RLH Dennis, P Eeles, ML Hall, M Hughes, N Hulme, S Jeffcoate, OT Lewis, T Melling, ML Munguira, M Oates, E Pollard, KJ Porter, NOM Ravenscroft, M Rothschild, K Schönrogge, DJ Simcox, MR Shaw, ALR Thomas, CD Thomas, H Van Dyck, M S Warren, I West, C Wiklund, KJ Willmott and MR Young. It is also a pleasure to thank G McGavin and D Carter, respectively, for access to specimens in the Hope Entomological Collections of Oxford, and those of the Natural History Museum. Livestock, and other references were kindly supplied by K Bailey, S Bradley, P Eeles, S Forster, A Gardner, M Hoare, R Lilley, J McFeely, C Raper, F Rayner, C Rivers, J Tucker, D Wedd, M Wilkins, K Willmott and C Wiskin.

We also thank Martin Harvey for many helpful comments on the text. Finally, we are extremely grateful to Andrew and Anne Branson for their sensitive and painstaking editing of this book, for its design and production, and for their enthusiastic support throughout.